Palgrave Frontiers in Philosophy of Religion

**Series Editors**
Yujin Nagasawa
Department of Philosophy
University of Birmingham
Birmingham, UK

Erik J. Wielenberg
Department of Philosophy
DePauw University
Greencastle, IN, USA

Palgrave Frontiers in Philosophy of Religion is a long overdue series which will provide a unique platform for the advancement of research in this area. Each book in the series aims to progress a debate in the philosophy of religion by (i) offering a novel argument to establish a strikingly original thesis, or (ii) approaching an ongoing dispute from a radically new point of view. Each title in the series contributes to this aim by utilising recent developments in empirical sciences or cutting-edge research in foundational areas of philosophy (such as metaphysics, epistemology and ethics).

More information about this series at
https://link.springer.com/bookseries/14700

Andrew Loke

# The Teleological and Kalam Cosmological Arguments Revisited

palgrave
macmillan

Andrew Loke
Department of Religion and Philosophy
Hong Kong Baptist University
Kowloon Tong, Hong Kong

ISSN 2634-6176 ISSN 2634-6184 (electronic)
Palgrave Frontiers in Philosophy of Religion
ISBN 978-3-030-94402-5 ISBN 978-3-030-94403-2 (eBook)
https://doi.org/10.1007/978-3-030-94403-2



Cover Illustration: Suchart Kuathan/Getty Images

This Palgrave Macmillan imprint is published by the registered company Springer Nature Switzerland AG.
The registered company address is: Gewerbestrasse 11, 6330 Cham, Switzerland

*To J.P. Moreland*
*For teaching me philosophy, and for being such an inspiration and encouragement over the years*

# Acknowledgements

The main idea for this book first came when I was studying at Biola University back in 2007. We were discussing the Teleological Argument during a philosophy of religion class, and I began to ponder how might one exclude all the alternative hypotheses to Design. An idea came to me in the form of a logical syllogism which I develop in this book.

This book is also a follow-up to my previous book *God and Ultimate Origins* (Springer Nature 2017). Since its publication I have had the privilege of debating philosophers Alex Malpass (2019) and Graham Oppy (2020) on the Kalām Cosmological Argument, and interacting with many people concerning the debates. My responses to their objections constitute a significant part of this book. I am grateful to them for the exchanges and have acknowledged their contributions in the footnotes of this book.

I have also presented my arguments at the World Congress of Philosophy (Peking University, 2018), the American Academy of Religion Annual Conference (2020), and conferences and seminars held at Oxford University, Birmingham University, National University of Singapore, Nanyang Technological University, and National Taiwan University, and I am grateful for the feedback I received.

The topics covered in this book are transdisciplinary in nature, and I am grateful for the interactions with experts in various disciplines and for the help I received. In particular, I would like to express my gratitude to scientists Aron Wall, Luke Barnes, Peter Bussey, Don Page, Chan Man Ho, mathematician Tung Shi Ping, theologians Mark Harris and Alister McGrath, and philosophers William Lane Craig, Graham Oppy, Wes Morriston, Alex Malpass, Josh Rasmussen, Wesley Wildman, Bruce Reichenbach, Garry DeWeese, Jacobus Erasmus, Andres Luco, Frederick Choo, Lee Wang Yen, Jonathan Chan, Mark Boone, Kwan Kai Man, Steve Palmquist, Andrew Brenner, Chan Kai Yan, Brian Wong, Rope Kojonen, Tim Maness, Phil Goff, and Zhang Jiji for very helpful correspondence.

I would like to thank Yujin Nagasawa for recommending my book proposal, the team at Springer Nature/Palgrave—in particular Brendan George, Rebecca Hinsley, and Md Saif—for their efficient assistance, and the anonymous reviewer for his/her helpful comments and recommendations for publication. I would also like to thank my research assistant Vincent Chan for his help with the bibliography.

The research for this book is funded by the Hong Kong Research Grants Council Early Career Scheme (ECS): 'A Reassessment of the Cosmological and Teleological Arguments for the Existence of God', project number 22603119, and partially supported by a grant from the Research Grants Council of the Hong Kong Special Administrative Region, China (project no. EdUHK 18606721).

31 July 2021
Hong Kong

# Contents

# List of Figures

# 1

# Introducing the Quest for an Explanation

## 1.1 Introduction

Throughout history, many have embarked on the quest to discover the answers to the fundamental mystery concerning the ultimate origins of the universe and the purpose of our existence. Some of the most brilliant thinkers in human history have contributed to this quest by formulating their answers to these Big Questions in the form of the Cosmological Argument—which attempts to demonstrate the existence of a Divine First Cause or Necessary Being—and the Teleological Argument (TA)—which attempts to demonstrate that our universe is the purposeful creation of a Divine Designer. Originating in Ancient Greece in the writings of Plato (e.g. *Laws*, 893–96) and found in other ancient philosophical traditions (e.g. the tenth-century Indian philosopher Udayana's *Nyāyakusumāñjali* I, 4), the Cosmological and Teleological Arguments have also been developed by scholars of the Abrahamic religions, since the 'First Cause' and Designer of the universe may be associated with Judaism, Christianity, or Islam.

Given that these arguments have been around for thousands of years, one might wonder what more can be said on behalf of them. Many have



A. Loke, *The Teleological and Kalam Cosmological Arguments Revisited*, Palgrave Frontiers in Philosophy of Religion, https://doi.org/10.1007/978-3-030-94403-2_1

thought that these arguments have been successfully rebutted by the extensive criticisms of David Hume and Immanuel Kant in the eighteenth century, and that they are now obsolete in light of modern science. A review of the literature however shows that—even in our present scientific period—these arguments are still being actively discussed in journals and monographs published by the world's leading academic publishers. One reason is that modern science itself has developed theories (e.g. the Big Bang theory, theories of fundamental physics) which seem to support the premises of these arguments. Moreover, the objections by Hume and Kant are answerable, as I shall explain in the rest of this book.

This book aims to expose the weaknesses in recent assessments of these arguments by their proponents and opponents, to offer a more compelling evaluation of alternative explanations, and to examine whether both arguments can be integrated in such a way that both are strengthened. It will move the discussion ahead in a new and significant way by providing original arguments in response to objections, including those found in leading academic publications within the last few years. These objections include (among others) the problem of ensuring that all the alternative hypotheses to Design have been considered and ruled out[1] (Ratzsch and Koperski 2019; see below), the problem of assigning prior probability for Design (Sober 2019), and the objection to the applicability of the Causal Principle to the beginning of the universe based on bounce cosmologies and the apparent challenge of fundamental physics to the directionality of causality and time (Linford 2020). Despite the huge amount of literature on the Cosmological Argument and Teleological Argument, I am not aware of any other publication which has provided the original arguments which I am going to offer in response to the objections, some of which have remained outstanding despite many years of intense discussions. To help the reader appreciate this point, I shall begin with a review of the background of discussion on the Cosmological Argument and Teleological Argument.

## 1.2 A Review of the Discussion

There are different versions of the Cosmological Argument (Craig 1980):

(1) The Leibnizian (named after Gottfried Leibniz 1646–1716; see *Monadology*, §32), which attempts to ground the existence of the contingent things of our universe in a Divine Necessary Being.
(2) The Thomist (named after Thomas Aquinas 1225–1274; see *Summa Theologica* I,q.2,a.3 and *Summa Contra Gentiles* I,13,a), which attempts to demonstrate that the universe is sustained in existence by a Divine First Cause.
(3) The Kalām Cosmological Argument (KCA), which attempts to demonstrate that the universe has a beginning of existence brought about by a Divine First Cause. The roots of the KCA can be traced to Plato's *Timaeus* sections 27 and 28 and the writings of the Christian philosopher John Philoponus (c.490–c.570), who argued against the possibility of an actual infinite number of earlier events in *Against Aristotle on the Eternity of the World*, frag. 132. Philoponus' work became an important source for Medieval Islamic and Jewish Proofs of Creation (Davidson 1969; Adamson 2007, chapter 4), and the KCA was developed by the Muslim philosopher al-Kindī' (805–873) and *mutakallimūm* (theologians who used argumentation to support their beliefs) such as al-Ghāzāli (1058–1111).

In this book, I shall focus on the Kalām version of the Cosmological Argument.

Concerning the KCA, the development of the Big Bang theory, which seem, to indicate that our spacetime manifold has a beginning, has led to renewed interest among philosophers and scientists concerning the question of First Cause of the beginning of the universe. Nevertheless, while the Big Bang is commonly understood as the beginning of spacetime, many cosmologists are now discussing pre-Big Bang scenarios in which the Big Bang is not the absolute beginning. While some cosmologists have proposed that entities such as a quantum vacuum or another universe existed before the Big Bang, others have asked where these came from. This question is related to whether everything that begins to exist has a cause (the Causal Principle, CP) and whether an infinite regress of causes and effects is possible. The KCA, as formulated by its most noteworthy recent proponent William Lane Craig, is as follows:

(1) Whatever begins to exist has a cause (Causal Principle).
(2) The universe began to exist.
(3) Therefore, the universe has a cause.

Craig argued that further analyses of the Cause of the universe show that this Cause possesses various theistic properties, such as being uncaused, beginningless, initially timeless and changeless, has libertarian freedom. and is enormously powerful (Craig and Sinclair 2009). Writing in *The Cambridge Companion to Atheism*, philosopher Quentin Smith noted that 'a count of the articles in the philosophy journals shows that more articles have been published about Craig's defence of KCA than have been published about any other philosopher's contemporary formulation of an argument for God's existence' (Smith 2007, p. 183). While many articles have argued in support of KCA, others have raised various objections.

With regard to premise (1), some philosophers have objected that we only have reason to suppose that the Causal Principle holds within our universe, but not with respect to the beginning of the universe itself (Oppy 2010, 2015).

With regard to premise (2) of KCA, Craig has defended two philosophical arguments for time having a beginning: the argument from the impossibility of concrete actual infinities and the argument from the impossibility of traversing an actual infinite. The first argument claims that the absurdities which result from paradoxes such as Hilbert's Hotel show that concrete infinities cannot exist, and since an infinite temporal regress of events is a concrete infinity, it follows that an infinite temporal regress of events cannot exist. The second argument claims that a collection formed by successive addition cannot be an actual infinite, and since the temporal series of events is a collection formed by successive addition, the temporal series of events cannot be an actual infinite. Others have raised various objections, such as claiming that actual infinite sequences are 'traversed' all the time in nature (e.g. whenever an object moves from one location in space to another) (see discussion in Puryear 2014), and arguing that Craig's defence of KCA depends on the highly controversial dynamic theory of time (according to which the members of a series of events come to be one after another) and begs the question against an actual infinite past (Oppy 2006). Stephen Hawking proposed that the initial state of the universe consisted of a timeless (no boundary) state (Hartle and Hawking 1983; Hawking 1988). This initial state can be understood as a beginningless impersonal First Cause from which all things came, and which avoids the need for a personal Creator.

I have addressed the objections noted above in my previous writings. For example, with regard to the objections noted above concerning premise 1 of KCA, I have proposed a new philosophical argument in Loke (2012b, 2017, chapter 5) which addresses the objections by Oppy (2010, 2015) and others, and which demonstrates that, if something (say, the universe) begins to exist uncaused, then many other kinds of things/events which begin to exist would also begin to exist uncaused, but the consequent is not the case; therefore, the antecedent is not the case. In this book (Chaps. 2 and 3), I shall further develop this Modus Tollens argument in response to more recent objections to the Causal Principle found in the writings of Rasmussen (2018), Almeida (2018), Linford (2020), and others, and in Chap. 6, I shall use it to respond to Hawking's objections to a Creator (including the objections found in his final book published in 2018).

With respect to the objections noted above concerning premise 2 of KCA, I have shown in Loke (2012a, 2014b, 2017, chapter 2) that the argument for a beginning of the universe based on the impossibility of concrete actual infinities does not beg the question against the existence of concrete actual infinities, by demonstrating that the argument can be shown to be based on the independent metaphysical fact that numbers are causally inert. With respect to the argument based on the impossibility of traversing an infinite, I have responded to the objection that actual infinite sequences are 'traversed' all the time in nature, by defending the view that time and space is a continuum with various parts but not having an actual infinite number of parts or points (Loke 2016; Loke 2017, chapter 2). Moreover, I have shown that this argument can be modified such that it does not need to presuppose the controversial dynamic theory of time (Loke 2014a, 2017, chapter 2; see further, chapter 5 of this book). Additionally, I have developed a new argument against an infinite causal regress which demonstrates that, if every prior entity in a causal chain has a beginning, then given the Causal Principle nothing would ever begin to exist; therefore, what is required is a beginningless First

Cause (Loke 2017, chapter 3). In this book (Chap. 5), I shall develop these arguments further in engagement with various pre-Big Bang scientific cosmologies and reply to the latest objections to these arguments (e.g. Almeida 2018; Linford 2020).

With respect to Hawking's conceptual challenge concerning the nature of First Cause noted above, I have argued in Loke (2017, chapter 6) that the First Cause is a libertarian free agent. Against this conclusion, it might be objected that one should not attribute libertarian freedom to the First Cause, because libertarian freedom is associated with a mind with the capacity for decision making, but it has not yet been shown that the First Cause has other properties of a mind with the capacity for decision making. In Chaps. 6 and 7, I shall show that this objection fails, and I shall also provide evidences that the First Cause has other properties of a mind with the capacity for decision making. The latter will be accomplished by developing the Teleological Argument and combine it with the KCA to demonstrate that the First Cause is an intelligent designer of the universe.

Concerning the Teleological Argument, 'according to many physicists, the fact that the universe is able to support life depends delicately on various of its fundamental characteristics, notably on the form of the laws of nature, on the values of some constants of nature, and on aspects of the universe's conditions in its very early stages' (Friederich 2018). Many scientists and philosophers have argued that this 'fine-tuning' is evidence for a Designer (Lewis and Barnes 2016). Others have cited the mathematically describable order of the universe (Polkinghorne 2011) as evidences for a Designer. Critics object that there could be alternative hypotheses which have yet to be considered. This problem beset various forms of design inference. For example, concerning 'inference to the best explanation' (IBE), which involves comparing explanations based on criteria such as explanatory power, explanatory scope, and so on, Ratzsch and Koperski (2019) state that substantive comparison between explanations 'can only involve known alternatives, which at any point represent a vanishingly small fraction of the possible alternatives … being the best (as humans see it) of the (humanly known) restricted group does not warrant ascription of truth, or anything like it'. Others have mentioned the problem of assigning prior probability for Design given that our inferences of intelligent design are based on our empirical knowledge of *human* intelligence, which may not carry over to hypotheses involving non-human designers (Sober 2003, p. 38). Additionally, many have insisted that we should try to find a scientific explanation for the apparent fine-tuning, for appealing to God can be used to solve any problem, so it is not helpful (Penrose and Craig 2019). Against Swinburne's (2004) formulation of the Teleological Argument, critics have also objected that the range of explanatory latitude is too wide: 'whatever the laws of nature turn out to be, the theist would explain these as brought about by God, hence … the supposed evidences [i.e. the laws of nature] provide no check on the validity of the explanatory premises' (Grünbaum 2004, p. 605).

This book will fill a gap in the literature by devising an original deductive argument (see Chap. 4) which demonstrates that the following are the only possible categories of hypotheses concerning fine-tuning and order: (i) Chance, (ii) Regularity, (iii) Combinations of Regularity and Chance, (iv) Uncaused, and (v) Design (The Designer may [or may not] use chancy, regular, or chancy + regular process; see the discussion on theistic multiverse scenarios in Chap. 4; given this clarification, it should be noted that (i)–(iv) are intended to be exclusive of Design).[2] My book collates a large variety of contemporary cosmological models and classifies them within the five categories. It demonstrates that there are essential features of each category such that, while the alternatives to design are unlikely, the Design hypothesis is not, and that one can thus argue for Design by exclusion without having to first assign a prior probability for Design. The exclusion of all the alternatives implies that the conclusion of design follows logically rather than being merely appealed to solve a problem; it also avoids Grünbaum's objection concerning the range of explanatory latitude. I shall show that KCA can be used to strengthen the TA by answering the question 'Who designed the Designer?' through demonstrating that there is a beginningless and hence un-designed First Cause, and by demonstrating that the ultimate explanation cannot be a scientific one, because the first event was brought about by a First Cause with libertarian freedom (a First Cause with libertarian freedom implies agent causation) and not by a mechanism describable by a law of nature (see Chap. 6). On the other hand, the TA strengthens the KCA by providing additional considerations for thinking that the First Cause is an (intelligent) Creator (see Chap. 7).

Finally, this book will provide an up-to-date discussion of various theories in scientific cosmology and fundamental physics that are relevant to the philosophy of religion debates concerning the ultimate origins of the universe. It responds to the God-of-the-gaps objection by demonstrating that the KCA-TA is not based on gaps in our understanding which can be filled by further progress in science, but is based on the analysis of the necessary conditions (e.g. what is required for a First Cause to bring about the first event) and follows from deduction and exclusion. It contributes to contemporary theological discussions concerning the relationship between God and time and the doctrine of creation, and responds to the theological objections to fine-tuning by Halvorson (2018) et al. It offers a superior form of design inference which avoids the problems that beset alternative forms. Additionally, it contributes to the discussions on issues of considerable philosophical interest such as time, causality, infinity, and libertarian freedom, and demonstrates the relevance of philosophical arguments for answering the question of ultimate origins against the Scientism of Hawking et al. and the New Verificationism of Ladyman et al. (2007). In these and other ways, this book promotes the dialogue between philosophers, scientists, and theologians concerning the Big Question of ultimate origins.

## 1.3 Problems with Scientism

Contemporary formulations of the Cosmological Argument and Teleological Argument involve considerations of both philosophy and modern scientific cosmology. Proponents of scientism have dismissed philosophy when considering the question about the ultimate origins of the cosmos, claiming that science is the only or the best way for understanding the nature of reality. Often an appeal is made to the predictive successes and technological applications of science, which metaphysics seems unable to offer. Against this sort of appeal, Feser (2017, p. 282) observes:

> A defender of scientism demands to know the predictive successes and technological applications of metaphysics or theology, and supposes he has won a great victory when his critic is unable to list any. This is about as impressive as demanding a list of the metal-detecting successes of gardening, cooking, and painting, and then concluding from the fact that no such list is forthcoming that spades, spatulas, and paintbrushes are all useless and ought to be discarded and replaced with metal detectors. The fallacy is the same in both cases. That a method is especially useful for certain purposes simply does not entail that there are no other purposes worth pursuing nor other methods more suitable to those other purposes. In particular, if a certain method affords us a high degree of predictive and technological power, what that shows is that the method is useful for dealing with those aspects of the world that are predictable and controllable. But it does not show us that those aspects exhaust nature, that there is nothing more to the natural world than what the method reveals.

On the other hand, scientism is susceptible to the objection that scientism cannot be proven by science itself (Loke 2014c). Indeed, its advocates 'rely in their argument not merely on scientific but also on philosophical premises' (Stenmark 2003). Additionally, science itself cannot answer the question 'Why scientific results should be valued?'; the answer to this question is philosophical rather than scientific. Likewise, the question 'Why is the testing of theories important for understanding how the natural world works?' cannot be answered by simply doing more testing; rather, the answer would require a philosophical explanation of how testing relates to our understanding of the workings of the natural world.

Moreover, philosophical conceptual analysis is evidently important for science itself. Cosmologist Sean Carroll quips that 'Physicists tend to express bafflement that philosophers care so much about the words. Philosophers, for their part, tend to express exasperation that physicists can use words all the time without knowing what they actually mean' (Carroll 2010, p. 396). The point here is that definitional issues are of fundamental importance and they underlie all our knowledge, including scientific knowledge. For example, if scientists do not define the terms in their scientific hypothesis carefully, then they do not even know what they are testing for, and their experiments would fail. It is a pity that some physicists like Lawrence Krauss are not careful enough about the concepts and words that they use, such as concerning 'nothing' (see Krauss 2012, cf. Bussey 2013).

Physicist Carlo Rovelli (2018) observes that philosophy has played an essential role in the development of science (in particular scientific methodology), and notes that

> Philosophers have tools and skills that physics needs, but do not belong to the physicists training: conceptual analysis, attention to ambiguity, accuracy of expression, the ability to detect gaps in standard arguments, to devise radically new perspectives, to spot conceptual weak points, and to seek out alternative conceptual explanations.

In his survey of the forms of reasoning and criteria of rationality that have characterized the production of knowledge across culture and history, McGrath (2018) observes the emergence and significance of the notion of multiple situated rationalities, which affirms the intellectual legitimacy of transdisciplinary dialogue. Noting the notion of multiple levels of reality, McGrath observes that the natural sciences themselves adopt a plurality of methods and criteria of rationality, making use of a range of conceptual tool-boxes that are adapted to specific tasks and situations, so as to give as complete an account as possible of our world (p. 2). For example, with regard to the scientific study of a frog jumping into a pond,

> The physiologist explains that the frog's leg muscles were stimulated by impulses from its brain. The biochemist supplements this by pointing out that the frog jumps because of the properties of fibrous proteins, which enabled them to slide past each other, once stimulated by ATP. The developmental biologist locates the frog's capacity to jump in the first place in the ontogenetic process which gave rise to its nervous system and muscles. The animal behaviourist locates the explanation for the frog's jumping in its attempt to escape from a lurking predatory snake. The evolutionary biologist adds that the process of natural selection ensures that only those ancestors of frogs which could detect and evade snakes would be able to survive and breed.

McGrath concludes that 'all five explanations are part of a bigger picture. All of them are right; they are, however, different' (pp. 59–60). Just as science itself brings together different explanations to help us see the bigger picture, there is a need to bring together different disciplines that would complement one other in our attempt to gain a fuller understanding of reality.

Contrary to Hawking, who infamously declared that 'philosophy is dead' (Hawking and Mlodinow 2010, pp. 1–2), cosmologist George Ellis observed that philosophy has an important role to play in scientific cosmology. He noted, with respect to the criteria for a good scientific theory (internal consistency, explanatory power, etc.), that 'these criteria are philosophical in nature in that they themselves cannot be proven to be correct by any experiment. Rather, their choice is based on past experience combined with philosophical reflection' (Ellis 2007, section 8.1). In view of the importance of philosophical considerations, cosmologists should not merely construct models of the universe without considering the philosophical problems associated with certain models, such as problems concerning the traversing of an actual infinite and the violation of Causal Principle, which have been highlighted by proponents of the Cosmological Argument. Indeed, scientists who are well-informed about the importance of philosophy have used philosophical arguments against an actual infinite number of earlier events to argue against cosmological models that postulate this. For example, cosmologists Ellis, Kirchner, and Stoeger write in an article published in the *Monthly Notices of the Royal Astronomical Society*: 'a realized past infinity in time is not considered possible from this standpoint—because it involves an infinite set of completed events or moments. There is no way of constructing such a realized set, or actualising it' (Ellis et al. 2004, p. 927). The proofs for the impossibility of a realized past infinity which Ellis et al. are referring to are two of the five philosophical proofs which I mention in Chap. 5, namely, the Hilbert Hotel Argument and the argument for the impossibility of traversing an actual infinite. This indicates that philosophical arguments are relevant for modern cosmology. This book will contribute to the discussion by developing some of these arguments in engagement with modern science.

## 1.4 Problems with Verificationism

Verificationism, which was popular in the early twentieth century, claims that only statements that are analytic or verifiable are meaningful. It has since been widely rejected, for the principle itself is neither analytic nor verifiable (Creath 2017). While its proponents claim that the principle could be regarded as a definition or axiom, this fails to meet the challenge of why we should adopt such a definition or axiom. The principle cannot meet its own demands (Trigg 1993, p. 20). Likewise, while confirmation by observation and repeated experiments is one way of knowing certain things, it would be wrong to think that this is the only way to know anything, for the view that 'confirmation by observation and repeated experiments is the only way to know anything' is a view which cannot be confirmed by observation and repeated experiments (for other ways of knowing, see below and Chap. 4). To equate factual (what is actually the case) with empirical (what is verifiable by observation) would be to commit the error of verificationism. Moreover, it begs the question against the existence of an immaterial timeless Creator who cannot be verified by observation given the limitation of the method (the method can only apply to observable material entities which exist in time).

While acknowledging 'we may no longer believe in the verificationist theory of meaning', Ladyman et al. (2007, p. 8) nevertheless propose a pragmatist New Verificationism which consists in two claims:

> First, no hypothesis that the approximately consensual current scientific picture declares to be beyond our capacity to investigate should be taken seriously. Second, any metaphysical hypothesis that is to be taken seriously should have some identifiable bearing on the relationship between at least two relatively specific hypotheses that are either regarded as confirmed by institutionally bona fide current science or are regarded as motivated and in principle confirmable by such science. (p. 29)

The main pragmatic motivation for adopting this principle is stated as follows: 'What we really want a verifiability criterion to capture is the pointlessness of merely putative domains of inquiry' (p. 308), such as inquiry concerning whether God is the cause of the Big Bang (p. 29). The reason why they think that such metaphysical inquiry is pointless is because, first, they claim that armchair intuitions about the nature of the universe ignore the fact that 'science, especially physics, has shown us that the universe is very strange to our inherited conception of what it is like'. Second, they claim that such metaphysical inquiry ignores 'central implications of evolutionary theory, and of the cognitive and behavioural sciences, concerning the nature of our minds' (p. 10). We shall now consider these two claims in turn.

## 1.5 In Defence of the Possibility of *a Priori* Metaphysical Knowledge

Concerning the first claim, Ladyman et al. state that 'much of what people find intuitive is not innate, but is rather a developmental and educational achievement. What counts as intuitive depends partly on our ontogenetic cognitive makeup and partly on culturally specific learning' (p. 10). Against the reliability of 'our common-sense image of the world' as an appropriate basis for metaphysical theorizing, they claim that 'modern science has consistently shown us that extrapolating our pinched perspective across unfamiliar scales, magnitudes, and spatial and temporal distances misleads us profoundly' (p. 11). For example, 'Casual inspection and measurement along scales we are used to suggest that we live in a Euclidean space; General Relativity says that we do not' (p. 11). Against the 'many examples of metaphysicians arguing against theories by pointing to unintuitive consequences' (p. 13), they ask: 'why should we think that the products of this sort of activity reveal anything about the deep structure of reality, rather than merely telling us about how some philosophers, or perhaps some larger reference class of people, think about and categorize reality?' (p. 16).

The warning to exercise caution when discussing matters that are far beyond our daily experiences is well taken. It is true that in the history of philosophy there has been a cascade of unduly optimistic estimates of the power of specifically philosophical reasoning, eventually corrected by empirically grounded insights.[3] Nevertheless, we need to distinguish these failures as well as 'common sense' and 'everyday intuitions' from philosophical principles of reasoning such as various forms of deductive and inductive reasoning which underlie the construction of scientific theories themselves, including General Relativity mentioned above. In other words, we need to distinguish 'common sense' and 'everyday intuitions' from philosophical principles of reasoning by which we show 'common sense' and 'everyday intuitions' to be highly unreliable and demonstrate those 'optimistic estimates' to be failures.

With regard to General Relativity, the idea that space itself can be curved may seem strange, but it does not violate deductive and inductive reasoning, properly understood. While quantum phenomena may appear foreign to our 'common sense' and 'everyday intuitions', it does not violate deductive reasoning[4] which assumes the laws of logic and which (together with inductive reasoning) is required for quantum physics itself. To illustrate, quantum physics is often heralded as a scientific theory that is well-confirmed by experiments, such as those that reveal quantum entanglement (an example cited against 'intuition' by Ladyman et al. on p. 19!). The confirmation would take the following form:

(1) If the experiment reveals quantum entanglement, then the prediction of quantum physics is confirmed.
(2) The experiment reveals quantum entanglement.
(3) Therefore, the prediction of quantum physics is confirmed.

This form of valid reasoning is known as modus ponens (1. If A, then B, 2. A. 3. Therefore, B), which is a form of deductive reasoning. Valid deductive reasoning can give a false result if the premise is false, but if the premise is true, then the conclusion which follows from valid deductive reasoning would be true as well. While De Cruz and Smedt (2016, p. 360) have complained that, unlike scientists who can often confirm or disconfirm their theories, philosophers 'do not have independent empirical techniques to confirm or disconfirm their intuitions', the above illustration shows that the confirmation or disconfirmation of scientific theories or intuitions itself would require the laws of logic. Kojonen notes that 'at least some compatibility between the human mind and the cosmos is required in order for the cosmos to be at all amenable to scientific discovery, and for human survival to have been possible in the first place' (Kojonen 2021, p. 42). Ladyman et al. would agree that scientific theories—and by implication, deductive reasoning which is assumed by scientific theories—is not 'merely telling us about how some philosophers, or perhaps some larger reference class of people, think about and categorize reality' (p. 16) but helping us understand 'the deep structure of reality' (ibid.). They wrote:

> Unlike Kant, we insist that science can discover fundamental structures of reality that are in no way constructions of our own cognitive dispositions .... As collective constructions, the institutional filters of science need not mirror or just be extensions of individual cognitive capacities and organizing heuristics. They have shown themselves to have a truth-tracking power—partly thanks to mathematics. (p. 300)

It is interesting to note that they acknowledge the role of mathematics which, similar to the laws of logic that underlie deductive reasoning, is both necessary for science and yet also knowable *a priori*. Ladyman et al. would acknowledge that mathematical equations such as 2 + 2 = 4, 4 × 4 = 16, and so on are not merely 'everyday intuitions' or 'common sense', but rather correspond with reality, such that they are able to confer 'truth-tracking power' to science. In the subsequent chapters, I will be using mathematical equations such as finite + finite = finite, 0 + 0 + 0 ... = 0 for some of my arguments.

Likewise, the laws of logic (e.g. A is A; it cannot be the case that A and not-A; either A or not-A) are not merely 'conceptual analysis', human psychology of reasoning, or human conventions. The laws of logic correspond with the way things are; indeed, they are necessarily true because a violation of the laws of logic would be non-existent. For example, consider a 'shapeless square': such a thing cannot exist because the existence of A implies it is not the case that not-A (the existence of a shape [e.g. square] implies that it is not shapeless). The fact that such things which violate the laws of logic cannot exist illustrates that the laws of logic are necessarily true. They do not merely exist in the human mind but they also apply to mind-independent concrete entities. For example, it remains the case that there cannot be shapeless squares billions of years ago even if there were no minds to think about them back then. While apparent contradictions can exist, a true contradiction (e.g. a shapeless square) cannot. Huemer (2018, p. 20) notes that

> If you think there is a situation in which both A and ~A hold, then you're confused, because it is just part of the meaning of 'not' that not-A fails in any case where A holds … Now, a contradiction is a statement of the form (A & ~A). So, by definition, any contradiction is false.

The above definition of 'not' (given which contradictions are impossible and the laws of logic are necessarily true) will be used for my arguments in this book. (One should not object to my arguments by using alternative definitions of these terms. To do that would be similar to someone objecting to 'All humans are mortal, Socrates is a human, therefore, Socrates is mortal' by using alternative definitions of human or Socrates, which of course misses the point of the argument by talking about something else. To rebut an argument one has to rebut the premise or the validity rather than use an alternative definition of the terms.)

One might think that the principle of superposition in quantum physics violates the laws of logic which underlie these reasonings. However, this is a misunderstanding. Superposition is the mathematical addition of probability densities of all of the possible states of a quantum system, and it is used to calculate the probability of observing the system in one of the states (e.g. a particle going through one slit or the other in the double-slit experiment). When the system is not being observed, it is not the case that a particle existing in contradictory states. Rather (according to the Copenhagen interpretation), the quantum of energy is spread across the possible states as a wave. It remains in that state until an observation collapses the wave to a particle. A wave has the potential to be observed at slit A or slit B, but it cannot be observed at both slits at the same time because an observation would cause it to be no longer be a wave, but a particle. Having the potential to be one thing or another does not violate the law of non-contradiction (Pratt 2012). (According to Bohm's interpretation, the system consists of a particle riding on a wave which follows the Schrodinger equation, and which guides the particle to only one position which is revealed when an observation is made. Again, there is no violation of the laws of logic.)

While multiple non-classical logics have been developed to meet specific tasks in knowledge production (McGrath 2018, p. 31), their proven utility has to do with definition, designation, proving, computability problem solving, and so on, i.e., they are helpful in situations where (say) the definitions are vague. On the other hand, classical logic applies to what is actually the case or can be the case (regardless of whether one can define it clearly, prove it, etc.). Thus, for example, it cannot be the case that shapeless squares exist.

Gödel Incompleteness Theorems do not entail the violation of the laws of logic, for there can be incomplete but consistent systems. Russell's paradox (which defines 'the Russell set' as the set of all things that are not members of themselves) can be resolved by arguing that the Russell set does not exist given that it has an inconsistent definition (Huemer 2018, pp. 42–43). Likewise, the liar paradox (Is the sentence 'this sentence is false' true or false?) does not entail the violation of the laws of logic, for one can argue that the liar sentence fails to express a proposition because the rules for interpreting the sentence are inconsistent; thus, it does not have the property of truth or falsehood (Huemer 2018, p. 29). Priest et al. (2018) mention the 'strengthened' liar paradox such as L: L is not true, and argue that, if this sentence is neither true nor false, it is not true; but this is precisely what it claims to be; therefore, it is true. Huemer (2018, pp. 34–36) replies by denying that L makes any claim at all. L does not make any claim because it fails to express a proposition. However, one can say that N: L is not true. Huemer explains 'N expresses the proposition that L is not true; yet L does not express that proposition, even though L is syntactically identical to N. Why is this? Because when we read L, we are invited to accept an inconsistent story about the proposition that it expresses; but when we read N, there is no inconsistent story about what N expresses' (p. 35).[5] Moreover, the claim that contradictions can exist in a self-referential paradox in linguistic games (which may be due to inadequacies of language) is in any case irrelevant to the claim that contradictions can exist in concrete entities such as the universe or ultimate reality.

Some religious mystical traditions (e.g. certain forms of Chinese Buddhism, Taoism, and apophatic theology) postulate a transcendent realm in which the laws of logic are violated (Capra 2010). However, this is impossible, for there cannot be shapeless squares in the transcendent realm either. This conclusion is not based on our inability to imagine it but based on what it would involve: the existence of A implies the non-existence of not-A.

Some might think that a solution to the Paradox of the Stone (If an omnipotent God exists, can He create a stone He cannot carry?) would require the claim that God can violate the laws of logic. However, this is not so. With regard to the Paradox, one can ask, 'If God exists, can God create a 'shapeless square'? The answer is no, because there is no such thing. Likewise, there is no such thing as 'a stone which God cannot carry'; thus, God cannot create such a stone. This does not mean that God's power is limited; rather, there is no such object ('shapeless square', 'a stone which God cannot carry') for God to bring about. Thus, the person who asks God to create a stone He cannot carry is asking God to do nothing, which poses no challenge to His power. Neither do the Christian doctrines of Trinity, Incarnation, and divine foreknowledge and freedom entail the violation of the laws of logic (see Moreland and Craig 2003; Loke 2014d).

One should note the distinction between the laws of logic and the laws of nature. There can be other universes with different properties and different laws of nature, but there cannot be other universes in which the laws of logic do not apply (as illustrated by the fact that there cannot be shapeless squares in other universes). As explained above, the laws of logic are necessary true and inviolable, and the impossibility of their inviolability can be known *a priori* with 100% epistemic certainty.

The 100% epistemic certainty concerning the inviolability of the laws of logic contrasts with the lack of 100% epistemic certainty in science because it is possible (no matter how improbable) that the observations based on which scientists infer the laws of nature are mistaken. A law of nature is derived from induction but—unlike deductive reasoning—inductive reasoning cannot yield 100% certainty because we cannot be 100% sure that there are no counterexample. Moreover, scientific theories, in their attempts to explain a connected sequence of phenomena by postulating an entity as a cause, face the difficulty that there may be other underlying causes for these phenomena which have not yet been discovered. While causes are necessary conditions for an event, many of them are yet unknown to us, and it is quite impossible for us to state all of them that would be sufficient for an event to obtain. In this way, scientific theories are underdetermined by the observations that purportedly supported them, and other theories for these observations remain possible (for classic discussions see Duhem 1954; Quine 1951; Laudan 1990). Given that there may be undiscovered causes for the phenomena we observe, science can never prove that the laws of logic can be violated or that something began to exist uncaused; on the contrary, as explained above, the laws of logic cannot be violated, and it will be shown in Chaps. 2 and 3 that the Causal Principle, that is, 'whatever begins to exist has a cause' is true as well. Given the Problem of Underdetermination, we should adopt an eclectic model of science whereby realist and anti-realist interpretations of scientific theories are adopted on a case-by-case basis, and adopt an anti-realist interpretation of a theory if a realist interpretation conflicts with well-established truths (Moreland and Craig 2003, pp. 314–318). For example, we should adopt an anti-realist view of a scientific theory if a realist interpretation would result in conflict with well-established understanding of the laws of logic (see above) and Causal Principle (see Chaps. 2 and 3).

While the laws of logic are limited in the sense that—by themselves—they cannot show us what exist, they can show us what cannot exist (e.g. a shapeless square cannot exist). Likewise, philosophical arguments (see Chap. 5) can show that an actual infinite number of prior events cannot exist, and therefore the universe (which we know does exist based on observation) cannot have an actual infinite number of prior events. Indeed, philosophical arguments are particular apt for proving negatives; just as one can prove that there cannot be shapeless squares, I shall show that there cannot be an infinite regress of events, and that it is not the case that something begins to exist uncaused.

In conclusion, I have shown that, contrary to popular misconceptions, quantum physics, Gödel Incompleteness Theorems, Russell's paradox, the liar paradox, and non-classical logics do not violate the laws of logic. Against the worry that how we think about the world may be very different from what the world itself really is, I have argued that the laws of logic correspond with what the world itself really is, and we can therefore use them to formulate various arguments concerning the world.

The laws of logic imply that the conclusion of a deductively valid argument from true premises must be true. Physics itself requires deductive and inductive reasoning the justification of which is philosophical, and one needs to distinguish between 'appearing weird' (e.g. superposition) from 'impossible' (e.g. it is impossible that 0 + 0 + 0 … be anything other than 0), which is what I shall demonstrate an infinite regress to be in later chapters. It should also be noted that, while what is mathematically impossible is metaphysically impossible (e.g. it is impossible that 0 + 0 + 0 … be anything other than 0), what is mathematically possible is not always metaphysically possible. For example, the quadratic equation $x^2-4 = 0$ can have two mathematically consistent and possible results for $x$: 2 or −2, but if the question is 'How many people carried the computer home?', the answer cannot be '−2', for in the concrete world it is metaphysically impossible that '−2 people' carried a computer home. Thus, the conclusion of '2 people' rather than '−2 people' is not derived from mathematical equations alone, but also from metaphysical considerations: '−2 people' lack the causal powers to carry a computer home. The metaphysically impossibility of '−2 people carrying the computer' would override the mathematical possibility in the quadratic equation. This shows that metaphysical considerations are more fundamental than mathematical considerations. The arguments against an infinite regress and against the violation of the Causal Principle which I discuss in the rest of this book are based on similar metaphysical considerations which are derived from understanding the nature of the world. This is not 'insisting that the physical world conform to some metaphysical principle'; rather, these metaphysical principles are based on understanding the nature of the world. The above conclusion implies that, even if a cosmological model is mathematically possible, it cannot be a correct model of the cosmos if it is metaphysically impossible.

It should be noted that the laws of logic would hold even at levels far beyond our daily experiences, such as at the beginning of time (there cannot be shapeless squares at such levels too). Likewise, we are able to know truths concerning relevance which hold even at levels far beyond our daily experiences. For example, the principle 'differences between prime numbers are irrelevant to the number's inability to give birth to a kitten' is clearly true, 'even though it certainly reaches far beyond ordinary experience; after all, it applies to infinitely many distinct numbers and infinitely many distinct ways to give birth to kittens'; we are able to recognize that 'the differences in the size of number make no categorical difference with respect to the ability to give birth to kittens' (Rasmussen and Leon 2018, p. 43).

As for the concern that causality and temporality may 'break down' at the beginning of the universe (Drees 2016, p. 199), following the laws of logic, the 'breaking down' of these would imply being *uncaused* and *timeless*. It will be shown in subsequent chapters that, using the laws of logic and undeniable experiences, one can formulate a Modus Tollens argument to show that the intuition 'all events have a cause' applies to the universe at large (see Chap. 3, contra De Cruz and Smedt 2016, p. 360), and that other arguments can be formulated to show that there is an *uncaused* and (initially) *timeless* First Cause of the universe.

## 1.6 Reply to the Evolutionary Objection Against Metaphysical Knowledge

Concerning the second claim by Ladyman et al. (2007) regarding the implication of evolutionary theory, they wrote:

> proficiency in inferring the large-scale and small-scale structure of our immediate environment, or any features of parts of the universe distant from our ancestral stomping grounds, was of no relevance to our ancestors' reproductive fitness. Hence, there is no reason to imagine that our habitual intuitions and inferential responses are well designed for science or for metaphysics. (2007, p. 2)

In their reply to why this would not undermine our scientific knowledge, they wrote 'even if one granted the tendentious claim that natural selection cannot explain how natural scientific knowledge is possible, we have plenty of good reasons for thinking that we do have such knowledge. On the other hand, we have no good reasons for thinking that *a priori* metaphysical knowledge is possible' (p. 7).

However, as explained above, science itself requires the correctness of *a priori* metaphysical knowledge of the laws of logic and mathematics; hence, the success of science in yielding scientific knowledge—which Ladyman et al. acknowledge despite our evolutionary history!—is one good reason for thinking that *a priori* metaphysical knowledge is possible. Moreover, regardless of the success of science, we do know that shapeless squares are not possible, and so on, which shows that we do have *a priori* metaphysical knowledge, and this is true regardless of how we might explain how we could have acquired such knowledge as well as scientific knowledge given evolution. (Plantinga 2011 famously argued that a theistically guided evolution would be able to explain this, whereas naturalism would not, but my argument here does not depend on Plantinga's argument, although I do think that it has plausibility. I have argued that evolution is compatible with Christian theism in Loke 2022.)

## 1.7 Reply to Empiricist Objections

Many who take a dismissive attitude towards metaphysics trace their view back to Carnap's influential paper 'Empiricism, Semantics, and Ontology' (1950). Carnap claims that 'If someone wishes to speak in his language about a new kind of entities, he has to introduce a system of new ways of speaking, subject to new rules; we shall call this procedure the construction of a linguistic framework for the new entities in question' (p. 21). It should be noted, however, that Carnap did not prove that there cannot be 'ways of speaking, subject to rules' (i.e. a linguistic framework) applied to speaking about a Creator of the universe which philosophers have been doing for thousands of years since the predecessors of Plato, who formulated the Cosmological Argument for a divine Creator. Carnap only offered illustrations of linguistic frameworks involving mathematical entities and material entities. However, these examples do not prove that there cannot be other kinds of linguistic framework involving other kinds of entities and following other rules. Bradley (2018, p. 2249) observes that 'Dismissivists have tended to assume that "Empiricism, Semantics and Ontology" provides an argument, but when we look there is little to be found'.

Bradley goes on to observe that the argument which Carnap had in mind is based on the Verificationism of his earlier writings, and he notes that Verificationism has long been rejected (see above). Bradley claims that, nevertheless, there is a lack of justification for metaphysical conclusions, but he fails to consider recent works on (say) the Cosmological Argument which shows that a Creator exists. As explained in later chapters, what deductively follows from the true premises of this valid argument is the existence of a Creator with libertarian freedom.

One might object that Carnap's main point is that theists have not specified under which conditions 'God exists' can be known to be true or false.[6] In reply, in this book, 'God' is understood to be referring to the Creator who brought about the beginning of the universe. One can specify the conditions under which the proposition 'A Creator brought about the beginning of the universe' can be shown to be true or false as follows: One can show this proposition to be false by proving that the universe has no beginning; one can show this proposition to be true by proving that the universe has a beginning and proving the Causal Principle using the Modus Tollens argument (see Chap. 2). Since the proposition 'A Creator brought about the beginning of the universe' is meaningful, it can be the conclusion of an argument. Given that this proposition follows as the conclusion of the Kalām Cosmological Argument, and given that the premises of the Kalām Cosmological Argument are true and that its deduction is valid (as argued in later chapters), this proposition is true.

Following Kant, it might be objected that we cannot know that the universe is an effect of God, for to know that A is causally related to B it seems that I must have sensory experience of both A and B so as to establish that they are regularly connected, but we don't have such sensory experiences of God (Evans 2010, pp. 151–152; citing Kant 1965, A603–14, B631–42).

To respond to this objection, it the distinction between affirming that there is a cause and identifying the properties of the cause should first be noted. Before scientists discover the cause of (say) an explosion of a certain chemical substance, did they think that the event has a cause? They sure did on the basis of Causal Principle, even though they have not specified the conditions or a universal law. Likewise, on the basis of Causal Principle, the beginning of matter-energy would have a cause understood as a necessary condition; how to identify the cause is a different issue.

With regard to the identification of cause, the specification of regular connection ('universal law') can be understood as one of the ways of identifying the properties of causes; this way is inductive. However, no proof has been offered to think that it is the only way. On the other hand, other than induction, deduction is also a method of inference. A deductive argument has already been provided previously to show that the necessary condition for the beginning of matter-energy is uncaused, beginningless, possessing libertarian freedom, and enormously powerful, that is, a transcendent Creator. Given this deductive argument, the inductive method is not required in this case to identify the properties of the Cause.

Sceptics might object that there is no empirical evidence that the universe or the singularity is created by God. Ladyman et al. (2007, p. 29) put it this way:

> Suppose that the Big Bang is a singular boundary across which no information can be recovered from the other side. Then, if someone were to say that 'The Big Bang was caused by Elvis', this would count, according to our principle, as a pointless speculation. There is no evidence against it—but only for the trivial reason that no evidence could bear on it at all.

However, direct empirical evidence is not the only way to find out the truth. On the contrary, for any evidence *x* to indicate that something else *y* is true, the laws of logic and various forms of reasoning are required to show how *y* follows from *x* or is supported by *x*. Scientists are able to conclude that the Big Bang happened, even though none of them have directly observed the Big Bang, because they are able to reason from the evidences (e.g. red shift, cosmic microwave background radiation) to the conclusion. Moreover, even though they do not have direct empirical evidence concerning how everything within our universe is formed, they can nevertheless deduce that that these things came from the Big Bang. Likewise, as I show in the rest of this book, we can conclude that the universe has a First Cause because there is evidence that there exists a series of causes and effects and it can be shown that an infinite regress is not possible, and it can be deduced that this First Cause has libertarian freedom, that is, a Creator. Even though we do not have direct empirical evidence concerning how the Big Bang singularity of the universe is formed, we can nevertheless conclude that it came from a Creator given that this Creator is the *First* Cause of the universe while the singularity cannot be the first cause because the singularity does not have libertarian freedom which (as I shall show in Chap. 6) the First Cause must have.

Many people today assume that for someone to claim that God created the universe would be to pretend to know what we cannot possibly know. This assumption is related to the Kantian assumption that we can only know the phenomena and that we cannot know the causes beyond the phenomena. However, on the one hand, the fact that something cannot be directly experienced does not imply that we are unable to have any knowledge of it. Experience is not the only source of knowledge. Introspection, rational insight, and moral insight are some other sources of knowledge. On the other hand, Oxford philosopher Richard Swinburne (2005, p.39) notes that the atomic theory of chemistry has shown 'in precise detail some of the unobservable causes of phenomena—the atoms whose combinations give rise to observable chemical phenomena'. Sudduth (2009, p. 206) likewise observes that the evolution of modern science and scientific methodology has made the Humean and Kantian crude empiricism no longer sensible. He elaborates:

> Neither Hume nor Kant envisioned the success of scientific reasoning from observable states of affairs to unobservable entities and causal processes on the grounds of the explanatory power of the latter. Extra-solar planetary science infers the existence, estimated mass, size, and orbital paths of unobservable planets from observable wobbles in the planet's parent star … Boltzmann utilized the atomic model to explain the behavior of gases and liquids. Eventually, the existence and behavior of atoms was explained in terms of yet smaller particles—protons, neutrons, and electrons.

The above scientific findings indicate that the unobservable causes of phenomena are not in principle unknowable; on the contrary, we can know many details about these unobservable causes through deductive and inductive reasoning. Starting from the phenomenon of the universe—an empirical premise!—one can likewise ask 'what caused this phenomenon?', and use deductive reasoning to arrive at the knowledge that there is a First Cause of the phenomenon which has the properties of being a Creator, as demonstrated by the KCA, which Kant has failed to refute (Craig 1979) and which I shall defend in the rest of this book.

One might object that the conclusion that unobserved causes such as atoms exist is arrived at via verification by experimentation, but one cannot use such a method to prove that an unobserved God exists. In reply, the above examples are only meant to show that unobserved does not imply unknowable. On the one hand, there is no argument which proves that verification by experimentation is the only way to know unobserved causes. On the other hand, as explained above using the example of quantum entanglement, verification itself requires deductive reasoning (in addition to inductive reasoning). As have been explained above, that the laws of logic on which deductive reasoning is based are necessarily true. This implies that the conclusion of a deductively valid argument from true premises must be true, and I shall show in the rest of this book that such an argument (viz. KCA) can be formulated to show that a Divine First Cause exist.

It should be noted that, unlike the Ontological Argument, which is an *a priori* argument that starts by defining God, the KCA does not start by defining God, and neither is it dependent on the Ontological Argument. Rather, as explained in the rest of this book, the KCA starts with the observation that a series of causes and events exist in the world and then demonstrates that an infinite regress of causes and events is impossible, before reasoning deductively to an independently existing First Cause with the capacities of libertarian freedom. The KCA is thus an *a posteriori* argument, and it uses deductive reasoning, which science itself requires. While the laws of logic cannot tell me that there is a square on my table and I need observational evidence to know that there is a square, by using the laws of logic I can know that the square on my table cannot be a square and shapeless at the same time. Likewise, while the laws of logic cannot tell me that there is a series of causes and effects and I need observational evidence to know that, by using the laws of logic I can know that the series of causes leading to me cannot be finite and infinite. Thus, the series either has a first member or it does not, and given the arguments against an actual infinite regress (see Chap. 5), it can be deduced that the series has a first member, that is, a first cause. Likewise, by using the laws of logic I can know that this first cause cannot be caused and uncaused. Thus, it is either caused or uncaused, and given that it is the first, it can be deduced that it is uncaused and that it has caused an effect (i.e. it has started a series of causes and effects resulting in my existence). The rest of the properties of the First Cause, that is, beginningless, timeless, has libertarian freedom, enormously powerful, and so on, can likewise be deduced similarly, as will be explained in Chap. 6.

A Kantian might object that 'this is just your way of thinking, you are thinking that that there is an actual object called a First Cause which corresponds with your idea'. However, it has been explained previously that the laws of logic correspond to reality, and that the conclusion of a sound argument (i.e. a deductively valid argument from true premises) must be true, it is not just a way of thinking or perspective. My great grandfather (a prior cause of my existence) is not just an idea; even though I have never seen him, it can be inferred that he really existed, for otherwise I would not have existed. Likewise, it will be argued in the rest of this book that it can be inferred that the First Cause really existed, for otherwise I would not have existed.

## 1.8 Conclusion and Overview of Following Chapters

Contemporary formulations of KCA and TA involve considerations of both philosophy and modern scientific cosmology. Contrary to Hawking, who infamously declared that 'philosophy is dead' (Hawking and Mlodinow 2010, pp. 1–2), cosmologist George Ellis observed that philosophy has an important role to play in cosmology. He noted, with respect to the criteria for a good scientific theory (internal consistency, explanatory power, etc.), that 'these criteria are philosophical in nature in that they themselves cannot be proven to be correct by any experiment.

Rather, their choice is based on past experience combined with philosophical reflection' (Ellis 2007, section 8.1). In view of the importance of philosophical considerations, cosmologists should not merely construct models of the universe without considering the philosophical problems associated with certain models, such as problems concerning the traversing of an actual infinite and the violation of Causal Principle, which have been highlighted by proponents of KCA (see below). Scientists who are well-informed about the importance of philosophy have used philosophical arguments against an actual infinite number of earlier events to argue against cosmological models that postulate this (Ellis et al. 2004, p. 927). This indicates that philosophical arguments are relevant for modern cosmology. This book will contribute to the discussion by developing these arguments in engagement with modern science

Against the New Verificationism proposed by Ladyman et al. (2007, p. 29) which claims that we have no good reasons for thinking that *a priori* metaphysical knowledge is possible (pp. 7, 29), I have shown that mathematics and the laws of logic are both necessary for science and yet also knowable *a priori*. I have also shown that the laws of logic are necessarily true; they would hold even at levels far beyond our daily experiences, such as at the beginning of time. I explain that, contrary to popular misconceptions, quantum physics, Russell's paradox, Gödel Incompleteness Theorems, and non-classical logics do not entail the violation of the laws of logic. As for the concern that temporality and causality may 'break down' at the beginning of the universe (Drees 2016, p. 199), following the laws of logic, the 'breaking down' of these would imply being timeless and uncaused. It will be shown in subsequent chapters that, using the laws of logic, various arguments lead to the conclusion that there is an uncaused and initially timeless First Cause of the universe. Against the worry that how we think about the world may be very different from what the world itself really is, I have argued that the laws of logic correspond with what the world itself really is, and we can therefore use them to formulate various arguments concerning the world. I replied to various empiricist and Kantian objections and note that the necessity of the laws of logic implies that the conclusion of a deductively valid argument from true premises must be true.

It has been noted previously that the KCA is traditionally formulated as follows:

(1) Whatever begins to exist has a cause.
(2) The universe began to exist.
(3) Therefore, the universe has a cause. (Craig and Sinclair 2009)

Craig argues that further analyses of the Cause of the universe show that this Cause possesses various theistic properties.

To make the deduction of the theistic properties explicit, I shall reformulate the KCA and combine it with the TA as follows (KCA-TA):

(1) There exists a series of causes and effects and changes (= events).
(2) The series either has an infinite regress that avoids a First Cause and a first change, or its members are joined together like a closed loop that avoids a First Cause and a first change, or its members are not so joined together and the series has a First Cause and a first change.
(3) It is not the case that the series has an infinite regress.
(4) It is not the case that its members are joined together like a closed loop.
(5) Therefore, the series has a First Cause and a first change (from 1 to 4).
(6) Since the First Cause is the first, it is uncaused.
(7) Since whatever begins to exist has a cause (Causal Principle), the First Cause is beginningless.
(8) Since every change is an event which has a beginning as something/part of a thing gains or loses a property, and since the first change (= first event) does not begin uncaused (given the Causal Principle), the first change (= first event) is caused by a First Cause which is initially changeless (from 5 and 7; here, 'initial' refers to the first in the series of states ordered causally, not first the series of changes/events/temporal series).
(9) Since the First Cause is initially changeless, it is transcendent and immaterial (i.e. it is distinct from the material universe and is the cause of the universe).
(10) In order to cause an event (Big Bang or whatever) from an initial changeless state, the First Cause must have

- the capacity to be the originator of the event in a way that is undetermined by prior event, since the First Cause is the first, and
- the capacity to prevent itself from changing, for otherwise the First Cause would not have been initially changeless and existing beginninglessly without the event/change.
- 10.1 and 10.2 imply that the First Cause has libertarian freedom.

(2) In order to bring about the entire universe, the First Cause is enormously powerful.
(3) (+ the Teleological Argument) In order to bring about a universe with its fine-tuning and order, the First Cause is highly intelligent.
(4) A First Cause that is uncaused, beginningless, initially changeless, transcendent, immaterial, has libertarian freedom, and is highly intelligent and enormously powerful is a Creator of the Universe.
(5) Therefore, a Creator of the universe exists.

The above argument is deductively valid; the key question is whether the premises are true. I shall defend premises 3 and 4 in Chap. 5, premise 7 in Chaps. 2 and 3, premises 8–11 in Chap. 6, and premise 12 in Chaps. 4 and 7. Here is an overview of the following chapters.

In Chaps. 2 and 3, I explain the notions of causality and the laws of nature which are fundamental for KCA-TA, defend the Causal Principle (premise 7 of KCA-TA) against various objections, and develop a Modus Tollens argument which shows that the Causal Principle is true.

In Chap. 4, I explain another notion which is fundamental for KCA-TA, namely, 'design'. I note that various properties of the universe have been suggested as indicative of the work of a designer. In this book, I focus on two such properties: 'fine-tuning' and 'order'. (The word 'order' refers to the arrangement of things in relation to each other [*Oxford English Dictionary*], and in the scientific literature it can be used in various ways such as 'low entropy', 'non-chaotic', or 'governed by laws'. I use the term to refer to patterns of events which can be described by advanced mathematics and which are characterized as 'laws of nature'; see Chap. 2.) I defend these two notions against various objections, and note that, while various forms of design inference have been suggested, the problem of unconsidered alternative explanations besets all of them. I address this concern by first devising an original deductive argument which demonstrates that the following are the only possible categories of hypotheses concerning 'fine-tuning and order': (i) Chance, (ii) Regularity, (iii) Combinations of Regularity and Chance, (iv) Uncaused, and (v) Design. I go on and demonstrate that there is an essential feature of (i), (ii), and (iii) which renders them unlikely as an explanation for the fine-tuning and order of the universe.

For categories (iv) and (v), I shall evaluate them by showing that an actual infinite regress of causes and events is not possible. I undertake this task in Chap. 5 and evaluate various cosmological models which postulate an actual infinite number of prior events. I note that these cosmologies face various scientific and philosophical problems. On the other hand, there are at least five arguments which demonstrate that an actual infinite regress of causes and events is not the case. I summarize some of these arguments and explain why there are good reasons for thinking that they are sound. I also explain that a closed causal loop involves a viciously circular setup which would not work, and it is contradicted by the Generalized Second Law of Thermodynamics (Wall 2013a, 2013b). Given the refutations of a closed loop (premise 4 of KCA-TA) and an actual infinite regress of causes and events (premise 3), there is a First Cause and a first event (premise 5).

In Chap. 6, I explain and defend premises 6–11 of KCA-TA which show that the First Cause is not part of the physical universe as postulated by Hawking's no-boundary proposal (which in any case has been shown to be scientifically flawed by other cosmologists). Rather, premises 6–11 show that the First Cause is uncaused, beginningless, initially changeless, has libertarian freedom, and is enormously powerful, that is, a transcendent immaterial Creator of the Universe. The conclusion that the First Cause is a Creator who brought about the first event purposefully rather than accidentally can be further strengthened by considering the evidences of fine-tuning and order of the universe.

In Chap. 7, I complete my comparison of categories (iv) Uncaused and (v) Design concerning the fine-tuning and order of the universe. I offer three considerations against (iv) and reply to various objections against the likelihood of Design, and conclude that, while the alternatives to design are unlikely, the Design hypothesis is not. I explain how my argument from exclusion avoids the problems which beset other design inferences, such as the difficulty of assigning prior probability for Design.

In the concluding chapter (Chap. 8), I summarize the conclusions and contributions of my book and explain how science, philosophy, and religion can continue to work together in our understanding of the Ultimate Designer.

## Notes

1. In this book, 'ruled out' does not require 'perfect' elimination understood as demonstrating that other possible hypotheses have zero probability. It only requires showing that their probability is so low that they can be eliminated as reasonable alternatives to Design even if we assign them very generous probability estimates (see Sect. 7.5).
2. I thank Chan Man Ho for clarification of this point.
3. I thank Wesley Wildman for this point.
4. Concerning quantum superposition, see below.
5. While Huemer gives no model to serve as proof of consistency, this does not invalidate his argument, which is simply intended to show that his 'solution to the liar paradox holds that the liar sentence fails to express a proposition due to an inconsistency built into our language' (p. 29).
6. I thank Jonathan Chan for raising this objection.

## Bibliography

Adamson, P. 2007. *Al-Kindī*. New York: Oxford University Press.

Almeida, Michael. 2018. *Cosmological Arguments*. Cambridge: Cambridge University Press.

Bradley, Darren. 2018. Carnap's Epistemological Critique of Metaphysics. *Synthese* 195, 2247–2265.

Bussey, Peter. 2013. God as First Cause – A Review of the Kalām Argument. *Science & Christian Belief* 25: 17–35.

Capra, Fritjof. 2010. *The Tao of Physics*. Boston, MA: Shambhala.

Carroll, Sean. 2010. *From Eternity to Here: The Quest for the Ultimate Theory of Time*. New York: Dutton.

Craig, William Lane. 1979. Kant's First Antinomy and the Beginning of the Universe. *Zeitschrift für Philosophische Forschung* 33: 553–567.

———. 1980. *The Cosmological Argument from Plato to Leibniz*. London: Macmillan.

Craig, William Lane, and James Sinclair. 2009. The Kalām Cosmological Argument. In *The Blackwell Companion to Natural Theology*, ed. William Lane Craig and J.P. Moreland. Chichester: Wiley-Blackwell.

Davidson, H.A. 1969. John Philoponus as a Source of Medieval Islamic and Jewish Proofs of Creation. *Journal of the American Oriental Society* 89 (2): 357–391.

Drees, Willem. 2016. The Divine as Ground of Existence and of Transcendental Values: An Exploration. In *Alternative Concepts of God: Essays on the Metaphysics of the Divine*, ed. Andrei Buckareff and Yujin Nagasawa. Oxford: Oxford University Press.

Duhem, P. 1954. *The Aim and Structure of Physical Theory*, Trans. P. Wiener. Princeton: Princeton University Press.

Ellis, George. 2007. Issues in the Philosophy of Cosmology. In *Philosophy of Physics*, ed. J. Butterfield and J. Earman. Amsterdam: Elsevier.

Ellis, George, U. Kirchner, and W. Stoeger. 2004. Multiverses and Physical Cosmology. *Monthly Notices of the Royal Astronomical Society* 347: 921–936.

Evans, C. Stephen. 2010. *Natural Signs and Knowledge of God: A New Look at Theistic Arguments*. Oxford: Oxford University Press.

Feser, Edward. 2017. *Five Proofs of the Existence of God*. San Francisco: Ignatius Press.

Friederich, Simon. 2018. Fine-Tuning. In *The Stanford Encyclopedia of Philosophy*, ed. Edward Zalta. https://plato.stanford.edu/archives/spr2018/entries/fine-tuning/.

Grünbaum, Adolf. 2004. The Poverty of Theistic Cosmology. *The British Journal for the Philosophy of Science* 55: 561–614.

Halvorson, Hans. 2018. A Theological Critique of the Fine-Tuning Argument. In *Knowledge, Belief, and God*, ed. Matthew Benton, John Hawthorne, and Dani Rabinowitz. Oxford: Oxford University Press.

Hartle, James, and Stephen Hawking. 1983. Wave Function of the Universe. *Physical Review D* 28: 2960–2975.

Hawking, Stephen. 1988. *A Brief History of Time*. London: Bantam.

Hawking, Stephen, and Leonard Mlodinow. 2010. *The Grand Design*. New York: Bantam Books.

Helen De Cruz, and Johan De Smedt. 2016. Naturalizing Natural Theology. *Religion, Brain & Behavior* 6 (4): 355–361.

Huemer, Michael. 2018. *Paradox Lost*. Cham: Springer.
Kant, Immanuel. 1965. *Critique of Pure Reason*, Trans. Norman Kemp-Smith. London: Macmillan.
Kojonen, Rope. 2021. *The Compatibility of Evolution and Design*. Palgrave Frontiers in Philosophy of Religion Series. Cham: Springer Nature.
Krauss, Lawrence. 2012. *A Universe from Nothing*. New York: Free Press.
Ladyman, James, Don Ross, David Spurrett, and John Collier. 2007. *Every Thing Must Go: Metaphysics Naturalized*. Oxford: Oxford University Press.
Laudan, Larry. 1990. Demystifying Underdetermination. In *Scientific Theories*, ed. C. Wade Savage. Minneapolis: University of Minnesota Press.
Lewis, Geraint, and Luke Barnes. 2016. *A Fortunate Universe: Life in a Finely Tuned Cosmos*. Cambridge: Cambridge University Press.
Linford, Dan. 2020. The *Kalām* Cosmological Argument Meets the Mentaculus. *The British Journal for the Philosophy of Science*. https://doi.org/10.1093/bjps/axaa005.
Loke, Andrew. 2012a. Is an Uncaused Beginning of the Universe Possible? A Response to Recent Naturalistic Metaphysical Theorizing. *Philosophia Christi* 14: 373–393.
———. 2012b. Is an Infinite Temporal Regress of Events Possible? *Think* 11: 105–122.
———. 2014a. A Modified Philosophical Argument for a Beginning of the Universe. *Think* 13: 71–83.
———. 2014b. No Heartbreak at Hilbert's Hotel: A Reply to Landon Hedrick. *Religious Studies* 50: 47–50.
———. 2014c. The Benefits of Studying Philosophy for Science Education. *Journal of the NUS Teaching Academy* 4: 27–35.
———. 2014d. *A Kryptic Model of the Incarnation*. London: Routledge.
———. 2016. On Finitism and the Beginning of the Universe: A Reply to Stephen Puryear. *Australasian Journal of Philosophy* 94: 591–595.
———. 2017. *God and Ultimate Origins: A Novel Cosmological Argument. Palgrave Frontiers in Philosophy of Religion Series*. Cham, Switzerland: Springer Nature.
———. 2022. *The Origin of Humanity and Evolution: Science and Scripture in Conversation*. London: T & T Clark.
McGrath, Alister. 2018. *The Territories of Human Reason: Science and Theology in an Age of Multiple Rationalities*. Oxford: Oxford University Press.
Moreland, J.P., and William Lane Craig. 2003. *Philosophical Foundations for a Christian Worldview*. Downers Grove: InterVarsity Press.

Oppy, Graham. 2006. *Philosophical Perspectives on Infinity*. Cambridge: Cambridge University Press.

Oppy, Graham. 2010. Uncaused Beginnings. *Faith and Philosophy* 27: 61–71.

———. 2015. Uncaused Beginnings Revisited. *Faith and Philosophy* 32: 205–210.

Penrose, Roger, and William Lane Craig. 2019. The Universe: How Did It Get Here & Why Are We Part of It? [Video File]. https://www.youtube.com/watch?v=9wLtCqm72-Y. Accessed 22 April 2020.

Plantinga, Alvin. 2011. *Where the Conflict Really Lies: Science, Religion, and Naturalism*. Oxford: Oxford University Press.

Polkinghorne, John. 2011. *Science and Religion in Quest of Truth*. New Haven: Yale University Press.

Pratt, Bill. 2012. Does Quantum Mechanics Invalidate the Law of Non-Contradiction. Part 2. https://www.toughquestionsanswered.org/2012/01/04/does-quantum-mechanics-invalidate-the-law-of-non-contradiction-part-2/.

Priest, Graham, Francesco Berto, and Zach Weber. 2018. Dialetheism. *The Stanford Encyclopedia of Philosophy* (Fall 2018 Edition), ed. Edward N. Zalta. https://plato.stanford.edu/archives/fall2018/entries/dialetheism/.

Puryear, Stephen. 2014. Finitism and the Beginning of the Universe. *Australasian Journal of Philosophy* 92: 619–629.

Quine, W.V.O. 1951. Two Dogmas of Empiricism. In *Reprinted in From a Logical Point of View*, 2nd ed., 20–46. Cambridge, MA: Harvard University Press.

Rasmussen, Joshua. 2018. Review of God and Ultimate Origins. *European Journal for Philosophy of Religion* 10: 189–194.

Rasmussen, Joshua, and Felipe Leon. 2018. *Is God the Best Explanation of Things: A Dialogue*. Cham: Springer Nature.

Ratzsch, Del, and Jeffrey Koperski. 2019. Teleological Arguments for God's Existence. In *The Stanford Encyclopedia of Philosophy*, ed. Edward N. Zalta. https://plato.stanford.edu/archives/sum2019/entries/teleological-arguments/.

Rovelli, C. 2018. Creath, Richard, "Logical Empiricism", The Stanford Encyclopedia of Philosophy (Summer 2017 Edition), Edward N. Zalta (ed.), https://plato.stanford.edu/archives/sum2017/entries/logical-empiricism/.

Smith, Quentin. 2007. Kalām Cosmological Arguments for Atheism. In *The Cambridge Companion to Atheism*, ed. Michael Martin. Cambridge: Cambridge University Press.

Sober, Elliott. 2019. *The Design Argument*. Cambridge: Cambridge University Press.

Sober, Elliot. 2003. The Argument from Design. In N. Manson (ed.), *God and Design*. Routledge, pp. 25–53. Reprinted in W. Mann (ed.), *The Blackwell Guide to Philosophy of Religion*, 2004, pp. 117–147.
Stenmark, Mikael. 2003. Scientism. In *Encyclopedia of Science and Religion*, ed. J. Wentzel Vrede van Huyssteen, 2nd ed. New York: Macmillan Reference.
Sudduth, Michael. 2009. *The Reformed Objection to Natural Theology*. London: Routledge.
Swinburne, Richard. 2004. *The Existence of God*. 2nd ed. Oxford: Clarendon.
———. 2005. The Value and Christian Roots of Analytical Philosophy of Religion. In *Faith and Philosophical Analysis*, ed. H. Harris and C. Insole. Aldershot: Ashgate.
Trigg, Roger. 1993. *Rationality and Science: Can Science Explain Everything?* Oxford: Blackwell.
Wall, Aron. 2013a. The Generalized Second Law Implies a Quantum Singularity Theorem. *Classical and Quantum Gravity* 30 (16): 165003. Preprint: arXiv:1010.5513v4 [gr-qc].
———. 2013b. Corrigendum: The Generalized Second Law Implies a Quantum Singularity Theorem. *Classical and Quantum Gravity* 30 (19): 199501.

# 2

# Causation and Laws of Nature

## 2.1 Introduction

In this chapter and the next, I shall explain the notions of causality and the laws of nature which are fundamental for KCA-TA (Kalām Cosmological Argument-Teleological Argument), and defend the Causal Principle 'whatever begins to exist has a cause'. The defence of the Causal Principle is very important for philosophy of religion debates and science and religion dialogues, as it provides the basis for a response to Hawking's claim that

> You can't get to a time before the Big Bang because there was no time before the Big Bang. We have finally found something that doesn't have a cause, because there was no time for a cause to exist in. For me this means that there is no possibility of a creator, because there is no time for a creator to have existed in. (Hawking 2018, p. 38)

I shall respond to Hawking's claim in Chap. 6 after establishing the Causal Principle in Chaps. 2 and 3.



A. Loke, *The Teleological and Kalam Cosmological Arguments Revisited*, Palgrave Frontiers in Philosophy of Religion, https://doi.org/10.1007/978-3-030-94403-2_2

In his *Inquiry and Essays*, the eighteenth-century philosopher Thomas Reid (1710–1796) declared 'that neither existence, nor any mode of existence, can begin without an efficient cause, is a principle that appears very early in the mind of man; and it is so universal, and so firmly rooted in human nature, that the most determined scepticism cannot eradicate it' (Reid 1983, p. 330). His contemporary and well-known sceptic David Hume had apparently raised an objection by claiming that the ideas of cause and effect are distinct and we can conceive of an uncaused beginning-to-be of an object (Hume 1739/1978, p. 79). However, Hume confessed in a letter written in 1754 that 'I never asserted so absurd a Proposition as that anything might arise without a cause: I only maintain'd that, our Certainty of the Falsehood of that Proposition proceeded neither from Intuition nor Demonstration; but from another Source' (Hume 1932, i., p. 187). Thus, it seems that Hume himself would agree that the mere conceivability of something beginning to exist uncaused does not provide sufficient grounds for rejecting the principle stated by Reid (Anscombe 1974). Others however have argued that (in the absence of arguments to the contrary) conceivability does entail possibility, and philosophers influenced by Hume have raised doubts about Reid's principle. For example, in the Preface to the Second Edition of his *Critique of Pure Reason*, Immanuel Kant argued that, while the principle of causality is valid for objects as phenomena, it may not be valid for objects as things in themselves (the noumenal world). In more recent years, some scientists and philosophers have claimed that quantum physics indicate that uncaused events happen all the time (Grünbaum 2009, p. 15). It has also been argued that, even if things do not begin to exist uncaused within our universe, it might be the case that our universe itself begun to exist uncaused (Oppy 2010, 2015; Almeida 2018).

The debate is fascinating and of importance to metaphysics, philosophy of science, philosophy of religion, and science and religion dialogues. In this book, instead of defending the stronger claim that 'neither existence, nor any mode of existence, *can* begin without an efficient cause' (Reid), I shall defend the weaker claim that 'neither existence, nor any mode of existence, begins without a cause', that is, 'whatever begins to exist has a cause' (here, the word 'cause' refers to either an efficient cause or a material cause; I shall explain this point below). For convenience of exposition, I shall henceforth refer to this weaker claim as *the* Causal Principle. I shall first define the key terms of the Causal Principle in the next section, and then respond to some objections to the Causal Principle. In the next chapter, I shall defend an argument in support of the Causal Principle. I shall show that the Causal Principle remains defensible not only on the dynamic (A-) theory of time but also on the static (B-) theory of time (which is widely accepted by cosmologists).

## 2.2 Defining the Key Terms of the Causal Principle

I shall begin by discussing the definitions of the key terms of the Causal Principle 'whatever begins to exist has a cause' and the related terms 'time', 'eternal', 'event', 'change', 'perdurantism', and 'uncaused'.

'Whatever' refers to all that exists (regardless of whether they are things, events, substances, states of affairs, arrangements, etc.). Some have objected to the Causal Principle by claiming that everything came from pre-existent materials (e.g. my body came from pre-existent molecules) and therefore there isn't anything which begins to exist. Those who affirm *creatio ex nihilo* (according to which God is the efficient cause who brought about the universe without material cause) would dispute the claim that everything came from pre-existent materials, but in any case the objection is based on a misunderstanding, since 'whatever' refers to events and arrangements as well. (Thus, for example, even though my body came from pre-existent molecules, there was a beginning to the event at which the molecules constituted the first cell of my body resulting in a new arrangement of the molecules. The event and new arrangement were *caused* by the fertilization of my mother's egg by my father's sperm.) Therefore, the Causal Principle does not require the demonstration of *creatio ex nihilo* (nor does it deny *creatio ex nihilo*; see below). Rather, the Causal Principle is claiming that, regardless of whether something begins from pre-existing materials or not, it has a cause.

'Begins to exist': something has a beginning if it has a temporal extension, the extension is finite,[1] and it has temporal edges/boundaries, that is, it does not have a static closed loop (see Chap. 5) or a changeless/timeless phase (see Chap. 6) that avoids an edge. Consider, for example, Oppy's defence of the claim (against Craig) that it is possible for the initial state of reality to *come into existence* uncaused *out of nothing* (Oppy 2015, section 4, italics mine). The terms in italics indicate a temporal boundary, that is, a beginning. Whereas on Craig's theistic hypothesis, God (the First Cause) does not come into existence uncaused out of nothing; rather, God is timeless sans creation and in time with creation (Craig and Sinclair 2009, p. 189). On this view, God's existence has a timeless phase which avoids a boundary and is therefore beginningless.

In relation to the definition of 'beginning', there are different views of time which need to be distinguished. A relational view of time defines time as an extended series of changes/events ordered by 'earlier than' and 'later than' relations, whereas a substantival view of time affirms that time can exist as an extended substance independently of change.

According to the dynamic (A-) theory of time, the members of a series of changes/events come to be one after another. Whereas on the static (B-) theory of time, our spacetime is a four-dimensional block and the series of events is a tenselessly existing manifold all of whose members are equally real and the 'flow' of time is regarded as illusory. By defining 'beginning to exist' in terms of 'temporal extension' and 'boundary', I am using a definition that is compatible with both static and dynamic theories of time.

Against some philosophers who have doubted the existence of time altogether (Pelczar 2015), Simon (2015) notes that 'it would suffice if we could know via a combination of introspection and memory that our experience changes. But this is commonplace: I remember that I was experiencing a sunrise, and I introspect that I no longer am.' Moreover, 'it would suffice if we could conclude that experiences *take time* … in the words of Ray Cummings (1922), 'time is what keeps everything from happening all at once' (ibid.). Thus, the fact that I do not hear all the notes of a Beethoven symphony all at once is evidence that events do not happen all at once; rather, there is a sequence. It has sometimes been claimed that a massless particle travelling at the speed of light is 'timeless'. However, what this means is that according to Special Relativity, something travelling at the speed of light would not 'experience' time passing.

One needs to note the distinction between experience and reality. Even though a massless particle travelling at the speed of light does not 'experience' time passing, in reality it still has a beginning in time at its point of origin from where the particle is emitted. (For the discussion on timelessness, see further, Chap. 6.)

There are also different uses of the word 'eternal' which need to be distinguished. 'Eternal' can mean (1) having no beginning and no end; however, 'eternal' has also been used in the literature to refer to (2) something that does not come into being or go out of being. On the static theory of time, the universe can have a beginning (in the sense explained above) and thus is not eternal in the first sense, and yet does not come into being or go out of being, and thus is eternal in the second sense. In line with the latter usage, 'eternalism' is used in the literature to refer to the view that our spacetime is a four-dimensional block and the series of events is a tenselessly existing manifold all of whose members are equally real. However, one must be careful to note that this does not imply that the universe has no beginning. (Moreover, 'eternal' has also been used to refer to (3) something that has no end but has a beginning; for example, Vilenkin affirms 'eternal inflation' and yet he argues that the universe has a beginning; see Chap. 4.)

An event is understood as a change. The existence of changes is undeniable. It is true that according to the B-theory of time, the 'moving present' (often called the 'flow of time') which we experience in our consciousness is regarded as illusory. (Because of this, the static theory of time is sometimes misleadingly regarded as timelessness or changelessness. The key issue concerns the definition of time and change; see below.) Nevertheless, no time-theorist (whether A- or B-theorist) would deny (for example) that he/she has undergone numerous changes since he/she was conceived (e.g. he/she has grown taller, heavier, etc.). Nathan Oaklander (2004, p. 39) observes, 'The rock-bottom feature of time that must be accepted on all sides is that there is change, and the different views concerning the nature of change constitute the difference between A- and B- theories of time.'

A change is understood here as involving a thing or part of a thing[2] gaining or losing one or more properties. On a dynamic (A) theory of time, the gaining/losing of properties involves a coming to be/passing away of properties. On a static (B) theory of time, the gaining/losing of properties does not involve a coming to be/passing away of properties; rather, it involves having different temporal parts at different times (perdurantism). The different parts have boundaries and hence beginnings (see the definition of 'beginning' above).

Thus, it is true that on a static theory, a four-dimensional block is 'unchanging' if this is understood as saying that there is no coming to be/passing away of properties, and that there is no 'earlier' event if this is understood as saying that there is no event that passes away before others. However, as Oaklander observes, there are still changes in the sense that the four-dimensional block has different temporal parts with different properties at different times. Moreover, some parts (e.g. those temporal parts in which there is water on earth) are posterior to ('later' in this sense) and dependent on prior ('earlier' in this sense) temporal parts (e.g. those temporal parts in which there is formation of hydrogen near the beginning of the Big Bang; scientists would say that the formation of water is dependent on the prior existence of hydrogen). In this sense later events are dependent on earlier events, and this remains true on the block theory. On a static theory of time, every event in the 'block' exists and is equally real, but nevertheless 'later' events are still dependent on 'earlier' events. Indeed, any theory of time which denies such a basic scientific fact as the formation of water in our universe is dependent on the prior existence of hydrogen would have to be rejected, and no B-theorist of time would deny that. (The dependence can be characterized using counterfactuals as follows: 'if there were no hydrogen formed earlier, there would not be water formed later'; I shall argue below that this dependence is causal.)

It might be objected that, while it makes sense to talk about things 'beginning to exist' within the spacetime block on B-theory of time, it makes no sense whatsoever to talk about the block itself beginning to exist.[3] But this is not true; if the spacetime block is finite in temporal extension etc. ('etc.' refers to 'does not have a static closed loop or a changeless phase that avoids an edge'), then that implies that the spacetime block has a beginning—the same sense of 'beginning to exist' is used. While the spacetime block does not 'come to be' on this B-theory view,[4] it still has a beginning in the sense of being finite in temporal extension etc., just as every part of it has a beginning in the sense of being finite in temporal extension etc. Craig and Sinclair (2009, p. 183) note that 'For B-Theorists deny that in beginning to exist the universe came into being or became actual'. Note that the concept of 'beginning to exist' is not absent in B-theory; indeed, scientists who are B-theorists (e.g. Carroll 2014) frequently speak about the beginning of universe. On B-theory 'beginning to exist' is not understood as 'came into being or became actual', but it is defined as 'exists tenselessly as a four-dimensional space-time block that is finitely extended' (Craig and Sinclair 2009, p. 184). The claim that 'the block does not exist in time thus to talk about a beginning is meaningless' is therefore false; regardless of whether the block exists in time or not, if it is finitely extended etc. then it has a beginning according to the static theory's definition of beginning. A block by definition has extension and an extension can be finite etc. One can say that the part of the spacetime block in which (say) Einstein exists is finite in the sense that it did not consist of an actual infinite moments but is finite etc. That is what it means to say that the block itself has a beginning.

One might object that there is a difference between the part of the block in which Einstein exists and the whole block itself, namely, the whole block itself does not exist in another time block whereas Einstein would exist in the time block. Nevertheless, I shall argue in Chap. 3 that, if the whole block has a beginning, it would have a cause just as the part of the block in which Einstein exists has a cause, the only difference is that, if the cause of the block is initially timeless (see Chap. 6), then it is not earlier than the block whereas the causes of Einstein (e.g. his parents) are earlier than Einstein. Both would still have causes, however.

One might ask how can the block have a cause if (according to static theory) it does not come into being or become actual, even though it has a beginning. In reply, the part of the block in which Einstein exists also does not come into being or become actual on the static theory, yet his existence is still causally dependent on his parents' existence in the sense that, if his parents had not existed, Einstein would not begin to exist. Likewise, I shall argue in Chap. 3 that the whole spacetime block has a cause in the sense that, if the cause does not exist, the spacetime block would not begin to exist.

Aristotle (*Physics* 2.3) famously identified four kinds of causes: efficient cause (the source of change, for example, the sculptor's act of bronze-casting the statue), material cause ('that out of which a thing comes to be and which persists', for example, the bronze of the statue), formal cause ('the form or the archetype', for example, the structure of the statue), and final cause ('in the sense of end [telos] or that for the sake of which', for example, the sculptor sculpting the statue for aesthetic purposes) (Mackie 2005). In this book, unless otherwise stated, 'cause' refers to either an efficient cause or a material cause, and which is either necessary or sufficient[5] for an effect,[6] understood as a change.[7] Weaver (2019, p. 261) notes that causation is multigrade, asymmetric (although not always temporally asymmetric), transitive, irreflexive and a dependence relation: 'when event *x* causes event *y*, *y* depends for its existence and contingent content on *x*.'[8]

Finally, there are two different senses of the phrase 'begins uncaused' which are often used in the literature and which should be distinguished:

(1) For any *x*, if *x* begins uncaused, then the beginning of *x* does not have a causally necessary condition understood as either an efficient cause or material cause. That is, either

(1.1) *x* begins without any causally necessary condition at all, or

(1.2) *x* begins without something that is known to be a causally necessary condition (under certain circumstances) for the beginning of *x*. For example, in the reality that we now inhabit, what is causally necessary for an increase in strength of a pre-existent electric field under certain circumstances would include (for example) the switching on of an electric field generator. If events such as the increase in strength of pre-existent electric fields happen without the switching on of electric field generators under the same circumstances, they would be regarded as uncaused and would entail a chaotic world e.g. I would suffer from electric shock even though nothing is switched on (see Chap. 3).

(2) Indeterministic events, such as (as many physicists would affirm) quantum events and (as many libertarians would affirm) a genuinely free act. It is controversial whether humans have libertarian freedom and whether quantum events are genuinely indeterministic. In any case, it should be noted that a libertarian free act does not imply that there is no causally necessary condition for the making of it; the pre-existence of the agent, for example, would be a causally necessary condition. Likewise, the

pre-existence of quantum field (for example) would be a causally necessary condition for quantum fluctuation while the pre-existence of atomic nuclei and the so-called weak nuclear force would be causally necessary conditions for beta-decay, in the absence of which the beta-decay would not occur (Bussey 2013, p. 20). The difference between supposed quantum indeterminism and (say) the supposed uncaused increase in strength of a pre-existent electric field in (1) above is that the former lacks a causally sufficient condition whereas the latter lacks a causally necessary condition.

In this book, unless otherwise specified, 'uncaused' is understood in the first sense, which is consistent with a key motivation for the Causal Principle, namely, *Ex Nihilo Nihil Fit* ('from nothing, nothing comes'). A genuinely free act would not be 'from nothing'; rather, it is from the agent (see further, Chaps. 3 and 6).

The conviction that 'from nothing, nothing comes' led Aristotle to insist that every state of the world must have come from a previous state of the world and hence the world must be everlasting (Cogliati 2010, p. 7)—this insistence resulted in the denial of the Christian doctrine of *creatio ex nihilo* among many ancient philosophers. However, such an insistence is unwarranted given the distinction between efficient cause and material cause. *Creatio ex nihilo* only denies that the world has a material cause; it does not deny that the world has an efficient cause. On the contrary, '*creatio*' implies that the Creator is the efficient cause who brought about the universe; in this sense, the world is from God and not from nothing.

Aristotle might object that 'from nothing, nothing comes' applies to material cause as well, and insist that 'from no material cause, nothing comes'. He might appeal to our daily experiences, which seem to support the inductive generalization that whatever begins to exist has a material cause. Craig replies that such an inductive generalization can be treated merely as an accidental generalization, 'akin to human beings have always lived on the Earth, which was true until 1968. The univocal concept of "cause" is the concept of something which brings its effects, and whether it involves transformation of already existing materials or creation out of nothing is an incidental question' (Craig and Sinclair 2009, pp. 188–9, 195). On the one hand, there has been no compelling argument offered to show that causes must involve the transformation of already existing materials. On the other hand, God as a causal agent could have causal powers that other entities (e.g. humans) do not have. While humans, for example, require pre-existing materials to work from in order to create (say) a table, God does not require that.[9] Moreover, there are independent arguments for the Causal Principle (see Chap. 3). Note, in particular, that the Modus Tollens argument for this principle explained in Chap. 3 is not dependent on inductive considerations, and because of this additional argument, the Causal Principle enjoys greater support than the principle that 'whatever begins to exist has a material cause', which, in any case, can be regarded as an accidental generalization, as Craig argues. In light of this, the affirmation that there is no physical entity prior to $t = 0$ only implies that the universe was not created out of pre-existent material; it does not imply that there cannot be an efficient cause which has the power to bring about the universe without requiring material cause. To insist otherwise would be to beg the question against *creatio ex nihilo* (see further, Chap. 6).

## 2.3 Causation, Fundamental Physics, and Laws of Nature

Causal eliminativists affirm that there are no obtaining causal relations in the mind-independent world (Weaver 2019, p. 24), while causal reductionists affirm that causation reduces to something else such as a law-governed physical history, where both the laws and physical history are non-causal (Weaver 2019, p. 62).

In favour of causal eliminativism, it might be thought that causes are merely human interpretations which involve concepts and modelling. However, if one takes up a piece of wood and hit one's head, one would realize that, while the application of the concept of cause to the wood may be a human interpretation, the wood does have real power to bring about the event of pain, and the correlation is real. Weaver (2019, p. 90) observes that instances of sensation and sense perception involve obtaining causal relations (the environment impressing itself upon the senses).

Moreover, the formation of beliefs implies that there are obtaining causal relations because formations are causal phenomena. 'When a cognizer forms a thought, they relate to the thought through causation. When a cognizer forms a desire, they cause (perhaps together with other factors) the desire' (p. 93).

While Bertrand Russell (1918) had declared causation to be a scientifically obsolete notion and logical positivists had tried to build philosophical systems without any reference to cause and effect, Koons and Pickavance (2015, p. 8) observe that

> Since then, causation has reclaimed its status as a central notion in philosophical theory. Edmund Gettier, in a famous article in 1963, challenged the traditional definition of knowledge as justified true belief, leading to new theories of knowledge that relied upon some kind of causal connection between states of knowledge and the world. Modern theories of sensory perception and memory, in particular, require reference to appropriate causal mechanisms. Work in the philosophy of language by Keith Donnellan, Saul Kripke, and Gareth Evans, among others, introduced causal theories of the meanings of words and the content of thought. Finally, the philosopher of science Nancy Cartwright demonstrated that causation is far from obsolete in the experimental sciences.

Causal reductionists such as cosmologist Sean Carroll (2014) claim that 'the notion of a "cause" isn't part of an appropriate vocabulary to use for discussing fundamental physics. Rather, modern physical models take the form of unbreakable patterns—laws of Nature—that persist without any external causes.' Carroll thinks that our construction of causal explanations for objects within the totality of physical reality is due to the fact that the objects obey the laws of physics, and that there is a low-entropy boundary condition in the past.[10] However, there is no physical law and no low-entropy boundary condition that apply to the totality of physical reality itself; hence, we have no 'right to demand some kind of external cause' (Carroll and Craig 2016, pp. 67–8).

In reply, it should be asked why the 'patterns' Carroll refers to are 'unbreakable'. While Carroll appeals to the so-called laws of nature, one should ask why the events described by fundamental physics follow those laws.

Now Hume famously stated that the laws of nature are simply regularities of events; there is no relationship of necessity between these events, nor are laws conceived of as something that govern the regularities. Hume also claims that 'we may define a cause to be an object, followed by another, and where all the objects similar to the first, are followed by objects similar to the second', and that 'all events seem entirely loose and separate' (*An Enquiry concerning Human Understanding* 1748, section VII). Following Hume, Regularity Theorists of Causation have analysed causation as regular patterns of succession and have regarded these regularities as 'brute facts' rather than as something in need of an explanation. Against this, others have argued that the question 'Why is the world regular (in the particular way that it is)?' needs to be answered by a deeper explanation, for otherwise the regularity of event *P* followed by event *Q* (rather than, say, event *R*, or *S*, or *T*, etc.) is just due to chance, which is highly improbable (Strawson 1989, pp. 205–6). I shall argue below that the deeper explanation is provided by the properties of the things which are involved in these regular patterns, and these properties can be called 'causal properties'.[11]

Regularity Theorists might object that the question 'What explains the regularity?' is merely pushed back on Strawson's strategy. For example, if the deeper explanation offered is 'Because of the nature of matter', they may ask 'what explains the nature of matter (or whatever)?' Since there must after all be some terminus of explanation, why not terminate with the regularities themselves (Psillos 2009, pp. 134–135)?

In reply, I would argue that terminating with regularities does not get rid of the problem of the improbability of one event following another regularly by chance. On the other hand, terminating with an alternative explanation such as 'because of causal properties grounded in the nature of matter', which, one might argue, is determined by a beginningless and uncaused First Cause (see Chap. 6) and therefore not the result of chance, would resolve this problem.

Carroll might insist that in fundamental physics, 'real patterns' described by laws explain causal regularities, but the question is, why the events described by fundamental physics follow those patterns/laws? A pattern/law of nature is not a concrete thing but merely a description of behaviour of concrete events/thing; thus, it is still the properties of those concrete event/things which ground the behaviour/law, and those properties can be called causal properties. As Feser (2013, p. 254) observes, the laws of nature are 'mere abstractions and thus cannot by themselves explain anything. What exist in the natural order are concrete material substances with certain essences, and talk of "laws of nature" is merely shorthand for the patterns of behavior they tend to exhibit given those essences.' Against Maudlin (2007), Dorato and Esfeld (2014) argue that the view that laws are grounded in properties (global properties rather than 'intrinsic' or local properties, in view of quantum entanglement) makes intelligible how laws can 'govern' the behaviour of objects. This is the decisive advantage of dispositionalism over primitivism (the view that laws are primitive).

Carroll might object that the equations of fundamental physics do not seem to specify which events are the causes and which events are the effects. Ladyman et al. (2007, p. 160) claim that 'matter has become increasingly ephemeral in modern physics, losing its connection with the impenetrable stuff that populates the everyday world … the ontology of modern physics seems to be increasingly abstract and mathematical'. Weaver (2019, p. 63) notes that the reason why causal eliminativism has been so prevalent in philosophy of physics 'is connected to a tendency in that sub-discipline to associate the substantial content of physical theories with the mathematical formalisms of those theories … because formalisms do not contain any causal notions … physical theories should not be understood causally'.

Nevertheless, Weaver also observes that many great physicists past and present, including the discoverers of relativity and quantum mechanics, 'adopted causal approaches to physics and conceived of their inquiry as a searching evaluation of the world that should uncover causes' (Weaver 2019, p. 71). The equations of fundamental physics do not specify causality because they do not provide an exhaustive description of reality. Consider the following example which illustrates that mathematical equations do not provide a complete account of the natural world and that an interpretative framework involving causal considerations is required: The quadratic equation $x^2 - 4 = 0$ can have two mathematically consistent results for '*x*': 2 or −2. Both answers are mathematically possible. However, if the question is 'How many people carried the computer home?', the answer cannot be '−2', because in the concrete world it is metaphysically impossible that '−2 people' carry a computer home, regardless of what the mathematical equation shows. The impossibility is metaphysical, not mathematical, and it illustrates that metaphysical issues are more fundamental than mathematics. The conclusion that '2 people' rather than '−2 people' carried the computer home is not derived from mathematical equations, but from causal considerations: '−2 people' lack the causal powers to carry a computer home.

Feser (2017, pp. 45–46) observes that 'since the equations of physics are, by themselves, mere equations, mere abstractions, we know that there must be something more to the world than what they describe. There must be something that makes it the case that the world actually operates in accordance with the equations, rather than some other equations or no equations at all.' In other words, the equations of physics merely provide an incomplete description of regularities without ruling out efficient causation and causal properties which (as explained above) operate at a more fundamental level as the ground of these regularities.

A number of concerns have been raised in the literature regarding the temporal order of events. It has been claimed that the Delayed Choice Quantum Eraser violates the notion that causes cannot be later than their effects. To elaborate on one version of this Eraser, according to the so-called Copenhagen interpretation of quantum mechanics, the photon either behaves as a wave or a particle when it passes through the double slit, and if scientists quickly place a detection device, the device would detect a particle, if not, a wave behaviour would be observed. Since the placement of the detection device happens after the photon passed through the double slit, it seems that the placement of the detection device determined what happened earlier (whether the photon would behave as wave or particle). However, this reasoning assumes the Copenhagen interpretation. According to Bohm's interpretation, the photon is always a particle guided by wave (the particle follows one path, while its associated wave goes through both paths); thus, the placement of the detection device did not determine what happened earlier but merely what happened to the photon at the moment of detection (Bricmont 2017, p. 145).

It has also been claimed that recent experiments in quantum mechanics (a photon prepared in a superposition with regard to its polarization hitting point A before point B on one route while hitting B before A on the other route; these two causal paths [A then B, or B then A] are in superposition) has indicated that, at the fundamental level, temporal order is not fixed (Indefinite Causal Order) (Qureshi-Hurst and Pearson 2020). However, the problem is that such claim assumes the Copenhagen interpretation, which (as explained previously) is unproven. Moreover, as explained previously in Chap. 1, instead of thinking of the superposed state as a photon existing in contradictory states, one can think of it as a quantum of energy spread across the possible states as a wave. Some parts of the wave reach A before B, while other (different) parts of the wave reach B before A; there is no contradiction and no violation of temporal order (it should also be noted that the emission of the photon happens before A or B: a definite temporal order!).

With regard to the so-called backward in time travelling positron in QED, this may be interpreted (in accordance with Paul Dirac's hole theory) as spacetime locations in the Dirac sea (a theoretical model of the vacuum as a sea of particles with negative energy) at which a negatively charged electron comes into being carrying the negative energy imputed to it by the Dirac sea (Greiner and Reinhardt 2009, p. 40), thus there is no violation of temporal order.

In any case, as I explain in response to Linford below, even if backward causation is possible and that it is the case that the future determines the past, given the arguments that the future is finite and that a closed loop is impossible (Chap. 5), the 'last' duration of the future would be the first, and the rest of my argument would still follow. Thus, in any case, the Cosmological Argument I defend is not affected by the above-mentioned concerns regarding the temporal order of events.

Ladyman et al. (2007, p. 160) claim that causation is problematic in the microscopic domain where, for example, 'the singlet state in the Einstein-Podolsky-Rosen (Bohm-EPR) experiment fails to screen off the correlations between the results in the two wings of the apparatus, and thus fails to satisfy the principle of the common cause'. In reply, Bohmian mechanics and the Ghirardi, Rimini and Weber (GRW) mass density theory are able to offer a causal explanation of the correlated outcomes of EPR-type experiments in terms of a non-local common cause (Egg and Esfeld 2014).

It might be objected that, 'from the point of view of microphysics, given an individual event, there is no objective distinction between which events make up that event's past and which its future. Therefore, there is no microphysical distinction between which are its causes and which its effects. Thus, there are no facts about microphysical causation' (Ney 2016, p. 146). Linford (2020) claims that 'efficient causation is a time asymmetric phenomenon' (p. 8)', but 'the direction of time does not appear in our best microphysical theories' (p. 4). He states that 'the distinction between the past and the future made in fundamental physics (if fundamental physics really does distinguish the past from the future) are unlikely to explain the distinction between causes and their effects or any of the other macrophysically observable temporal asymmetries' (n.4). Linford notes that 'the project of explaining all temporal asymmetry—including the asymmetry of efficient causation—in terms of the Mentaculus is ongoing', and if successful, 'efficient causation, qua macrophysical time asymmetry, will be given a reductive explanation in terms of the Mentaculus' (p. 8). Linford explains that the 'Mentaculus' hypothesis (which is part of what he calls the 'Albert–Loewer–Papineau reductive programme', or ALP) consists of the conjunction of three principles:

> First, whatever the fundamental dynamical laws happen to be. Second, the Past Hypothesis, that is, the hypothesis that the universe began in the low entropy macrophysical state … third, the Statistical Postulate, that is, the specification of a uniform probability measure over the portion of phase space consistent with whatever information we happen to have about the physical world. (pp. 7–8)

The implication of this project (if successful) is that

> Even if the coming into being of E requires explanatorily prior, physically necessary conditions C … the explanatorily prior, physically necessary conditions need not fall in any particular temporal direction with respect to E … the explanatorily prior and physically necessary conditions for the universe's 'beginning' can fall in the temporal direction away from the

> beginning … entities do not require explanatorily prior or simultaneous causes for their coming into being. (p. 11)

In reply, first, it does not follow from the fact that microphysics is not able to distinguish between past and future events that there are no facts about microphysical causation. The reason is that it might be the case that microphysics does not provide a complete explanation of microphysical reality, but only a certain aspect of it, and therefore what cannot be discerned from physics does not imply it does not exist.

Second, the underlying assumption of the above arguments is the Humean assumption that the direction of causation is parasitic on temporal direction, but this assumption can be challenged (see further, below and Chap. 3).

Third, an explicitly causal theory of quantum gravity has been proposed (Wall 2013a, b). While the correct framework for a truly quantum theory of gravity is far from settled, the current status of quantum gravity studies suggests that 'any case for the claim "quantum gravitational physics does not need causation" is at best uncertain and incomplete' (Weaver 2019, p. 274).

Fourth, Frisch points out that descriptions in scientific literature support the thesis that 'even at the level of fundamental research in physics, our conception of the world is ineliminably causal' (Frisch 2014, p. 66). He cites as an example a report from the Large Hadron Collider study group of CERN which mentions that

> There are various places in the machine where beams can be 'injected,' that other components allow 'suppression' of dispersion, and that others allow for the 'cleanup' of the beam. Finally, there is the 'beam dump' where the beam can be deposited with the help of 'kickers.' In the detector, when a photon passes through matter, it 'knocks out' electrons from the atoms 'disturbing the structure of the material' and 'creating' loose electrons. (Ibid., citing Pettersson and Lefèvre 1995)

Frisch rightly concludes that, although the word 'cause' is not used in these descriptions, the terms he quoted all describe what Nancy Cartwright would characterize as 'concretely fitted out' instances of 'causings' (Frisch 2014, p. 66). The fundamental particles described by nuclear physics clearly have dispositional properties, that is, tendencies to produce certain effects when they interact in certain ways (Martin 2008, p. 50).

Weaver (2019, p. 124) notes that 'the word interaction in scientific and physical research contexts is a causal term', citing the *Oxford Dictionary of Physics*, which gives the technical definition: An interaction is 'an effect involving a number of bodies, particles, or systems as a result of which some physical or chemical change takes place to one or more of them'. Weaver (2019, p. 234) observes that 'There are four fundamental types of interactions between fundamental entities in our best physical theories, viz., the strong, weak, electromagnetic, and gravitational interactions … No one (so far as I'm aware) in the physics literature denies that all four types of physical phenomena are interactive phenomena.'

Weaver also notes that, if there is causation in the physical base, then 'any attempt to reduce causal direction to the arrow of entropic increase, for example, will fail, for already within microphysical evolutions driving entropic increase are obtaining causal relations and therefore causal direction' (p. 131). Hence, it has not been shown that causal direction reduces to some direction in a non-causally interpreted physics given that what's fundamental in one of our currently best quantum theories should be interpreted causally (p. 143).

One might worry that the view that time-reversal invariant[12] entails that there are naturally possible worlds at which the imagined microdynamical causes are the effects whereas the effects are transmuted into the causes. In reply, Weaver (2019, p. 133) argues concerning the proposition 'every purely contingent event has a causal explanation featuring an obtaining irreflexive causal relation to back it' that a binary relation being necessarily asymmetric does not entail that the relation goes the same way in all possible worlds. It does not rule out the possibility that, if a gluon's activity causes a quark to take on certain properties in our world, the quark's beginning to exemplify those properties is the cause of the gluon's activity in another possible world. In other words, while the relationship between cause and effect is necessarily asymmetric, this does not imply that the kind of thing *x* which is the cause for an effect *y* in this world cannot be an effect *y* of cause *x* in another possible world. 'If at an arbitrary world *w*, the gluon's activity causes a quark to take on certain properties, then (at *w*) it is not the case that the quark's taking on those properties causes the gluon's activity' (ibid.). Additionally, there is a deductive argument for Causal Principle which shows that whatever begins to exist (this would include events at the level of fundamental physics) has a cause (see Chap. 3); therefore, causality is fundamental.

Concerning Norton (2003)'s 'mass on the dome' thought experiment, it does not pose a problem for my argument because the thought experiment (even if successful; this has been challenged by other philosophers) only goes to show that Newtonian mechanics is consistent with uncaused events. It does not show that uncaused events do happen. One can legitimately reply that, on the one hand, Newtonian mechanics is not a complete description of physical world (indeed, given quantum physics and relativity, we know it is not). On the other hand, given my Modus Tollens argument (see Chap. 3), we know that events do not happen without causally necessary condition(s). Additionally, Norton's thought experiment also assumes that time is composed of instants; but as Craig and others have argued, this view should be rejected because it results in paradoxes of motion (see Chap. 5).

Another problem with the Humean view of causation is that contingent relations between events would not support counterfactuals and warrant predictions in science (Mumford 2004, pp. 161–162). Thus, following Kripke (1980), who argues that there are metaphysical necessary truths discovered *a posteriori* (e.g. water is $H_2O$), many contemporary philosophers of science have argued that there are causally necessary connections between causal relata (such as events, substances, or states of affairs). The laws of nature have been regarded by them to be at least partly metaphysically necessary (necessitarian view; see, for example, Ellis 2001; Bird 2007), while other philosophers regard them as metaphysically contingent overall (contingentist view; see, for example, Fine 2002; Lowe 2002). Alternatively, one might deny that the laws of nature obtain with metaphysical necessity but argue that there is nevertheless a particular sense of necessity pertaining to natural laws (natural necessity) (Linnemann 2020, pp. 1–2). Fine (2002), for example, argues that metaphysical necessity is 'the sense of necessity that obtains in virtue of the identity of things' (Fine 2002, p. 254), and that not all natural necessities are metaphysical necessities. For example, 'light has a maximum velocity' is at most naturally necessary but not metaphysically necessary. Likewise, even though it is arguably naturally necessary that mass attracts mass with an inverse square law, this does not seem to render it metaphysically necessary (one would think that an inverse cube law for the attraction between masses is as such metaphysically possible). It might be objected that if an inverse cube law (rather than inverse square law) holds, we would not be dealing with 'mass' but with something else (e.g. 'schmass'). However, on the one hand, it is a natural necessity that there is no schmass, on the other hand, the objector is assuming the existence of schmass as a metaphysical possibility. This goes to underscore Fine's point that not all natural necessities are metaphysical necessities (Linnemann 2020).

Lange (2009, p. 45) contrasts the putative necessity of the laws of nature with other putative species of necessity, such as:

1. (Narrowly) logical necessity (e.g. either all emeralds are green or some emerald is not green)
2. Conceptual necessity (all sisters are female)
3. Mathematical necessity (there is no largest prime number)
4. Metaphysical necessity (water is $H_2O$)
5. Moral necessity (one ought not torture babies to death for fun)
6. Broadly logical necessity (as possessed by a truth in any of these categories)

Lange (2009, pp. xi–xii) notes that, while the laws of nature have traditionally been thought to possess a distinctive species of necessity (dubbed 'natural' necessity) an exception to which is (naturally) impossible, yet many have also regarded the laws of nature to be contingent; unlike the broadly logical truths listed above, the laws of nature could have been different from the way they actually are. Essentialists disagree; they characterize laws as possessing the same strong variety of necessity as broadly logical truths do (Ellis 2001). While one can imagine these laws to be false (e.g. one can imagine a different universe in which gravity does not exist), Bird (2007, p. 207) replies by claiming that

> imagination is a poor guide to the modality of laws, if one supposes that the power of imagination evolved to allow us to think about the sort of possibilities—concrete, perceptible states of affairs that we might actually come across (predators in the bushes)—rather than esoteric possibilities (if they really were such) we would never experience such as a world with different laws. It can be shown how Kripke's explanation for the illusion of contingency can be extended to laws.

While some have thought that the laws of nature break down at the Big Bang, physicist Paul Davies explains that there are still other versions of the laws of nature which hold at the Big Bang. Davies (2013) explains:

> Physicists have discovered that the laws of physics familiar in the laboratory may change form at very high temperatures, such as the ultra-hot environment of the Big Bang. As the universe expanded and cooled, various 'effective laws' crystallized out from the fundamental underlying laws, sometimes manifesting random features. It is the high-temperature versions of the laws, not their ordinary, lab-tested descendants, that are regarded as truly fundamental.

Nevertheless, there could still be alternative universes in which different properties and different laws of nature exist, whereas the laws of logic exist in all possible universe. Lange (2009, p. 77) argues that it is in this sense in which the contingency aspect of the laws of nature is to be understood, noting that the range of counterfactual suppositions under which the laws of nature must all be preserved, for the set of laws to qualify as stable, is narrower than the range of counterfactual suppositions under which the broadly logical truths must all be preserved, for the set of broadly logical truths to qualify as stable.

According to the dispositionalist view, the necessity aspect of the laws of nature is grounded in dispositional properties understood as natural clusters of powers (Mumford 2004, pp. 161, 170). On the dispositionalist view, apples regularly fall towards the earth because both apples and the earth have mass understood as a dispositional property, and the resulting regularities can be described by the abstract equations we call the laws of nature (Dumsday 2019, pp. 10–11). Dorato and Esfeld (2014) argue that the view that laws are grounded in properties (global properties rather than 'intrinsic' or local properties, in view of quantum entanglement) make intelligible how laws can 'govern' the behaviour of objects. This is the decisive advantage of dispositionalism over primitivism (the view that laws are primitive; see Maudlin 2007). According to the essentialist view, the causally necessary connections are explications of the essential properties of the natural kinds (Ellis 2001). Essentialists agree that some properties are essentially dispositional, but they argue that others (e.g. spatiotemporal properties) are not (Choi and Fara 2018; see further, Section 3.8.3).

Dumsday (2019, p. 119) has defended dispositionalism against Lange's attempt to reduce dispositions to subjunctive facts, by situating dispositionalism within robust natural-kind essentialism. 'Although the dispositions are real and irreducible (hence preserving dispositionalism), they are not ungrounded, but instead are rooted in the kind. Consequently, dispositions cannot be reducible to primitive subjunctives. And the kind in its turn is not reducible to a primitive subjunctive fact, as its explanatory role goes beyond that of such a fact' (p. 120). 'A primitive subjunctive fact is ordered to a possible future state of affairs, and cannot, in and of itself, explain the present instantiation of a categorical property like shape or size. By contrast, the kind-essence, as traditionally conceived, does exactly that' (ibid.; noting on p. 122 that kind-talk is utterly ubiquitous across all the natural sciences and the efforts of many physicists devoted to the classification of apparently fundamental types of particle in terms of kinds).

It has been objected that there are some laws of nature that could not be explained in terms of causal powers. For example, the law of conservation of energy indicates that interactions are constrained by the requirement of preserving the mass-energy, but that constraint does not seem to be the manifestation of a disposition (Chalmers 1999, pp. 12–13). Mumford (2004, p. 199) replies that what have been labelled as 'laws of nature' are actually a very diverse bunch: 'Some causal laws might be best explained in terms of causal powers but others might be better explained in terms of metaphysical connections between properties and others might merely describe the structure of space–time or the nature and limit of energy.' I shall argue in Chap. 3 that the law of conservation of energy should be explained in terms of the Causal Principle.

Mumford (2004) has gone further and argued that, given that the concept of a governing law of nature is no longer plausible, 'law of nature' should be discarded. Bird (2007, pp. 189–190) disagrees by appealing to the widespread usage and function of the term in science which indicate that the governing role is not essential to the definition. Bird defines a law of nature as follows: (L) The laws of a domain are the fundamental, general explanatory relationships between kinds, quantities, and qualities of that domain, that supervene upon the essential natures of those things (p. 201).[13]

Dumsday (2019, chapter 2) has replied to Mumford that at least some dispositions have CP clauses incorporating uninstantiated universals (which CP clauses help to delimit the range of manifestations of those dispositions), which imply that the laws of nature exist. Dumsday claims that a Platonist may argue that, while the disposition instances do the causal work, the Platonic universals set the rules by which they operate, and the laws of nature can be understood as relations between universals which govern the causal/dispositional roles that properties play as a matter of metaphysical necessity (Dumsday 2019, chapter 2). However, the notion of 'setting the rules' and 'govern' is misleading, since abstract objects do not have causal power to set or govern anything. As noted above, abstract objects such as the equations of physics are merely descriptive of behaviour; thus, there must be something concrete that 'makes it the case that the world actually operates in accordance with the equations, rather than some other equations or no equations at all' (Feser 2013, pp. 45–46).

Traditionally, this concrete entity is God. Bird (2007, pp. 189) notes the theologico-legal origins of the concept of the laws of nature as the decrees of God. Historically, the use of the term 'law of nature' is related to legislation by an intelligent deity (Brooke 1991, p. 26). Mumford (2004, pp. 202–203) objects to the use of this terminology, arguing that, while moral and legal laws are issued to conscious agents who can understanding them and decide whether to obey them or not, physical entities cannot understand and choose, and they could not have behaved other than the way they do because their behaviour are tied necessarily to their properties understood to be clusters of causal powers. He denies the existence of laws imposed on any things, which they then govern.

In reply, it can still be argued that the regularities indicate a Governor (God) who determined the properties of physical entities to be such that they move regularly according to equations in a law-like manner noted by Bird (see Chaps. 4 and 7). This conclusion does not require the views that (1) God has created a perfect world fine-tuned to his ends, (2) there is a universal and complete order in nature, and/or (3) what happens in nature can be described in universal and exceptionless laws.[14] On the contrary, the new groundbreaking view of nature as not universally law-governed (see above) fits well with the claim that, while God determined the properties of physical entities, He also judiciously intervenes in nature at key points to direct its ends (Gingerich 2006).

In any case, one still needs to ask where these 'laws of nature' come from. One might think that the 'laws of nature' express abstract relations between universals which physical things somehow 'participate in' (something like the way every tree participates in the Form of Tree; see Armstrong 1983). However, we would still need to know how it comes to be that there is a physical world that 'participates in' the laws in the first place, why it participates in these laws rather than others, and so on, and this indicates that the laws of nature cannot be ultimate explanations (Feser 2017, pp. 279–280). As I argue in the rest of this book, the answers to these questions are found in an intelligent First Cause.

In summary, fundamental physics does not provide a complete description of reality. It does not exclude efficient causation and causal properties which operate at a more fundamental level as the ground of the regularities described by fundamental physics. The latter point is further supported by the argument for the Causal Principle (see Chap. 3).

## 2.4 Considerations of Quantum Indeterminancy

The Causal Principle has been rejected in recent years by some philosophers due to considerations from quantum-mechanical indeterminacy (Grünbaum 2009, p. 15). However, others have responded that quantum particles emerge from the quantum vacuum which is not non-being, but something with vacuum fields (quantum particles are manifestations of fields) and which can be acted on by the relevant laws of nature (Bussey 2013, p. 33). Given that the pre-existent quantum fields and the capacities to be acted on by the relevant laws of nature are the necessary conditions for bringing about these quantum events, it is not the case that these quantum events are uncaused (see the definition of uncaused in Sect. 2.2).

It has sometimes been thought that Heisenberg's uncertainty principle violates the Causal Principle. This is a misunderstanding. The 'uncertainty' in question does not imply it is possible that energy comes from absolute nothing; it just means that the pre-existing energy (i.e. the vacuum energy which is already present) can (unpredictably) have a very high value in a very short period of time, such that the uncertainty of the energy measurement can be very large.

While some scientists have proposed theories according to which the universe began to exist from 'nothing' (e.g. Vilenkin 2006; Krauss 2012), cosmologist George Ellis objects that the efforts by these scientists cannot truly 'solve' the issue of creation, 'for they rely on some structures or other (e.g. the elaborate framework of quantum field theory and much of the standard model of particle physics) pre-existing the origin of the universe, and hence themselves requiring explanation' (Ellis 2007, section 2.7). Ellis' objection indicates that what these scientists mean by 'nothing' cannot be the absence of anything; rather, there needs to be something that can behave according to physics in order for their physical theories to work.

Moreover, it has already been explained in Chap. 1 that, in view of the importance of philosophical considerations for evaluating scientific theories, cosmologists should not merely construct models of the universe without considering the philosophical arguments against certain models. If what these scientists mean by 'nothing' is truly the absence of anything, then their theory would be refuted by philosophical arguments for the Causal Principle (see Chap. 3) which they have not successfully rebutted.

Even if it is the case that the negative gravitational energy of our universe exactly cancels the positive energy represented by matter so that the total energy of the universe is zero, as suggested by the Zero Energy Universe Theory (see Chap. 5), this does not imply that the positive and negative energy arose uncaused from zero energy. One can still ask what is the efficient cause which made the positive and negative energy to be the way they are. To conclude otherwise is to commit the logical fallacy of thinking that 'net zero imply no cause'. (This logical fallacy may be illustrated using the following analogy: the fact that my company's total expenses cancel the total revenue, such that the net profit is zero, does not imply that the expenses and revenue occurred without an efficient cause. We still need to ask what made the expenses and revenue to be the way they are.)

As for the radioactive disintegration of atomic nuclei, even if events such as the decay of a given atom of 235U at this instant rather than (say) two weeks from now do not have a sufficient cause, there is strong justification for maintaining that the phenomena (the decay and statistics they exhibit) themselves have underlying proportionate causal explanations, for they exhibit regularities that strongly indicate the existence of more fundamental ordered causes (Stoeger 2001, p. 87). These fundamental ordered causes would be entities that are causally antecedent to the radioactive disintegration of atomic nuclei. Physicist Peter Bussey (2013, p. 20) notes that 'beta-decay is due to the so-called "weak nuclear force", in whose absence the decay would not occur. So the cause of the new nuclear state is the weak force acting on the previous nuclear state.'

Additionally, many different interpretations of quantum physics exist, and some of them, such as Everett's Many Worlds interpretation and Bohm's pilot-wave model, are perfectly deterministic. A number of scientists and philosophers have argued that Bohm's theory is superior to the indeterministic Copenhagen interpretation (Towler 2009a, b; Goldstein 2013; I discuss this in Loke 2017, chapter 5), and that it can explain Heisenberg's uncertainty principle (Bricmont 2017, section 5.1.8). Contrary to popular opinion, physicist John Bell has not demonstrated the impossibility of hidden variables, but only the (apparent) inevitability of non-locality of quantum physics;[15] Bell himself defended Bohm's hidden variable theory (Bell 1987). Likewise, Alain Aspect (2002), the noted experimenter of quantum entanglement, agrees that his experiment does not violate determinism but only the locality condition. While it has been objected that Bohm's theory is incompatible with theory of relativity (Lewis 2016, p. 180), others have replied that, if Bohmian mechanics indeed cannot be made relativistic, it seems likely that quantum mechanics can't either (Dürr et al. 2014; Maudlin [2018] defends Bohmian mechanics by arguing that fundamental Lorentz invariance can be violated, and that observational Lorentz invariance can be explained by appealing to quantum equilibrium). With regard to quantum field theory, Bricmont (2017, p. 170) proposes Bohm-like quantum field theories in which dynamics are defined

> for the fields rather than for the particles, and the guidance equation would apply to the dynamics of field configurations. … One can also propose other Bohm-like quantum field theories, including theories of particles and their pair creation … all the predictions of the usual quantum field theories are also obtained in those Bohmian-type models and, to the extent that those models are rather ill-defined mathematically, the same thing is true for ordinary quantum field theories, which is not the case for non-relativistic quantum mechanics or the corresponding de Broglie–Bohm theory for particles.

Now Bohm's theory is not the only possible deterministic quantum theory; other deterministic quantum theories that are better than Bohm's (as well as better than the indeterministic Copenhagen interpretation) might well be discovered in the future. The inability to predict the appearance of the quantum particles in quantum vacuum may be due to our epistemological limitation and the incompleteness of current quantum physics. As Einstein [1949, p. 666] remarks, 'I am, in fact, firmly convinced that the essentially statistical character of contemporary quantum theory is solely to be ascribed to the fact that this (theory) operates with an incomplete description of physical systems.' Physicist John Wheeler notes that our current understanding of quantum mechanics is provisional, and that it is plausible to think that some deeper theory, waiting to be discovered, would explain in a clear and rational way all the oddities of the quantum world, and would, in turn, explain the apparent fuzziness in the quantum classical boundary (Ford 2011, p. 263). Given that our current understanding of the quantum world is provisional, it is false to claim that quantum physics has shown that events can begin to exist without necessary or sufficient conditions.

In conclusion, it has been argued that no compelling scientific evidence against the Causal Principle has been offered. In the next chapter, I shall discuss a number of arguments for the Causal Principle.

## Notes

1. Something can have a beginning even if its temporal extension is an actual infinite (e.g. if something begins to exist in the year 2020 and exists endlessly in the later-than direction on the static theory of time). However, if something is finite in temporal extension and has temporal edges, it would have a beginning.
2. Here, part of a thing refers to a temporal part. See perdurantism, below.
3. I thank Oners for raising this objection and for the discussion below.
4. Strictly speaking, the purported evidence for the B theory does not prove that the block never comes to be; see Chap. 6.
5. Proponents of probabilistic causation acknowledge that there are sufficient causes and necessary conditions, and they regard sufficient causes as constituting a limiting case of probabilistic causes, but they deny that this limiting case includes all bona fide cause–effect relations (Williamson 2009, p. 192). It should be noted that a cause can be causally sufficient but not causally necessary for an effect.
6. Weaver (2019, chapter 7) argues that there are no plausible metaphysical theories of omissions understood as absences that are causal relata, and that virtually all supposed cases of negative causation can be faithfully/accurately re-described without omissions/absences.
7. Hence, by uncaused First Cause, I mean the First Cause of change, and that this First Cause is not something that is brought into existence. However, such a First Cause might be something that is sustained in existence, and thus is caused in the sense of having a sustaining cause. See further, Chaps. 6 and 8.
8. Weaver goes on to explain that he agrees with David Lewis (1986) that causation should be understood in terms of causal dependence, but disagrees with Lewis' additional step of reducing causal dependence to counterfactual dependence. He argues on page 261 that the heart of the causal interpretation of General Theory of Relativity (GTR) is not a relation that is reducible to counterfactual dependence, probabilistic dependence, the transfer of energy or momentum, or some other reductive surrogate relation or process.

9. I thank Michael Dodds for this point.
10. Curiel (2019) notes that 'the Second Law of thermodynamics has long been connected to the seeming asymmetry of the arrow of time, that time seems to flow, so to speak, in only one direction for all systems'.
11. See the discussion on dispositionalism and essentialism below.
12. Collins (2009, p. 270) notes that the laws of physics are not strictly speaking time-reversal invariant—since time-reversal symmetry is broken in weak interactions, notably the decay of neutral kaons.
13. Cf. The *Oxford Dictionary of Physics*' definition of a law in science as 'a descriptive principle of nature that holds in all circumstances covered by the wording of the law' (Issacs 2000, p. 260).
14. Cartwright (2016) states that since at least the Scientific Revolution three theses have marched hand in hand.
15. Tim Mawson points out to me that Bell's results do not even show that non-locality is violated; this 'loophole' is sometimes discussed under the name 'Superdeterminism'.

## Bibliography

Almeida, Michael. 2018. *Cosmological Arguments*. Cambridge: Cambridge University Press.

Anscombe, Elizabeth. 1974. Whatever has a Beginning of Existence Must Have a Cause': Hume's Argument Exposed. *Analysis* 34: 145–151.

Armstrong, David. 1983. *What Is a Law of Nature?* Cambridge: Cambridge University Press.

Aspect, Alain. 2002. Bell's Theorem: The Naive View of an Experimentalist. In *Quantum (Un)speakables*, ed. J. Bell, Reinhold Bertlmann, and A. Zeilinger. Berlin: Springer.

Bell, John. 1987. *Speakable and Unspeakable in Quantum Mechanics*. Cambridge: Cambridge University Press.

Bird, Alexander. 2007. *Nature's Metaphysics*. Oxford: Oxford University Press.

Bricmont, Jean. 2017. *Making Sense of Quantum Mechanics*. Cham: Springer Nature.

Brooke, John. 1991. *Science and Religion: Some Historical Perspectives*. Cambridge: Cambridge University Press.

Bussey, Peter. 2013. God as First Cause – A Review of the Kalām Argument. *Science & Christian Belief* 25: 17–35.

Carroll, Sean. 2014. *Post Debate Reflections*. http://www.preposterousuniverse.com/blog/2014/02/24/post-debate-reflections/.

Cartwright, Nancy. 2016. The Dethronement of Laws in Science. In *Rethinking Order After the Laws of Nature*, 25–52. Cartwright, N. and Ward, K. London: Bloomsbury Academic.

Chalmers, A. 1999. Making Sense of Laws of Physics. In *Causation and Laws of Nature*, ed. H. Sankey. Dordrecht: Kluwer.

Choi, Sungho, and Michael Fara. 2018. Dispositions. In *The Stanford Encyclopedia of Philosophy*, ed. Edward N. Zalta. https://plato.stanford.edu/archives/fall2018/entries/dispositions/.

Cogliati, C. 2010. Introduction. In *Creation and the God of Abraham*, eds. D. B. Burrell, C. Cogliati, J. M. Soskice, and W. R. Stoeger. Cambridge: Cambridge University Press.

Collins, Robin. 2009. *The Teleological Argument: An Exploration of the Fine-tuning of the Universe. Blackwell Companion to Natural Theology*. Oxford: Wiley-Blackwell.

Craig, William Lane. 2016. *God Over All: Divine Aseity and the Challenge of Platonism*. Oxford: Oxford University Press.

Craig, William Lane, and James Sinclair. 2009. The Kalām Cosmological Argument. In *The Blackwell Companion to Natural Theology*, ed. William Lane Craig and J.P. Moreland. Chichester: Wiley-Blackwell.

Curiel, Erik. 2019. Singularities and Black Holes. In *The Stanford Encyclopedia of Philosophy*, ed. Edward Zalta. https://plato.stanford.edu/archives/spr2019/entries/spacetime-singularities/.

Davies, Paul. 2013. Frozen Accidents: Can the Laws of Physics Be Explained? http://www.pbs.org/wgbh/nova/blogs/physics/2013/12/frozen-accidents-can-the-laws-of-physics-be-explained.

Dorato, Mauro, and Michael Esfeld. 2014. The Metaphysics of laws: dispositionalism vs. Primitivism. In *Metaphysics in Contemporary Physics*, ed. Tomasz Bigaj and Christian Wüthrich. Dordrecht: Kluwer.

Dumsday, Travis. 2019. *Dispositionalism and the Metaphysics of Science*. Cambridge: Cambridge University Press.

Dürr, Detlef, Sheldon Goldstein, Travis Norsen, Ward Struyve, and Nino Zanghì. 2014. Can Bohmian Mechanics Be Made Relativistic? *Proceedings of the Royal Society A* 470. https://doi.org/10.1098/rspa.2013.0699.

Egg, Matthias, and Michael Esfeld. 2014. Non-local Common Cause Explanations for EPR. *European Journal for Philosophy of Science* 4: 181–196.

Einstein, Albert. 1949. Remarks Concerning the Essays Brought Together in this Co-operative Volume. In *Albert Einstein: Philosopher-Scientist*, ed., P. Schilpp. The Library of Living Philosophers.

Ellis, Brian. 2001. *Scientific Essentialism*. Cambridge: Cambridge University Press.
Ellis, George. 2007. Issues in the Philosophy of Cosmology. In *Philosophy of Physics*, ed. J. Butterfield and J. Earman. Amsterdam: Elsevier.
Feser, Edward. 2013. *Aristotle on Method and Metaphysics*. London: Palgrave.
———. 2017. *Five Proofs of the Existence of God*. San Francisco: Ignatius Press.
Fine, Kit. 2002. The Varieties of Necessity. In *Conceivability and Possibility*, ed. Tamar Gendler and John Hawthorne. Oxford: Clarendon.
Ford, Kenneth. 2011. *101 Quantum Questions*. Cambridge, MA: Harvard University Press.
Frisch, Mathias. 2014. *Causal Reasoning in Physics*. Cambridge: Cambridge University Press.
Gingerich, O. 2006. *God's Universe*. Cambridge, MA: Harvard University Press.
Goldstein, Sheldon. 2013. Bohmian Mechanics. In *The Stanford Encyclopedia of Philosophy*, ed. Edward Zalta. http://plato.stanford.edu/archives/spr2013/entries/qm-bohm/.
Greiner, Walter, and Joachim Reinhardt. 2009. *Quantum Electrodynamics*. Berlin: Springer.
Grünbaum, Adolf. 2009. Why Is There a Universe AT ALL, Rather Than Just Nothing? *Ontology Studies* 9: 7–19.
Hawking, Stephen. 2018. *Brief Answers to the Big Questions*. New York: Bantam Books.
Hume, David. 1739/1978. *A Treatise of Human Nature*, ed. L. A. Selby-Bigge. 2nd ed. Oxford: Oxford University Press.
Hume, David. 1932. In *The Letters of David Hume*, ed. J.Y.T. Greig, vol. 1. Oxford: Clarendon.
Issacs, A., ed. 2000. *Oxford Dictionary of Physics*. 4th ed. Oxford: Oxford University Press.
Koons, Robert, and Timothy Pickavance. 2015. *Metaphysics: The Fundamentals*. Chichester: Wiley-Blackwell.
Krauss, Lawrence. 2012. *A Universe from Nothing*. New York: Free Press.
Kripke, Saul. 1980. *Naming and Necessity*. Oxford: Blackwell.
Ladyman, James, Don Ross, David Spurrett, and John Collier. 2007. *Every Thing Must Go: Metaphysics Naturalized*. Oxford: Oxford University Press.
Lange, Marc. 2009. *Laws and Lawmakers. Science, Metaphysics, and the Laws of Nature*. Oxford: Oxford University Press.
Lewis, David. 1986. *On the Plurality of Worlds*. New York: Basil Blackwell.
Lewis, Peter. 2016. *Quantum Ontology*. Oxford: Oxford University Press.
Linford, Dan. 2020. The *Kalām* Cosmological Argument Meets the Mentaculus. *The British Journal for the Philosophy of Science*. https://doi.org/10.1093/bjps/axaa005.

Linnemann, N. 2020. On Metaphysically Necessary Laws from Physics. *European Journal for Philosophy of Science* 10: 23.
Loke, Andrew. 2017. *God and Ultimate Origins: A Novel Cosmological Argument. Palgrave Frontiers in Philosophy of Religion Series*. Cham, Switzerland: Springer Nature.
Lowe, E.J. 2002. Kinds, Essence, and Natural Necessity. In *Individuals, Essence and Identity*. Heidelberg: Springer.
Mackie, Penelope. 2005. Causality. In *The Oxford Companion to Philosophy*, 2nd ed. New York: Oxford University Press. http://www.oxfordreference.com/views/ENTRY.html?subview=Main&entry=t116.e356.
Martin, C.B. 2008. *The Mind in Nature*. Oxford: Clarendon Press.
Maudlin, Tim. 2007. *The Metaphysics Within Physics*. Oxford: Oxford University Press.
———. 2018. Philosophy Has Made Plenty of Progress. *Scientific American*. https://blogs.scientificamerican.com/cross-check/philosophy-has-made-plenty-of-progress/.
Mumford, Stephen. 2004. *Laws in Nature*. London: Routledge.
Ney, A. 2016. Microphysical Causation and the Case for Physicalism. *Analytic Philosophy* 57: 141–164.
Norton, John. 2003. Causation as Folk Science. *Philosophers' Imprint* 3 (4): 1–22. http://www.philosophersimprint.org/003004/3.
Oaklander, Nathan. 2004. *The Ontology of Time*. New York: Prometheus Books.
Oppy, Graham. 2010. Uncaused Beginnings. *Faith and Philosophy* 27: 61–71.
———. 2015. Uncaused Beginnings Revisited. *Faith and Philosophy* 32: 205–210.
Pelczar, Michael. 2015. *Sensorama: A Phenomenalist Analysis of Spacetime and Its Contents*. Oxford: Oxford University Press.
Pettersson, Thomas, and P. Lefèvre 1995. *The Large Hadron Collider. CERN Document Server*. http://cdsweb.cern.ch/record/291782.
Psillos, Stathis. 2009. Regularity Theories of Causation. In *Beebee, Hitchcock, and Menzies*, 131–157. Oxford: Oxford University Press.
Qureshi-Hurst, E., and A. Pearson. 2020. Quantum Mechanics, Time and Theology: Indefinite Causal Order and a New Approach to Salvation. *Zygon* 55: 663–684.
Reid, Thomas. 1983. In *Inquiry and Essays*, ed. Ronald E. Beanblossom and Keith Lehrer. Indianapolis: Hackett.
Russell, Bertrand. 1918. On the Notion of Cause. In *Mysticism and Logic andO-ther Essays*. New York: Longmans Green and Co.

Simon, Jonathan. 2015. Review of Michael Pelczar's *Sensorama. Notre Dame Philosophical Reviews: An Electronic Journal*. https://ndpr.nd.edu/news/sensorama-a-phenomenalist-analysis-of-spacetime-and-its-contents/.

Stoeger, William. 2001. Epistemological and Ontological Issues Arising from Quantum Theory. In *Quantum Mechanics: Scientific Perspectives on Divine Action*, ed. Robert Russell et al. Berkeley: Center for Theology and the Natural Sciences.

Strawson, Galen. 1989. *The Secret Connexion*. Oxford: Clarendon.

Towler, Mike. 2009a. Pilot Wave Teory, Bohmian Metaphysics, and the Foundations of Quantum Mechanics Lecture 7. www.tcm.phy.cam.ac.uk/~mdt26/PWT/lectures/bohm7.pdf. Accessed 20 Jan 2017.

———. 2009b. De Broglie-Bohm Pilot-Wave Teory and the Foundations of Quantum Mechanics. http://www.tcm.phy.cam. ac.uk/~mdt26/pilot_waves.html. Accessed 20 Jan 2017.

Vilenkin, Alexander. 2006. *Many Worlds in One*. New York: Hill and Wang.

Wall, Aron. 2013a. The Generalized Second Law Implies a Quantum SingularityTheorem. *Classical and Quantum Gravity* 30 (16): 165003. Preprint: arXiv:1010.5513v4 [gr-qc].

———. 2013b. Corrigendum: The Generalized Second Law Implies a Quantum Singularity Theorem. *Classical and Quantum Gravity* 30 (19): 199501.

Weaver, C. 2019. *Fundamental Causation: Physics, Metaphysics, and the Deep Structure of the World*. London: Routledge.

Williamson, Jon. 2009. Probabilistic Theories of Causality. In *The Oxford Handbook of Causation* ed. Helen Beebee, Christopher Hitchcock, and Peter Menzies. Oxford: Oxford University Press.

# 3

# Arguments for the Causal Principle

## 3.1 Introduction

As noted in Chap. 2, a number of scientists and philosophers have expressed scepticism concerning the Causal Principle 'whatever begins to exist has a cause'. In particular, they have objected that, even if the Causal Principle applies to things within the universe, it might not apply to the universe itself. A number of arguments have been offered in the literature in response to this objection. These include (1) an inductive argument, (2) an argument from the concept of non-being, (3) a Modus Tollens argument, (4) the rationality argument (if the universe began uncaused, an absurd universe is as likely to begin uncaused as a normal universe is; this generates serious scepticism about the reliability of our cognitive faculties, the truth of our sensory inputs, and our past knowledge, thus creating a reductio ad absurdum against the objection [Miksa 2020]), and (5) argument from fine-tuning and order. It should be noted that any one of these arguments would be sufficient for the purposes of the KCA. In other words, a proponent of the KCA does not have to rely on any single one of these arguments. Therefore, even if the objector of the KCA manage to find fallacies in one of these arguments, this does not



A. Loke, *The Teleological and Kalam Cosmological Arguments Revisited*, Palgrave Frontiers in Philosophy of Religion, https://doi.org/10.1007/978-3-030-94403-2_3

imply that the KCA has been rebutted. Rather, the objector would need to rebut all five of these multiple independent arguments.

I have defended the first three arguments in my previous work (Loke 2012, 2017, chapter 5). In this book, I shall develop my defence of (3) the Modus Tollens argument against objections which have been raised more recently since the publication of my previous work (I shall also respond to other relevant objections in older literature). In addition, I shall defend (5) the fine-tuning argument at the end of this chapter.

The Modus Tollens argument can be traced to the American philosopher-theologian Jonathan Edwards, who argues that 'if there be no absurdity or difficulty in supposing one thing to start out of non-existence into being, of itself without a Cause; then there is no absurdity or difficulty in supposing the same of millions of millions' (Edwards 1830, p. 53). Likewise, Arthur Prior reasons that 'if it is possible for objects to start existing without a cause, then it is incredible that they should all turn out to be objects of the same sort' (Prior 1968, p. 65). Craig and Sinclair explain the argument as follows:

> If things really could come into being uncaused out of nothing, then it becomes inexplicable why just anything or everything does not come into existence uncaused from nothing. Why do bicycles and Beethoven and root beer not pop into being from nothing? Why is it only universes that can come into being from nothing? What makes nothingness so discriminatory? There cannot be anything about nothingness that favors universes, for nothingness does not have any properties. Nothingness is the absence of anything whatsoever. As such, nothingness can have no properties, since there literally is not anything to have any properties. Nor can anything constrain nothingness, for there is not anything to be constrained. (Craig and Sinclair 2009, p. 186)

Craig and Sinclair's defence of this argument has been subjected to criticisms (e.g. Oppy 2010, 2015). In response to these criticisms (see below), I shall develop a version of the argument which can be formulated as follows:

1. If *x* (e.g. physical reality) begins uncaused, then *y* which begins to exist would also begin uncaused. (Here, y refers to event(s) or type of events that occurs/occurred/would have occurred.)
2. It is not the case that *y* begins uncaused.
3. Therefore, it is not the case that *x* begins uncaused.

As a rough analogy, consider the following story which illustrates the general principle that what begins to exist is brought about (and constrained by) the cause, and what would happen if this principle were violated.

Think about why my newly built house is the way it is, rather than a pile of rocks. The answer is simple: the house builder makes it that way; what begins to exist is brought about and constrained by the cause; if the cause were a huge explosion rather than a house builder, what began to exist would be rubble rather than a house.

However, if my house begins to exist uncaused, then that means that there is no cause, which makes it the case that only a house rather than other things (e.g. a rubble) begins to exist uncaused. That is, there would be no constraint on whatever begins to exist in the present circumstances from beginning to exist uncaused, in which case we would expect to see many other things (e.g. a rubble) begin to exist uncaused, but we do not.

I shall now elaborate on the argument.

Very briefly, the justification for premise 1 is that, (I) if something *x* begins to exist uncaused, then this means that there would not be any causally antecedent condition of *x* which would make it the case that *x* (rather than *y*) begins to exist uncaused, (II) the properties of *x* and the properties of *y* which differentiate between them would be had by them in the actual concrete world only when they had already begun to exist, and (III) the circumstance is compatible with the beginning of *y*. I shall argue in the following sections that (I), (II), and (III) jointly imply that there would be no difference between *x* and *y* where beginning to exist uncaused is concerned. To deny the consequent in premise 1, I only have to show that one event around me does not begin to exist uncaused. For example, I do not experience an event such as *y* = 'a rapid increasing in strength of electric fields around me' beginning to exist without causally antecedent conditions such as (say) having to switch on the electric field generator. Since the consequent in premise 1 is false, the antecedent is false; that is, it is not the case that something begins to exist uncaused.

It should be noted that the Causal Principle I am defending is not 'whatever begins to exist *must have* a cause'; but rather 'whatever begins to exist *has* a cause'. I shall *not* argue that there *could be no* exception, but there *is no* exception. In other words, I am defending a causal principle that is contingent, that is, true in the actual world, rather than in all possible worlds. The Modus Tollens argument I defend does not exclude possible worlds which are utterly chaotic and in which the initial state of reality (ISOR) begins uncaused, but such worlds are obviously not our actual world. As we shall see later, an objector to my argument might claim that ISOR begins uncaused and yet our world is not utterly chaotic because other things do not begin uncaused. In reply to this claim, I will *not* be arguing that it faces the problem of explaining how it *could be* the case that a thing of a certain kind (a kind to which ISOR belongs) begins to exist uncaused whereas things of other sorts do not. Rather, I will be arguing that the objector's claim faces the problem of explaining why (ex hypothesi) ISOR does in fact begin to exist uncaused while other things do not.[1]

The above clarifications address Almeida's (2018, pp. 87–90) objections to the Causal Principle. Almeida appeals to the argument that lawless possible worlds exist—one can imagine chaotic worlds in which events such as a raging tiger suddenly come into existence in the room uncaused happen often and unpredictably. While such events do not happen often and unpredictably in our world, there is a small chance they can happen. For example,

> there is some small chance, further, that the particles that compose the hand of the statue of David all move together upward and then downward and 'wave' at you … for any actual object whatsoever, there is a small chance that it spontaneously disappears and an intrinsic duplicate of the object appears on Mars. (Ibid.)

My Modus Tollens argument is not susceptible to the above objection because it does not deny that there are possible worlds which are chaotic, or that some things can begin to exist uncaused. Rather, what the argument would show is that, if something does begin to exist without any causally necessary condition whatsoever, our actual world would be very different because some uncaused events (e.g. electric fields increasing in strength under certain circumstances uncaused) would happen *all the time* in our world,[2] which is not the case.

Almeida (2018) also suggests that, for all we know, 'our world is one in which what we believe are the causal laws are statistical generalizations whose probabilities oscillate imperceptibly every 100 million years' (p. 38), and we happen to live in an epoch in which we do not perceive uncaused events, but this does not imply that the beginning of the universe is caused. However, given my argument, the statistical generalizations suggested by Almeida would not hold because, as I shall explain below, if something begins to exist uncaused, some uncaused events would happen *all the time* and such events would not have been preventable.

Finally, it is important to remind the reader of the definition of the key term 'beginning' which (as explained in Chap. 2) is understood as follows: something has a beginning if it has a temporal extension, the extension is finite, and it has temporal edges/boundaries (e.g. on the model of spacetime which exists for all $t > 0$ but not at $t = 0$, $t = 0$ is a boundary). The above definition for 'beginning' holds regardless of whether time is emergent property of our universe or whether there may not be any definite time order to events when the universe is so small that quantum gravity is important. The definition does not require 'time earlier than the universe'; it also does not require the universe to begin to exist in time. As long as the universe has a temporal extension, the extension is finite, and it has temporal edges/boundaries, it has a beginning (regardless of whether the beginning is 'in time' or not), and therefore susceptible to my Modus Tollens argument. In other words, my Modus Tollens argument does not require the existence of time before the existence of the universe. One must also be careful not to confound two distinct models of the world: one with temporal boundary and one without temporal boundary. If one thinks that the universe is timeless and has no temporal boundary (as postulated by Hawking's no boundary proposal), that would be a separate issue which I address separately using other arguments explained in Chap. 6.

In the following sections I shall explain the argument in greater detail as I respond to various objections. I shall show that my argument would work on a static theory of time as well. At this point it should be noted that, even if the static theory of time is true, there is still something unique about time which makes it different from spatial order. For example, within the spacetime block, the durations next to each human, including myself, are occupied by his/her parents such that, if they had not existed, him/her would not exist. In this sense there is still a certain dependence and ordering among things/events, which I shall call 'causal dependence' and 'causal order'. Whereas this is not so with spatial order: if my parents do not stand on my left or right, I would still exist. This indicates that time is different from space even if the static theory of time is true, and that there is still causal order in any case. The argument of this chapter is that, if the causal principle is false, we would not observe the temporal/causal order which we do observe.

## 3.2 Objection: The Initial State of Reality (ISOR) is the Only Thing That Begins Uncaused

To begin, against premise 1, Oppy has suggested that the initial state of reality (ISOR) is the only thing that begins uncaused, while later things/events begins caused (Oppy 2010, 2015). Oppy claims that this view is supported by a branching view of modality according to which all possible worlds share the initial state (the First Cause) of the history of the actual world which is necessary (such an initial state exists given that [as I shall argue in Chap. 5] an infinite causal regress and a causal loop is not the case). Oppy explains:

> My favourite theory of modality has the evident advantage of theoretical frugality. On the one hand, if there are objective chances, then any theory of modality is surely committed to the possibility of the outcomes that lie in the relevant objective chance distributions. On the other hand, it is not clear that we have good reason to commit ourselves to any possibilities beyond those that are required by whatever objective chances there might

> be; at the very least, any expansion of the range of possibilities clearly requires some kind of justification. (Oppy 2013b, p. 47)

Oppy would also argue that, in comparison with theism which likewise affirms an uncaused and necessarily existent First Cause, his view has the best trade-off between simplicity and explanatory power because he thinks that it is able to explain everything that theism explains without needing to posit an 'extra' entity, that is, God (Oppy 2013a).

In reply, it is trivially true that, if there is an initial state of reality and that this state is the First Cause, then such a First Cause would be uncaused. The key question is, what kind of thing is the initial state that is uncaused? There are two possible answers:

1. A First Cause with a beginning
2. A First Cause without a beginning (I shall argue in Chap. 6 that such a First Cause would also be transcendent, immaterial, has libertarian freedom, that is, a Creator.)

Now both Oppy and I affirm that the First Cause is factually necessary and also metaphysical necessary. A factually necessary being is one which is not causally dependent on other things; in this sense, it is not contingent (dependent on) anything else. A metaphysical necessary being is one which exists in all metaphysically possible worlds. The difference between our views is that on Oppy's view the initial state (i.e. the First Cause) is metaphysical necessary *and has a beginning* (whereas on my view the First Cause is beginningless), and he thinks that this metaphysical necessity would entail factual necessity, which he also affirms.

In the rest of this chapter, I shall argue against (1) using the Modus Tollens argument; that is, I shall argue it is not the case that something that begins to exist is factually necessary; thus, this Modus Tollens argument refutes Oppy's view, which entails the opposite. While ultimately there must be brute necessity where explanation stop, my Modus Tollens argument implies that this brute necessity is not something that has a beginning; therefore, a brute necessity is something that is beginningless. At this point, I would just like to note that Oppy should not reply to my Modus Tollens argument by claiming that the initial state begins to exist as a brute necessity given his theory of modality, since this would be begging the question by assuming (1) instead of (2) (his theory of modality by itself does not imply either (1) or (2)). Neither should Oppy reply by claiming that ISOR would be uncaused if it begins to exist. The reason is because it is trivially true that ISOR would be uncaused (since it is supposed to be the initial state); whether it *begins* uncaused is precisely the issue under dispute. Oppy might object by arguing that (1) is justified by his claim that it has the best trade-off between simplicity and explanatory power. He thinks that premises of arguments are evaluated based on theories which are evaluated based on certain virtues (e.g. simplicity, explanatory power), and he claims that his view has equal explanatory power and is simpler compared to theism (Oppy 2013b). However, a view that entails a contradiction cannot be true, even if it is simpler. Thus, simplicity cannot help his view since his view entails a contradiction as I shall explain using the Modus Tollens argument. Basically, the argument shows that, if ISOR begins uncaused, I should expect to see other things beginning uncaused around me, but I do not; hence, the antecedent is false. I shall elaborate on this argument and defend it against objections in the rest of this chapter.

Let us begin by considering another scenario. Suppose someone postulates that the circumstances are such that it is metaphysically possible for *x* and *y* to begin to exist, and that only *x* begins to exist *caused*. If he/she were asked 'what makes it the case that it is *x* rather than *y* that begins to exist *caused*', the answer would be simple: in these circumstances, the causally antecedent condition(s) makes it the case that it is *x* rather than *y* that begins to exist.

Now suppose someone postulates that the circumstances are such that it is metaphysically possible for *x* and *y* to begin to exist, and that only *x* begins to exist *uncaused*. The important question to ask is, 'What makes it the case that it is *x* rather than *y* that begins to exist *uncaused*?' (By 'makes it the case', I mean 'provides metaphysical grounding'.)

Note that the term 'metaphysical ground' is standardly used in the philosophical literature to mean something distinct from a cause [although it can also be a cause]. 'For instance, we might say that the members of a set are prior to the set itself; the existence of the set is grounded in its members. Or to take a more concrete example, the existence of any given composite object is grounded in the existence of its parts' [Tahko and Lowe 2020, section 5].) Using the laws of logic (the law of excluded middle), the following are the only possible answers:

1.1. If *x* rather than *y* begins to exist uncaused, then either
1.1.1. nothing makes this the case (brute fact) or
1.1.2. something abstract makes this the case, or
1.1.3. something concrete with property[3] S makes this the case, in which case either
1.1.3.1. S is a property of x which makes it different from y (i.e. S is a property which *x* has but *y* does not have), or
1.1.3.2. S is a property of something other than *x*: either
1.1.3.2.1. S is a property of the circumstances of *x*, or
1.1.3.2.2. S is a property of *y*, or
1.1.3.2.3. S is a property of the circumstances of *y*.

(Note: circumstance is defined as 'a fact or condition connected with or relevant to an event' [*Oxford English Dictionary*, OED]. Thus, anything other than *x, y*, or their circumstance would be irrelevant).

In the rest of this chapter, I shall argue:

1.2. It is not the case that 1.1.1, 1.1.2, 1.1.3.1, 1.1.3.2.1, 1.1.3.2.2, or 1.1.3.2.3.
1.3. Therefore, it is not the case that *x* rather than *y* begins to exist uncaused.

In particular, it should be noted that, given (as I shall show later) that the objector to my argument cannot appeal to 'brute fact' (1.1.1) or 'abstract entities' (1.1.2), it turns out that the objector has to affirm (1.1.3) something concrete with property S makes it the case that *x* rather than *y* begins to exist uncaused, in which case either 1.1.3.1, 1.1.3.2.1, 1.1.3.2.2, or 1.1.3.2.3.

As I shall explain later in this chapter,

- the falsity of 1.1.3.2.1 'S is a property of the circumstances of *x*' is entailed by (I) there would not be any causally antecedent condition which would make it the case that *x* rather than *y* begins to exist uncaused;

- the falsity of 1.1.3.1. 'S is a property of *x* which makes it different from *y*' and 1.1.3.2.2. 'S is a property of *y*' is entailed by (II) the properties of *x* and the properties of *y* which differentiate between them would be had by them only when they had already begun to exist; and
- the falsity of 1.1.3.2.3 'S is a property of the circumstances of *y*' is entailed by (III) the circumstance is compatible with the beginning of *y*.

Therefore, (I), (II), and (III) jointly imply that there would be no essential difference between *x* and *y* where beginning to exist uncaused is concerned, and this implies that Oppy's theory that '*x* (ISOR) begins uncaused but *y* begins caused' is false. Oppy should not respond by claiming that theories determine whether arguments are sound and that given his theory, (I), (II), and/or (III) is false.[4] The reason is because (I), (II), and (III) are implied by Oppy's own theory. To elaborate, (I) Oppy's theory that *x* (ISOR) begins *uncaused* but *y* begins caused implies that *no cause* making it the case that *x* (rather than y) begins uncaused. Moreover, (II) *x* and *y* having a beginning implies that the properties of *x* and the properties of *y* which differentiate between them would be had by them only when they had already begun to exist, and (III) *y* begins caused implies that the circumstances are compatible with the beginning of *y*. In other words, Oppy's theory implies (I), (II), and (III), which imply that his theory is false; that is, Oppy's theory entails a contradiction.[5]

I shall now discuss 1.1.1, 1.1.2, 1.1.3.1, 1.1.3.2.1, 1.1.3.2.2, and 1.1.3.2.3 in detail.

## 3.3 Against 1.1.1. Brute Fact

Concerning 1.1.1, the objector to my argument might suggest it is a brute fact that only ISOR (suppose *x* = ISOR) but not *y* begins to exist uncaused (Rasmussen 2018). While there is a difference between *x* and *y* (viz. that *x* begins uncaused but *y* begins caused), there is no difference between *x* and *y* that explains why *x* begins uncaused but *y* begins caused. In this case, *x* and *y* have a difference that has no further metaphysical ground. The objector might argue that this is analogous to the situation in which two carbon-14 atoms which are qualitatively identical with respect to beta-decay and as a result have the same objective chance of decay, and purely as a matter of chance, one but not the other decays in the next moment as a brute fact (i.e. without a metaphysical ground that differentiates between them with respect to decay). In this scenario, it is stipulated that no relevant (hidden) variable is left out, such that the chances in question are irreducible; that is, the event happened as a brute fact without any further explanation. In this case of genuine indeterminism, there is no difference between the atoms that explains why beta-decay happens to one but not the other.[6]

Now the following two brute fact claims should be distinguished:

1. 'x begins uncaused' is a brute fact.
2. 'x begins uncaused but y begins caused' is a brute fact.

The possibility of (1) (see Sect. 3.1) does not imply the possibility or actuality of (2). (2) is refuted by the following three independent arguments (while (1) is refuted by the arguments in this chapter):

Argument 1

It is ad hoc and special pleading to claim that 'unlike other things which begins caused, *x* begins uncaused' without any ground or justification for claiming that. (For why the special pleading objection does not apply to God, see Chap. 6.)

Argument 2

There is a difference between radioactive decay scenario and the uncaused beginning ('x begins uncaused but y begins caused') scenario. In the former there are two atoms which pre-exist before one of them decays. Whereas in the latter nothing pre-exists the uncaused beginning of *x*. This difference is significant in light of the following argument:

1. 'The possession of the property of "beginning uncaused" by *x*' requires the existence of *x*, and 'the possession of the property of "beginning caused" by *y*' requires the existence of *y*.
2. Therefore, the existence of *x* and *y* (with their numerical distinction) is required for '*x* has the property of beginning uncaused and *y* has the property of beginning caused' (From 1).

Since '*x* begins uncaused but *y* begins caused' requires (i.e. depends on) the numerically distinct existence of *x* and *y*, it cannot just be a brute fact. Therefore, 1.1.1 is false.

The difference between the radioactive decay scenario and the uncaused beginning scenario is significant because, in the radioactive decay scenario, the numerically distinct carbon-14 atoms *x* and *y* already exist before the decay. Now the objector might claim that there is no metaphysical ground for '*why x* but not *y* decays'. I argue against this under the third argument below; but the point here is that, in any case, there is still a metaphysical ground for '*x* but not *y* decays'; namely, *x* and *y* are not numerically identical, even though they are supposed to be qualitatively identical with respect to beta-decay.[7] Thus, it is the case that the numerical distinction between *x* and *y* is possessed by *x* and *y*, which concretely pre-exist 'the radioactive decay of *x* but not y'. On the other hand, I shall argue in Sects. 3.4 and 3.5 that the numerical distinction between x and y (call this property S) cannot be a metaphysical ground for '*x* begins uncaused but *y* begins caused'. The reason is because it is not the case that S is possessed by *x* and *y*, which concretely pre-exist 'the uncaused beginning of *x* but not y', and I shall argue in the rest of this chapter that nothing else provides the required metaphysical ground. Hence, this argument refutes uncaused beginning, but it is compatible with indeterministic radioactive decay.

Argument 3

If '*x* begins uncaused but *y* begins caused' is a brute fact, this implies that there is no metaphysical ground which restricts uncaused beginnings to only *x* or makes *x* different from other things/events such as *y* with regard to uncaused beginnings; thus (contrary to the supposition), *y* would also begin uncaused unrestricted. The argument can be formulated as follows:

1. If 'x begins uncaused but y begins caused' is a brute fact, then there is no metaphysical ground which makes it the case that *only x* (and not y) begins uncaused.
2. If there is no metaphysical ground which makes it the case that *only x* (and not y) begins uncaused, then there is no restriction of uncaused beginnings to *only x* (and not y).
3. If there is no restriction of uncaused beginnings to *only x* (and not y), then y would begin uncaused.
4. It is not the case that y begins uncaused. (From 1)
5. Therefore, it is not the case that 'x begins uncaused but y begins caused' is a brute fact.

Against premise 3, it might be objected that 'just because *y could* begin uncaused does not imply that it *would* begin uncaused'.[8] However, this objection is based on a misunderstanding. I agree that 'could' does not imply 'would'. But I didn't argue or assume otherwise. 'Could' concerns possibility, but premise 3 is not referring to possible events. Rather, premise 3 is referring to (supposed) actual events. In other words, premise 3 is not referring to what could happen but what does happen. It states that the absence of restriction of 'what does happen uncaused' to only x implies that other events (e.g. y) which do happen also happen uncaused. For example, consider the scenario in which something (say) the universe began to exist and there was also a rapid increasing in strength of electric fields under certain circumstances around me. In this scenario these are not just possible events (i.e. it is not merely the case that the universe *could* begin to exist and electric field *could* increase in strength), but actual events; that is, the universe did begin to exist and electric field did increase in strength. Now suppose the former is x and the latter is y. Since premises 1 and 2 refer to what actually happens rather than merely what could happen, what follows from premises 1 and 2 is 'there is no restriction that uncaused beginning *would* only occur for x and not for y'. This provides the justification for concluding that y would begin uncaused.

It might be objected that 'just because there is no restriction to prevent a thing from behaving in a certain way does not entail that the thing would behave in that way'. For example, there was no restriction to prevent Peter from going for a walk today, but in fact Peter chose not to go.

In reply, there is a distinction between (A) 'no restriction to prevent a thing from behaving in a certain way' and (B) 'no restriction of a kind of event (uncaused happening) to only one thing'. Concerning (A), I am not claiming that 'there is no restriction to prevent a thing from behaving in a certain way' entails that 'the thing would behave in that way'. For in this case the thing can have the capacity to behave otherwise; for example, Peter has a capacity to choose not to go for a walk. Thus, the antecedent condition does not entail that the thing will behave in that way.

The case is different concerning (B) 'no restriction of a kind of event to only one thing'. For example, if there is no restriction of 'falling to the ground' to a particular thing, then falling would happen to other things because of the nature of reality as described by the law of gravity. (The hot air above a fire rises because its density is lower than the surrounding air; in this case, its lower density serves as a restriction to prevent it from falling.) Likewise, the absence of metaphysical restriction in premise 3 implies that the nature of reality would be such that there is no limitation of uncaused beginnings to only one particular thing x, which implies that uncaused beginnings would be unlimited and would also happen to other things such as y, which implies that y would also begin uncaused.

The third argument is different from the second argument because it is arguably incompatible with indeterministic radioactive decay. To see this, one can simply substitute 'undergoes radioactive decay' for 'begins uncaused' and 'does not undergo radioactive decay' for 'begins caused' into the above argument, and one gets the conclusion 'Therefore, it is not the case that *x* undergoes radioactive decay but y does not undergo radioactive decay is *a brute fact*'. This conclusion is compatible with scientific evidence, for as noted in Chap. 2 that there is insufficient scientific evidence for thinking that quantum events are genuinely indeterministic. Concerning probabilistic causation, there is strong justification for maintaining that the phenomena (the decay and statistics they exhibit) themselves have underlying proportionate causal explanations, for they exhibit regularities that strongly indicate the existence of more fundamental ordered causes (Stoeger 2001, p. 87).

The argument is compatible with indeterministic libertarian free choice, according to which 'A is chosen by person P but not-A is not-chosen by P' is a brute fact. It does not follow from this that there is no metaphysical ground for *only A* (and not not-A) is chosen, because person P *is* the metaphysical ground which makes it the case that *only A* is chosen.[9] The argument is also compatible with the uncaused existence of God as a brute fact. It does not follow from this that 'God exists uncaused beginninglessly but y begins to exist caused' is a brute fact, because God's existence having no beginning whereas y's existence having a beginning is the metaphysical ground in this case (see Sect. 3.5). Neither does it follow from this that 'God exists uncaused beginninglessly but an eternal Quadrinity does not exist uncaused beginninglessly' is a brute fact, because there could be preventive conditions in the beginningless state which makes such a state incompatible with a Quadrinity existing (see Sect. 3.8.4).

Finally, even if the above three argument fail, one may offer a probabilistic version as follows: If there is no metaphysical grounding which restricts uncaused beginnings to *only x*, then some other event would occur uncaused at some point with an overwhelmingly high probability. Consider the decay example again as an analogy. The chances being the same entails that the other, yet undecayed atom would decay at some point (during a sufficiently long time) with an overwhelmingly high probability (or some other atom in a sufficiently big ensemble would also decay in the next moment with an overwhelmingly high probability.) Therefore, if only one carbon atom decayed after a sufficiently long time (or in a sufficiently big ensemble), that would be overwhelming evidence that the chance of that carbon atom decaying is after all different from the chance for other carbon atoms, which would then require a special or differentiating metaphysical ground with respect to decay. Likewise, if only x began uncaused after a sufficiently long time (or in a sufficiently big ensemble), that would be overwhelming evidence that the chance of x beginning uncaused is after all different from the chance for other events, which would then require a special or differentiating metaphysical ground with respect to uncaused beginnings.[10]

In summary, the restriction to 'only *x*' (but not other things) begins uncaused must be grounded in something such that there is a relevant difference between *x* and other things—one should not simply say that it is brute fact. For if there is no relevant difference between *x* and other things (say) *y*, this would imply that *x* and *y* are the same where beginning to exist uncaused is concerned. The argument can be formulated in this way:

1.1.1.1. If it is not the case that there is a metaphysical ground which differentiates between *x* and *y* with respect to uncaused beginning, then *x* and *y* would be the same with respect to uncaused beginning.
1.1.1.2. (According to the objector) *x* and *y* are not the same with respect to uncaused beginning (given that the objector claims that *x* begins uncaused but not *y*).
1.1.1.3. Therefore, (the objector would require the claim that) there is a metaphysical ground which differentiates between *x* and *y* with respect to uncaused beginning.

It should be noted that premise 1.1.1.1 is based on arguments 1, 2, and 3 (either one of which is sufficient) and none of the premises of these arguments assumes the truth of the Causal Principle, while premise 1.1.1.2 is what the objector claims. Thus, there is no circularity in my argument; that is, my argument does not beg the question against the objector by assuming the truth of the Causal Principle. What follows from premises 1.1.1.1 and 1.1.1.2 is that the objector would require the claim that there is a relevant difference between x and y. Since 'x begins uncaused but y begins caused' requires a metaphysical ground which differentiates between x and y in order for it to be possible that *only* x begins uncaused (but y begins caused), it cannot just be a brute fact (a brute fact by definition has no further metaphysical ground). Therefore, 1.1.1 is false. The objector to my Modus Tollens argument would need to appeal to metaphysical grounding in some form or another, and this will be discussed below.

## 3.4 Against 1.1.2. Abstract Entities

A Platonist objector might modify Oppy's objection by claiming that the initial state of reality (ISOR) refers to the initial state of *concrete* reality. The objector might then claim that this leaves open the possibility that there could be beginningless abstract objects which exist timelessly. An abstract object, by definition, cannot cause something. Nevertheless, a Platonist objector to my argument by claim that abstract objects might provide metaphysical grounding for why only ISOR begins to exist uncaused. For example, the objector might say that the relevant difference between *x* (ISOR) and *y* is that ISOR and *y* are different logical possibilities (call this Difference D), and that Difference D exists in the abstract world. Alternatively, the objector might suggest that objects *x* and *y* subsist in the abstract world before they begin to exist, the subsisting objects already have haecceities before they begin to exist, and these haecceities differentiate between *x* and *y*. Another alternative is:

> A Platonist might suppose that there are brute necessary truths about uninstantiated properties, including truths about which properties can begin to be instantiated uncaused. On this theory, perhaps (contra Loke) there are things—abstract things—prior to an uncaused beginning that could explain why that beginning has its particular properties. (Rasmussen 2018)

There are a number of problems with the Platonist objection.

First, the objector's claim is that '*x* (ISOR) begins uncaused and *y* begins caused'. This claim refers to concrete entities *x* and *y*, and requires something concrete (whether qualitative or non-qualitative) to differentiate between two different things *x* and *y* in the concrete world. Hence, merely appealing to the haecceities of subsisting objects which exist in the abstract world is inadequate. (Contrast with Scenario 1 in Sect. 3.3 in which H is possessed by *p*, which concretely pre-exists the decay of *p*.)

Second, abstract objects (if they exist) merely describe relations (e.g. mathematical relations such as 2 + 2 = 4; logical relations such as 'if, then', relations between events) or possibilities/necessities (e.g. shapeless square cannot exist but shaped square can exist), or are merely exemplifiable by things (e.g. the property of redness are exemplified by red things).

Abstract objects by themselves (i.e. apart from concrete objects) do not make it the case that things/events happen in one way rather than the other in the concrete world. Indeed, there are disputes concerning whether abstract objects even exist (Gould 2014). But even if they do, they do not make a difference to the concrete world.[11] Thus, Difference D in the abstract world would not make a difference concerning ISOR beginning uncaused in the concrete world but *y* does not, and therefore Difference D is not a relevant difference. It is a difference that makes no difference, which means it is not a relevant difference.

Third, to claim that an abstract object X would make a difference in the concrete world such that 'only ISOR begins uncaused but *y* does not' would be to think of X as something similar to a concrete entity which exists prior to ISOR. In that case 'ISOR' would not be the initial state (of reality) anymore; rather, X would be the initial state. In any case, even if one insists that X is still an abstract entity, X would have to be beginningless, because if it has a beginning it would (like 'ISOR') require something to make it the case that only X begins uncaused but *y* does not. In that case, one needs to ask how could a beginninglessly existing X make a difference that has a beginning of existence (viz. the beginning of 'ISOR') rather than a difference that coexist beginninglessly with X. As argued in Chap. 6, the answer to this question indicates that X has libertarian freedom, which means that X is a Creator, which is the conclusion of the Cosmological Argument! Hence, this objection to the Cosmological Argument would fail.

Against my view that abstract entities do not make a difference to the concrete world, an objector might insist that difference in possibilities can make a difference in whether possibility *x* rather than possibility *y* is realized in the concrete world without causing its realization, and that instantiation of possibility does not have to be seen as an effect of some cause.[12] To illustrate, consider possibility *x*: It is possible that Peter exists, and possibility *y*: it is possible that Peter does not exist. Given that Peter's parents chose to conceive, possibility *x* is realized but not possibility *y*. Peter's parents cause the realization of *x*. One might say that the difference in possibility *x* and possibility *y* makes a difference in whether possibility *x* rather than possibility *y* is realized in the sense that it explains why Peter's parents causing the realization of possibility *x* does not also cause the realization of possibility *y* (answer: because possibility *x* and possibility *y* are different and *y* is not realizable given that *x* is realized).

However, such an explanation is not really making a difference in the concrete world, but merely explaining the difference made. In other words, it is not the difference in possibilities that makes a difference in the concrete world, but Peter's parents (concrete entities) who make a difference in the concrete world. This illustrates that differences in possibilities in the abstract world do not really make a difference in whether possibility *x* rather than possibility *y* is realized in the concrete world; on the contrary, the instantiation of possibility in the concrete world is made by concrete entities.

Recall premise 1.1.1.1, which states:

> 1.1.1.1: If it is not the case that there is a metaphysical ground which differentiates between *x* and *y* with respect to uncaused beginning, then *x* and *y* would be the same with respect to uncaused beginning.

Premise 1.1.1.1 can be rephrased as:

> 1.1.1.1′: If there is no relevant difference between the possibility A that x begins to exist uncaused and the possibility B that y begins to exist uncaused, then both possibility A and possibility B would be instantiated if possibility A or possibility B was instantiated.

This belongs to a more general principle:

If there is no relevant difference between a possibility A and a possibility B, then both possibility A and possibility B would be instantiated if possibility A or possibility B is instantiated.

Premise 1.1.1.1 involves uncaused beginning and the general principle may not involve that, and that is a difference between them. One might object that this difference between them seems to contradict my earlier conclusion that a difference in abstracta cannot be a relevant difference.[13]

It is true that there is a difference between premise 1.1.1.1 and the general principle, but that difference is not relevant to making a difference in the concrete world and hence does not contradict my earlier conclusion. Rather, the difference is relevant in the sense that it merely describes the different consequences that would follow *if a difference is made in the concrete world.* By my reasoning, the difference between my premise 1.1.1.1 and the general principle is a difference in abstracta and can be a difference that is relevant for describing the different consequences that would follow if a difference is made in the concrete world. However, this difference is not relevant for making a difference in the concrete world in the first place; in particular, it is not relevant for making a difference such that ISOR begins uncaused in the concrete world but *y* does not.

To be a relevant difference, the difference has to make a difference in the concrete world, because making a difference in the concrete world such that ISOR begins to exist uncaused in the concrete world but *y* does not is what my sceptical opponent (not me!) is claiming here for the abstract realm to do, but that is nonsensical since abstract realm does not do this kind of thing. By saying this I am not assuming that concrete instantiation of a possibility has to be causal (hence, I am not begging the question); rather, I am merely explaining what my opponent is claiming in order for his/her objection to work and why he/she has failed to meet this claim; therefore, the objection fails.

The Platonist might attempt to support the objection by citing the argument that at least some dispositions have ceteris paribus (CP) clauses incorporating uninstantiated abstract universals, and that these CP clauses help to delimit the range of manifestations of those dispositions (Dumsday 2019, p. 22). To elaborate, Dumsday argues:

> Take some value of mass, and a second value of mass, specify the distance relation, and a physicist could tell us what the resulting attractive force would be, ceteris paribus. We can then specify that the two masses belong to two objects which have a particular value of positive charge actually instantiated in our world, and a physicist could again calculate what the attraction would be, or whether instead it would be trumped by the repulsive force between the two like charges. Now do so for a value of positive charge that is not and has never been instantiated in our world. Once again, a physicist could calculate the results. The uninstantiated value is

just as legitimate a part of the set of CP clauses of mass as are the instantiated values. (p. 13)

Dumsday concludes that with the CP clauses

> we have abstracta determining that certain events can or cannot take place under particular circumstances. If an uninstantiated value of positive charge were instantiated in entities possessing mass, then where those entities would normally undergo a gravitational attraction of a certain force, they might instead be repelled … Even in their uninstantiated state, these universals serve as truthmakers for counterfactuals involving actual, instantiated dispositions. This counts as playing a governing role in the physical universe. (p. 14)

Nevertheless, the above argument cannot be used to support the Platonist's objection to my argument[14] because the uninstantiated abstracta Dumsday mentions merely describes what would be the case if certain things were to exist concretely alongside other *pre-existing* concrete things, as well as whether certain events can or cannot take place under particular *pre-existing concrete* circumstances. For example, if a positive charge of a certain value were to exist concretely, its repulsive force would trump the attraction of the pre-existing concrete masses. The uninstantiated abstracta do not make a difference as to which metaphysically possible set of properties are actually instantiated in the concrete world. In the case of $Mass_1$ attracting $Mass_2$ with a force F ('if no positive charge P …'), the force F in the concrete world is determined by the concrete entities (the masses and the distances between their centres). The CP clause 'if no positive charge P' does not determine the force in the concrete world but merely indicates that if P were to exist concretely, F would be different in the concrete world.

In conclusion, abstract objects by themselves would not be able to provide metaphysical grounding for *x* (ISOR) rather than *y* begins uncaused. Hence, Oppy cannot simply claim that there are metaphysical principles 'initial thing begins uncaused' and 'everything non-initial things have causes' which explain why x begins uncaused but not y.[15] The reason is because metaphysical principles/laws of nature are not concrete entities like tables or chairs; rather, they are supposed to be abstract entities. Thus, no abstract metaphysical principle/law of nature could (by itself) make it the case that only ISOR rather than other things begins uncaused. What makes things happen one way or the other are concrete entities and their properties, and I shall argue in the rest of this chapter that no such concrete entities and their properties can make it the case that only ISOR rather than other things begins uncaused.

Against my view that abstract entities by themselves do not make a difference to the concrete world, Malpass cites the Archimedes principle 'any object, totally or partially immersed in a fluid, is buoyed up by a force equal to the weight of the fluid displaced by the object'. Given some actual facts about a concrete entity and the Archimedes principle, it follows that the body would float.[16]

In reply, the Archimedes principle is abstract and it merely describes the relation between the object and the fluid. What makes the body float are the properties of the fluid and the body, and the Archimedes principle merely describes the relation. As Feser (2013, p. 254) observes, the laws of nature are 'mere abstractions and thus cannot by themselves explain anything. What exist in the natural order are concrete material substances with certain essences, and talk of "laws of nature" is merely shorthand for the patterns of behaviour they tend to exhibit given those essences.' When we use the Archimedes principle as part of an explanation we are only using it as a shorthand. In other words, a principle/law of nature is part of the explanation only because of the essential properties of the concrete entities. Likewise, in order to metaphysically ground 'x (ISOR) rather than y begins uncaused', the properties of concrete entities are required, but as I shall show in the rest of this chapter, there are no such properties. Hence, Oppy's theory fails.

It might be asked how would I answer the question 'Why uncaused events do not begin around us?' Do I not appeal to the abstract causal principle 'whatever begins to exist has a cause?[17]

In response, I do affirm the Causal Principle, but this principle is just the consequence of my view that, in our actual world, things happen one way rather than the other as a result of concrete entities and their properties, which implies that without concrete causes doing the work nothing would begin to exist. That is why uncaused events do not happen.

It might be objected that 'in our actual world, things happen one way rather than the other as a result of concrete entities and their properties' is also an abstract principle.[18] This is true. My view is that abstract principle *by themselves* do not make a difference in the concrete world. In this case, the abstract principle is not 'by itself'; rather, it involves concrete entities. Moreover, in this case, 'no event begins uncaused' is merely descriptive of the *absence* of a difference in the concrete world concerning *x* and *y* where beginning uncaused is concerned; hence, it is consistent with my claim that abstract principle by themselves do not make a difference in the concrete world.

In summary, the objector who affirms 1.1.2 is claiming that abstract objects provide metaphysical grounding which differentiates between concrete events *x* and *y* with respect to uncaused beginnings. I argue that abstract objects by themselves do not ground a difference concerning concrete events. Moreover, if abstract objects by themselves do ground a difference with respect to concrete events such as *x* (but not *y*) begins uncaused, given that abstract objects are beginningless (more specifically, timeless) the uncaused event they ground should be timeless as well, but that is not the case.

In the following sections, we shall consider 1.1.3 something concrete with property S makes it the case that *x* (but not *y*) begins uncaused.

## 3.5 Against 1.1.3.1 S is a Property of *x*

Let us now consider 1.1.3.1. S is a property of *x* which makes it different from *y* (i.e. S is a property which *x* has but *y* does not have). 1.1.3.1 implies that x (ISOR) has some special and unique property S which makes it different from other things/events, and the possession of S by ISOR would be required to do the work of providing the metaphysically grounding which differentiates between ISOR and other things/events in order for it to be possible that only ISOR (but not other things/events) begins to exist uncaused.

However, the first time S is possessed by ISOR is at time $t_{isor}$, where $t_{isor}$ is the first time at which ISOR exists. Given that S is stipulated to be an essential property of ISOR and possessed by ISOR only, and given that

ISOR only begins to exist at $t_{isor}$, S would only begin to exist at $t_{isor}$. It follows that 'the possession of S by ISOR' can make it the case that 'it is only ISOR that begins to exist uncaused' only when ISOR has already begun to exist at $t_{isor}$. But what this means is that 'the possession of S by ISOR' cannot metaphysically ground the uncaused beginning of ISOR (but not other things/events) in the first place, since it is required that ISOR has already begun to exist in order that 'the possession of S by ISOR' can metaphysically ground the uncaused beginning of ISOR (but not other things/events).

Now there are causal account and non-causal account of something existing (Bliss and Trogdon 2014), and the objector is referring to a non-causal account here given that ISOR is supposed to be uncaused. But the point here is that, regardless of whether the account referred to here is causal or non-causal, any account/explanation/grounding in the form of S's possession by ISOR would only begin to exist when ISOR already exists. Any such account/explanation/grounding would *not* exist without ISOR having already begun its existence; thus, there is *nothing* for accounting/explaining/grounding the uncaused beginning of ISOR (but not other things/events). For a thing has to exist in order to have any properties at all. And a thing that begins to exist must have begun to exist in order to have any properties. Thus, no matter what S is, ISOR having S cannot make it the case that ISOR begins to exist, make it the case that ISOR begins to exist *uncaused*, and make it the case that nothing lacking S begins to exist uncaused.[19]

Hence, 'the property S is possessed by *x* only when *x* has already begun' challenges 'S grounds why only *x* begins uncaused', not on account of a temporal order, but on account of what needs to be grounded, namely, the uncaused beginning of *x* (but not y). While non-causal ontological dependence between simultaneous events is widely regarded as commonplace, in no case of such examples of non-causal metaphysical grounding is there a case of making it the case why only something having some properties begins to exist. (As I explained previously using the analogy of a house and house builder, we know that the work of making it the case that *x* rather than *y* begins to exist is usually done by causes, that is, causal grounding.) On the other hand, any such property S of *x* would not be able to metaphysically ground the uncaused beginning of *x* (but not y), because the uncaused beginning of $x$ (but not y) which supposedly needs to be grounded by S needs to happen in order that S can ground its happening—this violates the irreflexivity of metaphysical grounding. The uncaused beginning of existence of $x$ is supposed to be explanatorily *prior* to $x$'s possession of the property; something must exist in order to possess a property.[20] Thus, the conjunction of $x$'s uncaused beginning and $x$'s possession of the property cannot provide metaphysical grounding for the first conjunct (rather than the beginning of y).

Now I do not deny that there are some other unique (perhaps unknown) properties possessed by ISOR—of course there must be such properties, since ISOR is different from (say) a tiger, a dinosaur, and so on! Rather, my argument is that there cannot be any unique property (regardless of whether this unique property is known or unknown) *which can do the work of making it the case that only ISOR but not other things begins to exist uncaused if ISOR began uncaused.* As Oderberg (2002, p. 330) notes,

> it is no use saying that the nature or essence of some things (such as the universe itself) is to begin to exist uncaused, and of others not to, since before anything begins to exist how can the essence of a thing regulate the conditions under which *it* begins to exist? But since we know some things begin to exist only if caused, the Causal Principle cannot be false since its falsity would not allow of such a distinction.

More seriously, any attempt to provide such an account/grounding/explanation would entail a contradiction. For:

1. the uncaused (or 'uncorrelated'; likewise, below) beginning of ISOR (instead of other entities) is supposed to require the possession of S by ISOR in order to account/ground/explain only ISOR begins to exist uncaused: this implies that the possession of S by ISOR is explanatory prior to (i.e. metaphysically grounds) the uncaused beginning of ISOR.

However,

2. the possession of S by ISOR is supposed to require the uncaused beginning of ISOR (instead of other entities), for S would not be possessed by ISOR if ISOR does not already (begun to) exist: this implies that the uncaused beginning of ISOR is explanatory prior to (i.e. metaphysically grounds) the possession of S by ISOR.

(1) and (2) entail a contradiction and violate the irreflexivity of metaphysical grounding.

The following points should be noted.

First, when philosophers talk about metaphysical ground, they are not talking about their subjective mental states or their use of language. Rather, they are talking about objective properties of the world. Just as when we talk about the foundation of a house grounding the roof (at a particular position *instead of* another), we are talking about grounding as an objective property of the world.

Second, by 'explanatorily prior', I am not referring to 'how humans choose to explain things'. Rather, I am referring to the relationship between things/properties in the actual world. Just as '*x* (e.g. foundation) grounds *y* (e.g. the roof)' implies that it is not the case that 'the roof grounds the foundation' (this illustrates the irreflexivity of metaphysical grounding), likewise '*p* is explanatory prior to *q*' implies that it is not the case that '*q* is explanatory prior to *p*'. To illustrate the notion of 'priority' and to explain the distinction between 'require' and 'imply': I require the possession of money in order to pay for the house. The possession of money is prior. To say that *q* requires *p* is to convey that *p* is prior to *q*. However, to say that *q* implies *p* does not convey that *p* is prior to *q*.

Third, the above argument against 1.1.3.1 remains valid on a static theory of time. On static theory of time, it remains the case that there are also other events (e.g. *y* = increasing in strength of electric field) which have the beginning of existence in the same sense as the supposed beginning of ISOR, that is, being finite in temporal extent in whatever dimensions and having 'edges' (i.e. does not have a static closed loop or a changeless phase that avoids an edge). According to 1.1.3.1, S is supposed to metaphysically ground why ISOR begins uncaused but y (which lacks S) does not begin to exist uncaused. However, S is possessed when

ISOR already begins, which makes S unable to do this work of grounding, as argued previously.

An objector might ask, 'why couldn't the property of necessary existence or being the initial state be the special property S that metaphysically ground the uncaused beginning of ISOR?'

In response, on the one hand, as noted previously in Sect. 3.2, whether initial state begins necessarily (rather than existing beginninglessly necessarily) is precisely the issue under dispute, and the objector should not beg the question by assuming that this is the case. On the other hand, there is an independent reason for thinking that there cannot be any special property S which metaphysically grounds such an uncaused beginning, namely, the demonstration (explained above) that any such property would be unable to do the necessary work of metaphysical grounding. As Malpass puts it, 'the conjunction of $x$ existing and $x$ having a special property cannot explain the conjunct that $x$ exists. Conjunctions don't explain their own conjuncts.'[21]

The objector would ask how could a theist explain God's uncaused existence without falling into similar problems. After all, whatever special property *S* that is used to explain why God is uncaused will *already* have to be had by God.[22]

In reply, it should be noted that I am not claiming that anything that exists requires a special property to explain why it exists. Such a principle is obviously false. For example, my existence does not require a special property to explain why I exist. Rather, my existence is (at least partly) explained by my already-existing (i.e. pre-existing) parents who brought me into existence and I am not required to have a special property S. However, if instead of my already-existing parents I have already existed and always-already existed at all earlier durations and that my existence has no temporal boundary (i.e. beginningless) and supposing that I am not being sustained in existence, then no already-existing parents would be required and I am not required to have a special property. In this case, my beginninglessness is not a special property that explains why I exist; rather, my beginninglessness is merely a way of describing my always-already existence that has no temporal boundary, which also implies that no parents are required; that is, I would be uncaused (supposing that I am also not being sustained in existence).

The same reasoning applies to God. God is supposed to have always-already existed at all earlier durations and has no temporal boundary (i.e. beginningless), and He is not being sustained in existence; hence (unlike things with beginnings), no already-existing pre-existing causes are required and He is not required to have a special property S. In this case, His beginninglessness is not a special property that explains why He exists. Rather, His beginninglessness is merely a way of describing His always-already existence that has no temporal boundary (i.e. beginningless), which (together with the fact that He is not being sustained in existence) also implies that no causes are required; that is, He would be uncaused.

There are two distinct senses of explanation which need to be clarified: (1) a statement or account that makes something clear (OED); (2) to provide a metaphysical grounding for. In the case of ISOR beginning uncaused, I was arguing that there needs to be a special property S that not only makes something clear but also provides a metaphysical grounding for why ISOR begins uncaused but B begins caused (but there cannot be such an S). In the case of God existing beginninglessly, God's beginninglessness merely makes clear why is it the case that no cause or special property is needed (it doesn't provide any metaphysical grounding which S is supposed to provide). In particular, by explicating the meaning of beginninglessness, we can see why it implies that God would be uncaused (assuming for the sake of parity that both God and Oppy's ISOR are unsustained). Thus, 'beginningless' itself is not a special property S.

Hence, God existing uncaused doesn't need to be explained by S. Whereas according to 1.1.3.1, Oppy's view that '*x* (ISOR) but not y begins uncaused' needs to be explained by special property S, because (1) ISOR does not always-already exist in the same sense as God, but is finite in temporal extent and has a temporal edge, just like *y* (2) S is not supposed to be merely a way of describing ISOR's beginning of existence (since *y* also begins to exist); rather, S is also supposed to 'make a difference' by explaining why ISOR begins uncaused but *y* does not begin uncaused. But as argued previously, there cannot be any property S that can do the required work of explaining why ISOR begins uncaused but *y* begins caused. Therefore, 1.1.3.1 is false.

## 3.6 Concerning 1.1.3.2.1

Concerning 1.1.3.2.1, one cannot appeal to the concrete circumstance of *x* making it the case that only *x* begins uncaused, since this would amount to saying that the circumstances causes *x* and yet *x* is supposed to have begun uncaused, that is, without causally antecedent conditions. Indeed, on Oppy's theory there isn't any concrete circumstances in which *x* (ISOR) begins uncaused, since ISOR is supposed to be the very initial state which begins to exist.

## 3.7 Concerning 1.1.3.2.2. S is a Property of y

Oppy might suggest the possibility that, once ISOR begins uncaused, ISOR brings about caused entities (e.g. *y*) with properties which would make it the case that they would begin caused (rather than uncaused). Thus, even though ISOR begins uncaused, this does not imply that later entities would also begin uncaused, since later entities would be causally dependent on earlier entities for their beginning.

The problem with this view is that the property of *y* which (according to 1.1.3.2.2) is supposed to make it the case that '*y* would begin to exist caused' would be had by *y* when *y* has already begun to exist (caused). This implies that (contrary to 1.1.3.2.2) this property is unable to ground the caused beginning of *y*.

Therefore, the reason why *y* does not begin uncaused is not because of the property of *y*. Rather, as explained previously in Sect. 3.4, it is just the consequence of my view that what makes things happen are concrete entities and their properties, which implies that without concrete causes nothing would begin to exist. That is why uncaused events do not happen.

## 3.8 Concerning 1.1.3.2.3. S is a Property of the Circumstances of y

### 3.8.1 Objection: Current Spatial Considerations Prevent Things from Beginning to Exist Uncaused Now

Oppy has offered another argument for why (say) a tiger does not begin to exist uncaused now if the initial state of reality began to exist uncaused. Focusing on the well-established features of the part of reality that we now inhabit, Oppy argues that causally prior to some concrete object occupying the space currently occupied by another concrete object, the current occupant must vacate the space to make room for the new object. Thus, the former occupant's ceasing to occupy the space is a cause (but not the sole cause) of the new object's coming into being. Generalizing this line of thought, Oppy writes,

> Pick any tiger shaped space in the room. In order for a tiger to occupy that space, that space must have appropriate internal and boundary properties: there are after all, lots of ways that the boundary and interior of that space could be that are simply inconsistent with the occupation of that space by a tiger. But, if that's right, then it seems to me that we should allow that … the coming about of the consistency of the boundary and interior of the space with occupation by a tiger—is a cause of the coming into existence of the tiger. And as before, if this is a cause of the coming into existence of the tiger, then it … isn't true that the tiger 'comes into existence uncaused out of nothing. (p. 67)

Oppy then goes on to make a few remarks concerning possible objections to his argument, claiming, among other things, that even if there is no cause of (say) a table ceasing to occupy the location that it currently occupies, it will still be the case that the table's ceasing to occupy the location is a cause of the coming to existence of the tiger in that location. Additionally, even if most of the universe consists of tiger-shaped spaces the interior and boundary conditions of which are consistent with occupation by a tiger, it would still be the case that the coming about of such spaces would be a cause of the popping into existence of a tiger in that space (pp. 66–67).

In summary, Oppy's suggestion that, since in the reality that we now inhabit, the spaces the interior and boundary conditions of which are consistent with occupation by any entity *y* would be a causally necessary condition for the beginning of *y* in that space, *y* could not begin uncaused now. In a more recent paper, Oppy suggests the reason why it is impossible for a raging tiger to 'suddenly come into existence uncaused out of nothing' in the room in which you are reading this is that there is no place in that room for a tiger to come to occupy uncaused. He writes:

> In the causal order, the displacing activity of the displacing object—the object 'popping into existence'—would have to be both (causally) prior to the displacement of the displaced object (in order to cause the displacement) and (causally) posterior to the displacement of the displaced object (in order that the displacing object exists and hence is able to bring about the displacement). But that's impossible. (Oppy 2015, p. 4)

He also argues that,

> if—per impossible—something did 'pop into existence' at a particular location, we would properly regard the vacation of the space now occupied by the thing that comes into existence by the thing(s) that previously occupied that space as a cause—i.e., a necessary causal condition—of the existence of the new occupant of that space. Thus, even in this case, we would not have something popping into existence uncaused. (Ibid.)

An argument against Oppy's objection based on spatial considerations has been offered by Erasmus. Erasmus notes that spatial substantivalism (the view that space is a substance that is able to exist by itself independently of material objects) is a metaphysically possible view, and indeed many scientists and philosophers hold to this view. Erasmus concludes that, if the Causal Principle is false, then 'a space is a thing that can itself come into existence uncaused. Accordingly, we should observe empty spaces, and objects of the related shape, constantly coming into being all around us' (Erasmus 2018, p. 166).

Even if the possibility of spatial substantivalism and Erasmus argument is rejected, Oppy's objection can be replied to in various other ways. For example, one can think of 'uncaused beginnings' which do not involve displacement of objects or vacation of spaces, and hence are compatible with the fact that the spaces of our universe are occupied. An example would be 'a pre-existing electric field increasing in strength uncaused under certain circumstances', where 'field' is understood as a region of space in which there is a force. It is observed that the space that is occupied by the electric field is compatible with their existing and increasing. Furthermore, it is observed that different strengths of electric fields (as well as other fields) can occupy the same amount of space, unlike Oppy's example of tigers where an increased number of tigers would require more space.

Moreover, our experiences indicate that in the reality that we inhabit, pre-existent fields (e.g. electric and magnetic fields) and 'spaces the interior and boundary conditions of which are consistent with an increasing in strength of a pre-existent electric field' are found around us. This can be seen from the fact that, when we switch on an electric or magnetic field generator around us, there will be an increase in strength in the relevant field around it. Indeed, in our experiences we have observed many instances of such events beginning to exist. This shows that our circumstances are compatible with such events beginning to exist around us, and that such events do begin to exist (call one such event *y*).

Concerning *y*: an increase in strength of a pre-existent electric field, Oppy would object that, since an increase in strength of a pre-existent electric field requires spaces the interior and boundary conditions of which are consistent with this increase, this event cannot begin uncaused. In reply, in the reality that we now inhabit, what is causally necessary for an increase in strength of a pre-existent electric field is not merely the presence of such spaces. Rather, it would also include (for example) the switching on of an electric field generator under certain circumstances. In this situation, the switching on of the electric field generator would be a cause. Recall that (I) *x* (ISOR) begins *uncaused* but *y* begins caused implies that *no cause* making it the case that *x* (rather than y) begins uncaused, (II) *x* and *y* having a beginning implies that the properties of *x* and the properties of *y* which differentiate between them would be had by them only when they had already begun to exist, and (III) the circumstances are compatible with the beginning of *y*. Now it has been explained in Section 4 that (I), (II), and (III) jointly imply that there would be no difference between *x* and *y* where beginning uncaused is concerned. Hence, if *x* (ISOR) begins uncaused, there would be no difference between 'beginning of ISOR without cause' and 'beginning of increase in strength of electric field without [having to switch on the generator as a] cause' (see the definition of 'uncaused' in Chap. 2). This implies *y* would begin without having to switch on the generator. The consequent is not the case; therefore, it is not the case that ISOR begins uncaused.

An objector might suggest reinforcing Oppy's argument by developing further hypotheses about how existing things might place causal conditions with respect to any new state of affairs, such that only the first state of reality could begin without a causal condition (Rasmussen 2018). However, I have explained above that saying that '*y* begins to exist uncaused' would mean '*y* begins to exist without having to switch on the electric field generator under certain circumstances'. The point is that there can be no further hypotheses about how existing things would place causal conditions with respect to increasing in strength in electric field in this case, given that I define uncaused in this case as 'without having to switch on the electric field generator under certain circumstances'.

### 3.8.2 Objection Based on the Distinction between Different Senses of Beginning to Exist

Against my argument using the case of electric field increasing in strength under certain circumstances, it might be objected that there is a distinction between 'there was no entity *E* and then there was an entity *E* with property *p*' (i.e. an entity beginning to exist)—such as the beginning of existence of ISOR—and 'an entity *F* which already exists coming to possess property *q* it did not previously have' (i.e. an event beginning to exist)—such as an electric field increasing in strength. One might then claim that this distinction would be the relevant difference that explains why ISOR begins uncaused but an electric field increasing in strength under certain circumstances does not.

In response, both cases are compatible with the definition of beginning of existence which I used for my argument; namely, *x* has a beginning of existence if x has a temporal extension, the extension is finite, and it has temporal edges. Here, the beginning of *x* can refer to 'then there was an entity *E* with property *p*' (such as the beginning of existence of ISOR), and the beginning of *x* can also refer to 'an entity *F* which already exists coming to possess property *q* it did not previously have' (such as an electric field increasing in strength under certain circumstances). In both cases there is a beginning of possession of a property, and my argument would still apply. That is, if the event 'then there was *E* with *p*' begins to exist uncaused, then

I. this event would not have any causally antecedent condition which would make it the case that only this event rather than '*F* coming to possess *q*' begins to exist uncaused;
II. the properties of the events 'then there was *E* with *p*' and '*F* coming to possess *q*' would be had by them only when they had already begun to exist; and
III. the circumstances are compatible with the beginning of existence of these events.

As explained above, (I), (II), and (III) imply that there would be no essential difference between these events where beginning to exist uncaused is concerned. Thus, the distinction between different senses of 'beginning to exist uncaused' would not be a relevant difference that accounts for why only ISOR begins to exist uncaused but an electric field increasing in strength under certain circumstances does not begin to exist uncaused.

Likewise, there is no relevant difference between 'beginning to exist within time' and 'beginning to exist with time' simpliciter where my Modus Tollens argument is concerned. The reason is because both cases involve being finite in temporal extension and having temporal edges, which is the definition of beginning of existence used in my argument, and my argument would still apply to both cases, since they both fit the definition. None of the premises nor the justification for the premises requires the assumption that there is earlier time or no earlier time before something begins to exist. Therefore, whether there is time or no time before something begins to exist does not affect my Modus Tollens argument.

In summary, the Modus Tollens argument applies equally well to objects and events. For even though there is a distinction between events beginning and objects beginning, yet both of them are finite in temporal extension and have temporal edges; hence, my Modus Tollens argument applies to both cases. To rebut my argument, the objector would have to rebut the premises of my argument rather than merely redefining the problem by saying that he/she is merely referring to objects, which is merely dodging the problem facing his/her view.

### 3.8.3 Objection: Pre-existing Things Such as a Pre-existent Law of Nature Might Prevent Things from Beginning to Exist Uncaused Now

The objector might suggest that the relevant difference between *x* and *y* is that there is nothing prior to *x* whereas there is something prior to *y*, and that once *x* begins uncaused, *x* causes 'the circumstances of *y*' which causes *y*. That is why (*x* begins uncaused but) *y* begins caused, that is, because of the causal powers of 'the circumstances of *y*' which brought about *y*.

In reply, the above answer is inadequate, because it does not answer why doesn't y begin without a cause. If *y* begins caused, *y* is caused by its circumstances (circumstance is defined as 'a fact or condition connected with or relevant to an event or action', OED). To claim that '*y* begins caused because the circumstances of *y* causes *y*' does not answer 'why does the beginning of y (but not the beginning of x) needs a cause' but merely restates the fact that '*y* begins caused (by the circumstances)'. It also fails to answer why doesn't y-type events begin uncaused but are observed to be correlated in an orderly manner with (say) the switching on of the electric field generator. While such circumstances as the switching on of the electric field generator may causally explain those y-type events brought about by them, these circumstances do not explain why y-type events (but not x) do not also begin without such circumstances resulting in a lack of orderly correlation.

The objector might suggest an alternative explanation that, once the universe existed, the universe would be the antecedent condition which makes it the case that other things do not begin to exist uncaused. For example, once the universe began to exist uncaused, there would be laws of nature which prevent other things from beginning to exist uncaused and therefore we do not observe things beginning to exist uncaused now. As Oppy (1991, p. 196) argues, perhaps just any and everything can come into existence uncaused, but our universe is governed by certain conservation laws which ensure that such events do not actually happen. Perhaps there is a true subjunctive conditional to the effect that, were *y* to come into existence uncaused, then *y* would possess some property P*, and it is the truth of this subjunctive conditional which accounts for the fact that *y* does not come into existence uncaused.[23] One might say that *y* would possess the property of violating the law of conservation of energy if *y* begins uncaused, and *y* does not possess this property if *y* begins caused, and that the consequence of violating the law of conservation of energy would prevent *y* from beginning uncaused. With regard to my example of increasing in strength of electric field under certain circumstances, one might object that, since an uncaused change in electric charge would create an unbalanced surplus of charge, the so-called uncaused beginning of increase in strength of electric field has the property of being susceptible to prevention by conservation laws.[24] Alternatively, one might suggest that perhaps the Causal Principle only began to exist with the beginning of existence of our universe.

Oderberg (2002, p. 331) replies that Oppy's appeal to the laws of nature begs the question. He writes:

> On a regularity view of laws it is just a general description of what happens, not an explanation. On a necessitarian view of laws, it still does not explain why things require a cause of their existence, since the necessity inherent in the law derives from the powers of existing things, and so appeal to the law merely invokes their existence rather than explains it. Why should the laws be as they are? Because of the powers of existing things. But why can't those things come into existence uncaused? Because that's how the laws are.

Against Oderberg, some might object by claiming it has not been shown that all the laws of nature derive from the powers of existing things. Others have cited the law of conservation of energy as a counterexample to dispositionalism, arguing that the law indicates that interactions are constrained by the requirement of preserving the mass-energy, but that constraint does not seem to be the manifestation of a disposition (Chalmers 1999, pp. 12–13). It has been suggested that the law of conservation may derive from symmetries. For example, Lange (2016, p. 64) proposes that

> A conservation law ... may have an explanation. In fact, one way for a conservation law to be a constraint is for it to arise from a symmetry principle ... As is well known, various classical conservation laws follow from various spacetime symmetries within a Hamiltonian dynamical framework: energy conservation follows from the laws' invariance under arbitrary temporal displacement, linear momentum conservation from their invariance under arbitrary spatial displacement, and angular momentum conservation from their invariance under arbitrary rotations. If these derivations explain why the conservation laws hold (as they are often said to do), then the conservation laws are constraints, not coincidences.

However, one should ask why the symmetries hold. Lange (2016, p. 82) claims that a symmetry principle is 'a 'metalaw': a law that governs the laws that are expressed by subnomic claims (the 'first-order' laws). But why should the metalaw hold? Bird (2007, pp. 213–214) states that it is a mystery why symmetries and conservation laws hold, and suggests that 'the dispositional essentialist ought to regard symmetry principles as pseudo-laws ... it may be that symmetry principles and conservation laws will be eliminated as being features of our form of representation rather than features of the world requiring to be accommodated within our metaphysics'. Lange (2016, p. 94) acknowledges that Bird may be proved right, but objects that 'in any event, a metaphysics that cannot do justice to explanations by constraint is at a serious disadvantage'.

In reply, one can do justice to explanations by constraint by arguing that the representation mentioned by Bird holds because of the Causal Principle, such that if the Causal Principle is violated, the law of conservation would be violated as well. It is interesting to note Chalmers' (1999, pp. 12–13) observation that 'a characteristic feature, and a major strength, of thermodynamics is that it applies at the phenomenological level whatever the details of the causal process. It is precisely this feature of the laws of thermodynamics that makes it difficult to portray them as causal laws.' Similar to the law of conservation of energy, the Causal Principle 'whatever begins to exist has a cause' also does not depend on the details of the causal process. The hypothesis that the law of conservation of energy is based on the Causal Principle would explain this observation.

More importantly, one should ask how would the existing universe, law of nature, and so on be able to prevent other things from beginning to exist uncaused if ISOR began to exist uncaused. Now Bigelow et al. (1992) had postulated that

> the world has an essence, and that essence requires that mass-energy, charge, lepton number, etc. are conserved in all interactions … perhaps there is a property corresponding to the kind, the property of being a world, and this property has as its essence the disposition to conserve energy, etc. in response to any event.

On the other hand, Fine argues that, while the proposition that electrons have negative charge is metaphysically necessary in virtue of the identity/definition of electrons, 'energy is conserved' is at most naturally necessary but not metaphysically necessary, because it is hard to see how it could be partly definitive of energy that it should be conserved (Fine 2002, p. 261).[25] Wolff (2013) objects that a certain conservation law closely tied to symmetry principles via Noether's second theorem is an instance of a metaphysically necessary physical law, because it supposedly follows from the interdependence of matter and gauge fields, and this interdependence can seem to look like the result of a mere mathematical identity (p. 904). Against Wolff, Linnemann (2020, p. 7) argues that, even in the derivation of the conserved current via Noether's second theorem, some particular equations of motion (which on Wolff's view would count as examples of physical laws that are naturally but not metaphysically necessary) were used. Linnemann claims instead that the conservation of topological currents holds in virtue of the identity of the fields, and thus, on Fine's account of necessity, with metaphysical necessity (p. 10). On the other hand, Wilson (2020) claims that the conservation laws are metaphysically necessary because it occurs in every branch of the wavefunction.

However, given that multiple interpretations of the basic mathematical formulation of quantum mechanics are possible, any metaphysical claim of the form 'quantum mechanics entails *x*' is likely to be false (Lewis 2016, p. xi). Whether Linnemann's or Wilson's account can withstand further scientific scrutiny remains to be seen. I shall now argue below that, in any case, our world would have been very different if ISOR began to exist uncaused, and that the subjunctive condition concerning *y* mentioned above would be false because the uncaused beginning of *y* would not have been preventable if something begins to exist uncaused.

To begin, the problem with the view that once ISOR began it imposed a metaphysical principle or natural law that ensures subsequent entities begin caused is that (as argued previously) metaphysical principle or natural law is not concrete but abstract. Abstract objects do not make things happen in one way rather than the other. Thus, no such principle or law by itself could make it the case that only ISOR rather than other things begins uncaused. What makes things happen one way or the other are concrete entities and their properties. Thus, the objector should not simply suggest a principle or law of nature and say that suffices to explain why things do not begin uncaused now if ISOR began uncaused. Neither should the objector simply suggest that the Causal Principle only begins to exist with the beginning of existence of our universe. These suggestions would not work in the absence of concrete entities and their properties doing the metaphysical work of grounding why x (ISOR) rather than y begins uncaused. (As explained in previous sections, the abstract Causal Principle is just the consequence of my view that what makes things happen are concrete entities and their properties, which implies that without concrete causes doing the work nothing would begin to exist. For the objector's hypothesis that 'the Causal Principle only begins to exist with the beginning of existence of our universe', there would need to be concrete entities and their properties doing the metaphysical work of grounding the restriction of the Causal Principle to *y* but not *x*.)

Hence, one should ask how would the conservation law (or any other law of nature) be able to prevent (say) the uncaused beginning of an unbalanced surplus of charge if ISOR begins uncaused. Now in order for *x* (e.g. law of conservation of energy) to prevent *y* from beginning to exist, *x* would have to either remove the causally necessary conditions or act on the circumstances to make them incompatible with *y* beginning to exist. (For example, in order to prevent a moving battery-operated toy car from entering a room, I would have to either remove the causally necessary conditions, such as by removing the batteries, or act on the circumstance to make it incompatible with the event occurring, such as by filling the room with hard objects such that there is no space for the car to enter into; see Sect. 3.8.1.) However, if ISOR began to exist uncaused, what this implies is that

I. there is no causally necessary condition which makes it the case that only ISOR rather than other things begins to exist. In particular, any time *t* and any location *l* would not be such a causally necessary condition.
II. Additionally, as explained previously, any difference between ISOR and other things would be had by them only when they had already begun to exist. (I) and (II) imply that there would be no difference between them where the requirement for causally necessary condition is concerned. Moreover,
III. it has been explained previously that the circumstances around us have been shown to be such that they are compatible with the beginning of existence of an increase in strength of electric field under certain circumstances.

(In addition to (I), (II), and (III), the three problems noted below are relevant as well.)

Against the above, an objector might argue that once x (ISOR) begins uncaused, *x* causes the concrete circumstance of *y* to have properties which ground certain laws of nature and make it incompatible for events to begin uncaused within it. As Carrier (2018) argues, 'the very reason we do not observe a violation of *ex nihilo nihil* is that those extant properties and laws now prevent "just anything" from happening. The only *nihil* we observe is actually *a thing*: propertied spacetime. And that thing, being existent, now limits what can happen.' For example, one might suggest the possibility that some concrete substance (say, spacetime [this assumes a substantival view of space and time])[26] began to exist uncaused with the beginning of the universe, and the properties of this concrete substance determine (causes) the total mass-energy of the universe to remain constant throughout time by making the circumstances incompatible with increase in total mass-energy. Therefore, even if our present circumstances are compatible with the beginning of increase in strength of electric field under certain circumstances, we would *not* observe such events beginning to exist *uncaused*, and the law of conservation of mass-energy would hold. In this sense, $y$ beginning in spacetime but $x$ does not would be a differentiator between $y$ and $x$ (ISOR), which explains why the former requires a cause and the latter does not.[27]

However, there are at least three arguments against the above objection; these arguments are independent and any one of them would suffice (I shall discuss the first two in this section and the third in the next section).

First, consider the above-mentioned example of a moving battery-operated toy car entering the room. One can make the circumstances incompatible with this event by filling the room with hard objects, because the car necessarily occupies space. However, an increase in strength of electric field does not necessarily occupy more space; in fact, the spaces around us are compatible with different levels of strengths of electric fields. This indicates that there is in fact no existing thing which makes it the case that the circumstances of our universe are such that it is incompatible with increase in total level of strength of electric field and hence increase in total mass-energy if something can begin uncaused. Rather, as I have argued, the reason why the law of conservation holds is because the causal principle (something does not begin to exist uncaused) holds. My argument is that, if this principle is false, as the sceptic suggests, then the law of conservation would not hold, but the consequent is not the case.

Second, the law of conservation of energy would not prevent energy-conserving events from *beginning uncaused* (e.g. without causal interaction) if ISOR begins to exist uncaused. It should be noted that there is no law or spatial considerations that now prevent energy-conserving changes from *beginning*. Indeed, such events happen frequently in the present, which indicates that the present circumstances are compatible with such changes happening. For example, hydrogen is currently being fused into helium in the sun, and in the process of the causal interaction some of the mass of the hydrogen atoms is converted directly to energy according to $E = mc^2$. Now if ISOR begun to exist *uncaused*, then given (I) the beginning of ISOR does not have causally necessary condition which makes it the case that only the beginning of ISOR rather than certain other events *y* begins uncaused, (II) any difference between the beginning of ISOR and the beginning of *y* would be had by them only when they had already begun to exist, and (III) the compatibility of the circumstances with the beginning of y, we can likewise expect these energy-conserving events (let these = *y*) to begin to exist uncaused now. These events would not be preventable by the law of conservation of energy given that they are compatible with it. But we do not observe these energy-conserving events/changes beginning uncaused now; on the contrary, scientists have described the causal interactions that brought about events such as the fusion of hydrogen into helium. Thus, the antecedent is false.

### 3.8.4 A Second Form of Modus Tollens Argument

Third, there is a second form of Modus Tollens argument which is immune to the objection that after the initial state has begun uncaused, some concrete substance would prevent things/events from beginning uncaused. It is immune to the objection because it implies that these would be other spacetime blocks which would begin uncaused initially (not after the initial state has begun) and massively disrupt our universe in an unconstrained manner, in which case our universe would have been very different. The argument can be formulated as follows:

1. If *x* (ISOR) begins uncaused, then some other possible spacetime blocks *y* and *z* would also begin uncaused.
2. It is not the case that *y* and *z* begin uncaused.
3. Therefore, it is not the case that *x* begins uncaused.

The justification for premise 1 is that,

I. if x (ISOR) begins uncaused, this implies that the beginning of *x* would not have causally necessary condition which makes it the case that only *x* (rather than other possible spacetime blocks *y* and *z*) begins to exist.
II. Additionally, any difference between the possibilities *x*, *y*, and *z* would be had by them in the concrete only when they had already begun to exist. Moreover,
III. there would be no pre-existing entity that makes it incompatible for *y* and *z* from beginning to exist initially.

(I), (II), and (III) jointly imply that there would be no difference between the possibilities *x*, *y*, and *z* where beginning to exist uncaused is concerned. Hence, if *x* begins uncaused, *y* and *z*—as well as other things with other possible physical laws or metaphysical principles—would all begin uncaused initially.

This consequence has surprisingly been accepted by atheist Richard Carrier (2018) as a way to account for the fine-tuning of our universe. His reasoning is that, since 'nothing (other than what is logically necessary) prevents anything from happening to that Nothing' (ibid.), anything that is logically possible—including an actual infinite multiverse ensemble which includes our 'fine-tuned' universe—would begin to exist uncaused from nothing. Carrier concludes:

> This entails that the assertion *ex nihilo nihil*, 'from nothing, comes [only] nothing,' is false. Because that is a rule, and Nothing contains no rules. No such rule can therefore exist when there is Nothing, so as to govern that Nothing. Therefore it cannot be the case that *only* nothing comes from Nothing. In fact we cannot even establish that it is *likely* that only nothing will come from Nothing. (Ibid.)

Nevertheless, there are several problems which Carrier fails to note.

First, Carrier's hypothesis faces Mawson's objection (2011) that on such a hypothesis in which every possibility is actualized, the probability of any universe in which we can more or less continually and consistently understand through induction would have been infinitesimally small, but that is not the case (see the discussion on multiverse in Chap. 4).

Second, for any spacetime block *x*, it is possible that there are *y*, *z*, … and so on which expand and collide with *x* and leave behind detectable effects. Cosmologists have been discussing the possibility of collision of multiple universes. Some have claimed that there are 'scars' of the collision detected on the Cosmic Microwave Background Radiation (CMB) as 'cold spots', while others have disputed the existence of the these 'scars' or argued that these 'scars' could have alternative explanations such as inhomogeneous reheating associated with non-standard inflation (Mackenzie et al. 2017). In any case, it is possible (though perhaps not yet proven) that there had been collision of universes leaving behind scars on the CMB. It has been objected that multiverses are brought about by an early inflationary phase and the process of inflation would have prevented the collision (Siegel 2018). However, if (instead of being caused by inflation) our universe began to exist *uncaused* alongside an actual infinite number of universes which also begun uncaused as Carrier suggests, a huge number of universes would collide with one another and with our universe in an *unconstrained* manner, and the huge number of collisions would generate huge amounts of radiations and would leave behind much more obvious traces rather than a few disputed 'scars' on the CMB. (The fact that there could also be other possible spacetime blocks that begin uncaused but do not affect our universe does not deny this consequence.) It would not help to say that the universes could have merged together, leading to the formation of our present universe, since given the lack of a cause, the beginnings would be unconstrained. In this case, not only 'a small universe' would begin uncaused, rather all kinds of possible universes (including those as big as ours!) with opposing properties would begin uncaused, since there would be no difference between them where beginning to exist uncaused is concerned, which implies that there would be no constraint whatsoever and the CMB would be totally disrupted. The opposing effects of the opposing properties would also cancel one another and no order would emerge from it. In short, the result would be totally disruptive rather than resulting in the fine-tuned and highly mathematically ordered universe which we see. It would also massively disrupt the spacetime substance of our universe, and our universe would have been very different such that the law of conservation of energy would not hold and would therefore not prevent events such as the *uncaused* beginning of increase in strength of electric field under certain circumstances.

Against this second form of Modus Tollens argument, the objector might utilize the branching view of modality affirmed by Oppy which claims that all possible worlds share the initial state of the history of the actual world (Oppy 2013b holds this view because he thinks it is more parsimonious). Given this theory, the initial state of our spacetime block is necessary and there is no other possible spacetime block at the initial state. In other words, based on this branching theory of modality, there is no other *metaphysically possible* alternatives concerning the initial state, even though there are other *logically possible* alternatives such that we are able to tell alternative logically consistent stories about the initial state. Thus, when one asks, 'Why is the initial state A rather than B (e.g. one who is totally disrupted by other spacetime blocks)?' Oppy could answer, 'Because A is necessary, and it follows from my branching view on modality.'[28] (It should be noted that the first form of Modus Tollens is not affected by this theory of modality because it is not referring to possible things or events, but to *actual* events such as increasing of electric field strength under certain conditions. As argued previously, there would be no difference between the beginning of ISOR and later actual events such as increasing in strength of electric fields [which we know happens frequently!] where beginning to exist uncaused is concerned, if it were actually the case that ISOR begins uncaused. However, this theory of modality is relevant to the second form of Modus Tollens which concerns *possible* initial states.)

In reply, I shall argue below that the branching theory of modality is unproblematic if the initial state is a beginningless First Cause. However, it is problematic if the initial state is a First Cause with beginning. In other words, my argument below is perfectly consistent with the branching theory of modality and does not require a rejection of that theory of modality; it only requires the rejection of the view that the First Cause has a beginning which is contradicted by the Modus Tollens argument and result in the unscientific denial of the fine-tuning problem.

To elaborate, empirical evidences show that it is metaphysically possible for physical entities with beginnings to have different arrangements. For example, we know that the tables and chairs in this room can be arranged differently; likewise, we know that atoms can be arranged differently. That is the reason why it is problematic to think (as Oppy claims) that the physical things of Oppy's ISOR could not be arranged differently, that the arrangement of physical entities at the beginning of the initial state of our physical spacetime block is the only metaphysically possible arrangement, and that there are no other metaphysically possible arrangements or metaphysically possible spacetime blocks. Moreover, we know that, even though physical entities with beginnings can have different arrangements, the actual arrangement is usually dependent on the cause. For example, in the newly built house analogy mentioned at the beginning of this chapter, the house is the way it is (rather than a pile of rocks) because the cause (i.e. the house builder) makes it that way; what begins to exist is brought about and constrained by the cause. However, if ISOR begins uncaused as Oppy suggests, then there would be no cause and no pre-existing preventing conditions which constrain what begins to exist. That is the reason for thinking that other spacetime blocks would also begin uncaused initially, since any property which differentiate between them and our spacetime block would be had by them only when they had already begun to exist.

Oppy might raise a tu quoque style objection by claiming that the theory that there is an immaterial, beginningless First Cause (call this 'God') would face the same problem. For example, if God exists uncaused beginninglessly and initially timelessly (see Chap. 6), why wouldn't other timeless concrete entities exist uncaused alongside God? Alternatively, why is God three persons (as Christians claim) rather than four persons (a Quadrinity)?

In reply, since the objector raised the tu quoque objection by claiming that an immaterial, beginningless First Cause suffer the same problem as a First Cause with beginning, the burden of proof would be on the objector to justify his/her claim by showing that (say) multiple timeless concrete entities or a Quadrinity is metaphysically possible. The objector might reply that it is possible because it is conceivable. However, what is conceivable is not always metaphysically possible. To elaborate, I have explained above that, on my view, what exists is constrained by pre-existing conditions (e.g. the arrangement of the house is constrained by the pre-existing material and house builder). If God exists uncaused beginninglessly, then there would be a beginningless pre-existing condition which constrains what exists. This beginningless condition might make the beginningless state incompatible with other timeless concrete entities or a Quadrinity existing. In other words, it might be the case that those circumstances are such that it is not metaphysically possible for those entities to exist timelessly. In order for the objector's objection to work, the objector would have to bear the burden of proof to rule out this possibility, but they have not done so; hence, their objection fails.

The objector might ask why cannot Oppy also appeal to preventive conditions to explain why other spacetime blocks do not begin uncaused. For if the static theory of time is true and the initial state of our spacetime block is tenselessly existing at time $t_1$, it might have some property located at $t_1$ which prevented any other blocks existing alongside it.

In reply, while a concrete entity existing at $t_1$ might have properties that are incompatible with some other things existing at $t_1$ and hence prevent their existence at $t_1$, the problem is that on Oppy's view there is no entity or condition pre-existing $t_1$ which can prevent or constrain what begins to exist uncaused at $t_1$ in the first place. For in order to constrain what begins at $t_1$, condition C must act prior (in some sense of prior, see below) to $t_1$, for otherwise if (say) z has begun at $t_1$ it would be too late to constrain or prevent z's beginning. However, on Oppy's view, there is no such condition. In particular, on his view, there is nothing concrete that exist timelessly sans the initial state of the universe, since on his view that initial state is supposed to be initial state of all reality and has a beginning. If there were a timeless state, then *that* timeless state would be beginningless and would be the initial state of reality in terms of order of being (not temporal order) and hence would be prior to the beginning of the spacetime block in that sense. Thus, on Oppy's theory there simply isn't a state prior to the beginning of universe for any constraining or preventing condition to do the required work, not even a timeless state of existence which exists sans ISOR (this contrasts with Craig's view according to which God exists timelessly sans the universe; see Chap. 6).

Moreover, many physicists do think that the initial conditions at the beginning of the universe could have been different. So why did our universe begin in such a 'fine-tuned' way which allowed life to exist? This is related to fine-tuning problem which I shall elaborate in Chap. 4. At this point it suffice to note that it is unreasonable to say, 'Hey, there's no fine-tuning problem, the initial state begins necessarily!' Most scientists would rightly find this response implausible. For given the observation of our universe with its highly ordered systems (quantum systems, biological systems, solar systems, galactic systems, etc.) and highly ordered laws of physics as well as 'fine-tuning', what is the best explanation for its beginning? Something or nothing? It is unreasonable to think that our universe with its billions of stars and highly ordered laws of physics ultimately begun uncaused; that it just happened to be like that without anything determining it to be like that. For why should the initial conditions of the universe begin in such a way that the universe can allow for existence of life if there was nothing that determines the conditions to begin that way rather than other way and having other properties? As I shall explain further in Chap. 4, the fundamental principles or laws of nature do not uniquely determine a fine-tuned universe (and avoid the Boltzmann Brain problem, etc.). 'Physics is blind to what life needs. And yet, here we are' (Lewis and Barnes 2016, p. 181). That is one reason why many scientists recognize that there is a 'fine-tuning problem' which cannot be resolved simply by claiming that the initial state begins uncaused necessarily and therefore there is no fine-tuning problem. To reply this way is unscientific and irrational, that is, appealing to a magical event without a magician.

Because many scientists recognize the implausibility of simply saying that the universe began 'fine-tuned' and uncaused, they have proposed the multiverse hypothesis in an attempt to address this implausibility. However, as I shall explain in greater detail in Chap. 4, one problem with the multiverse hypothesis is that, even if there are many universes, whatever led to their formation would itself require fine-tuning (i.e. must be highly ordered) in order to generate so many different kinds of universes (whether deterministically or indeterministically) such that eventually one that is 'fine-tuned' is generated by chance. As Collins (2018, p. 90) notes, 'anything that produces such a multiverse itself appears to require significant fine-tuning'. Thus, Oppy's initial state would still need to be fine-tuned—but to say that such a fine-tuned state begins uncaused is precisely the problem that we started with!

Oppy objects that if the fine-tuning of the initial state is required for naturalism, it would be required for theism as well (Oppy 2013b, p. 59). He might therefore be motivated to raise a tu quoque style objection by arguing that both theories (Oppy and mine) suffer the same problem in the sense that, on both theories, there is no state prior to the initial state at which it is possible for some property or condition to do the constraining or preventive work concerning the initial state.[29]

In response, my argument against uncaused beginning is not special pleading because it is based on evidence. As explained earlier, we have evidence based on observation and science that it is metaphysically possible for physical entities with beginnings to have different arrangements. Given this, we should ask why the initial arrangement of the universe began in such a way (rather than other ways) that allows for existence of life (fine-tuning problem). On the other hand, the objector has not provided evidence that it is metaphysically possible for timeless things without beginning to have different arrangements (e.g. a Quadrinity) in order to substantiate the tu quoque objection that theism suffers from the fine-tuning problem. Additionally, as explained earlier, we have evidence based on observation and science that constraining/preventive work is needed for things/events with beginning (e.g. so as to explain why is it not the case that a rubble rather than a house begins to exist). On the other hand, the objector has not provided evidence to substantiate the tu quoque objection that a similar constraining/preventive work is also needed for timeless things without beginning. Given the lack of evidence that a beginningless state would suffer from the same problem, the tu-quoque style objection fails.

Moreover, consider another analogy: Suppose that fire begins uncaused. In that case the beginning of fire would be unconstrained; in particular, the lack of oxygen would not constrain the beginning of fire. However, suppose that oxygen is causally necessary for fire. If there is an eternal lack of oxygen, then no fire can ever exist in that eternal state; such a state would be incompatible with the existence of fire. This scenario is analogous to my view according to which there is a beginningless state which can do the work of prevention. In order to (say) prevent other timeless entities/Quadrinity from existing uncaused beginninglessly alongside God, the preventive conditions do not need to act in any sense prior, but can be part of the beginningless state which makes such a state incompatible with those other timeless entities/Quadrinity existing. Such a beginningless state is not limited by a temporal edge, whereas 'our spacetime block begins at $t_1$' is limited by a temporal edge and hence comes too late to constrain what begins to exist at the same edge, as explained above.

It might be objected that, on Oppy's theory of modality, the lack of a previous state to the beginning of the universe is precisely why it is metaphysically necessary. For according to his theory, in order for something to have been otherwise, it would have to be caused to be otherwise by a previous state, yet there is no previous state to the beginning of the universe according to his view.

In reply, the point that 'in order for it to have been otherwise, it would have to be caused to be otherwise' may be applicable to later events, but the fact that on Oppy's theory there is nothing prior to the initial state of our spacetime block which has a beginning to constrain what begins to exist is precisely the problem. For the lack of a previous state to the beginning of universe on Oppy's theory—which implies the lack of constraining and preventive conditions—is precisely one reason why he cannot say what makes it metaphysically impossible from beginning in a different way, and thus undermines his claim that the initial state of our universe has a metaphysically necessary beginning. It would be fallacious to reply to my argument that 'without a prior state, other uncaused events cannot be prevented' by saying that this is not applicable to the initial part of universe which is just necessary. For a theory of modality is merely abstract, which merely describes what is necessary and possible. What does the required metaphysical work are concrete things and their properties. However, the lack of a previous state to the beginning of the universe implies that there is no concrete previous state which can do the required metaphysical work of constraining or prevention, that is precisely why it cannot be metaphysically necessary.

On the other hand, I have explained above that, while the branching theory of modality is problematic for an initial state with a beginning, it is unproblematic for initial state that is beginningless. Being beginningless implies that the initial state would not face the problem posed by my Modus Tollens argument, which only works against the view of uncaused beginnings (such as that held by Oppy). Moreover, something that is beginningless and unsustained will be initial, since there cannot be another thing existing prior (whether temporally prior or causally prior) to an entity that is beginningless and unsustained, and given the branching theory of modality, such an initial entity would be necessary. Therefore, his properties could not have been different. Additionally, given the branching theory of modality which implies that First Cause is metaphysically necessary, it isn't the case that multiple beginningless timeless concrete entities could have existed initially if there is only one such entity initially, and it isn't the case that the beginningless First Cause could have been a Quadrinity if it is a Trinity. There just isn't any possible alternative initial state, since all possibilities share that initial state. This conclusion provides another response to the tu-quoque style objection.

## 3.9 Objection Concerning the Distinction between Could and Would

Back to the first form of the Modus Tollens argument, sceptics might object by claiming that my argument only shows that, if ISOR began to exist uncaused, other entities could also begin to exist uncaused; it does not show that they would also begin to exist uncaused.

In reply, 'could' concerns possibility, but I am not referring to possible events here. Rather, I am referring to actual events, and arguing that there would be no difference between them where beginning to exist uncaused is concerned if one of them begins uncaused. For example, consider the scenario in which something (say) the universe began to exist and there was also a rapid increasing in strength of electric fields under certain circumstances around me. In this scenario these are not just possible events (i.e. it is not merely the case that the universe *could* begin to exist and electric field *could* increase in strength), but actual events; that is, the universe did begin to exist and electric field did increase in strength. Since (as explained above) there would be no difference between these events where beginning to exist uncaused is concerned if the universe did begin to exist *uncaused*, the increasing in strength of the electric field under certain circumstances around me would also begin to exist uncaused (e.g. without requiring the switching on of an electric field generator).

An objector might appeal to Oppy's branching theory of modality according to which what is *possible* for these later events will be set by the causal powers of the previous event, and then argue that according to this theory it is not possible for later events to begin uncaused.[30]

However, this objection misses the point of my argument which refers to actual later events (and not possible events). That is, we know that electric fields actually increase in strength from time to time, and given what I have argued above concerning I, II, and III, there would be no difference between the beginning of these actual events and the actual beginning of our universe where beginning to exist uncaused is concerned if (as Oppy says) our universe actually began uncaused, and that these actual later events would have been uncaused rather than caused as Oppy thinks.

Another objector might appeal to a different theory of modality according to which it is conceivable and possible that many different possible strengths could begin to exist uncaused. He/she might then argue that, since none of these possibilities is privileged over the other, none of them would begin uncaused even if ISOR begins uncaused.[31]

There are two steps to my response.

First, in our actual world, increasing in strength of electric field happens often. For example, I just observed an event y: 'an increased in strength E under certain circumstances' begins to exist at time ty after I switched on a generator. My argument is that, if ISOR begins uncaused, then given I, II, and III (explained above), there would be no difference between beginning of ISOR and beginning of y where beginning to exist uncaused is concerned. The fact that y could have been of different strengths (e.g. 2E, 3E) does not deny either of I, II, and III, which jointly imply my conclusion. That is, the fact that there is no privilege over which strength begins does not imply that none would begin. On the contrary, given I, II, and III, y would have begun uncaused, since there would have been no difference between the beginning of ISOR and the beginning of y where beginning to exist uncaused is concerned. Therefore, the beginning of y at ty should have been uncaused, that is, without having to switched on a generator. But that consequent was not the case. Thus, it is not the case that ISOR began uncaused.

Second, what about the fact that y could have been of different strengths and that there is no privilege over which strength begins? What does this fact imply? One can argue that this fact implies that, not only would the increasing in strength of the electric field under certain circumstances around me began to exist uncaused, the increasing in strength would be of very different values as well. The reason is because,

I. since increasing in strength E would have begun to exist uncaused, as shown under the first step, what this implies is that the beginning of E would not have causally necessary condition which makes it the case that only E (rather than other possible strengths, for example, 2E, 3E) begins to exist.
II. Additionally, as explained previously, any difference between the possibilities E, 2E, 3E, and so on would be had by them only when they had already begun to exist. Moreover,
III. the circumstances around us are compatible with different levels of increasing in strengths of electric fields up to a certain physical limit.

Given (I), (II), and (III), there would be no relevant difference between these different increasing in strengths where beginning to exist uncaused is concerned. Hence, if increasing in strength E began uncaused, the other possible strengths would all begin uncaused up to the physical limit (e.g. suppose 6E is the limit; increasing in strengths of E, 2E, and 3E would all begin to exist and they add up to 6E), and our universe would have been very different.

With regard to the second step, one may argue more directly from the uncaused beginning of our spacetime block (if that happens) to the uncaused beginning of other possible strengths of electric field, as follows:

1. If possibility *x* (e.g. our spacetime block) is actualized and begins uncaused, then some other possibilities (e.g. *y: increasing in electric*

*field strength of E, 2E, and 3E*) would also be actualized and begins uncaused.

2. It is not the case that *y* is actualized and begins uncaused.
3. Therefore, it is not the case that *x* is actualized and begins uncaused.

The justification for premise 1 is that,

I. if possibility *x* is actualized and begins uncaused, this implies that the beginning of *x* would not have causally necessary condition which makes it the case that only possibility *x* (rather than some other possibilities, for example, y) begins to exist.
II. Additionally, any difference between the possibilities *x* and *y* would be had by them in the concrete only when they had already begun to exist. Moreover,
III. the circumstances are compatible with the beginning of y (as explained under step two above).

I, II, and III jointly imply that there would be no difference between possibility *x* and possibility y where being actualized and beginning uncaused in the concrete world is concerned, which implies premise 1.

## 3.10 Objection: The Causal Principle is Inconsistent with Libertarian Freedom

Almeida (2018, pp. 38–39) objects that the Causal Principle is inconsistent with libertarian freedom, which he understood as implying that, in the case whereby agent S freely chooses to do A, 'the cause of A is S, and there is nothing that causes S to cause A. S's causing A is an event that comes into existence uncaused. So according to source libertarianism it is perfectly possible that some things come into existence uncaused.'

In reply, an agent causing some effect is not itself an event, but just a way of describing an agent causing an event (Craig and Sinclair 2009 p. 194n. 101). My argument does not rule out libertarian free choice, since one can understand libertarian free choices as indeterministic but not uncaused (see further, Chap. 6).

## 3.11 Objection Based on Lack of Directionality

Don Page has raised the concern that

> We have learned that the laws of physics are CPT invariant (essentially the same in each direction of time), so in a fundamental sense the future determines the past just as much as the past determines the future … the effective unidirectional causation we commonly experience is something just within the universe and need not be extrapolated to a putative cause for the universe as a whole.[32]

Page's worry is related to Linford's objection to the KCA that 'the explanatorily prior and physically necessary conditions for the universe's "beginning" can fall in the temporal direction away from the beginning' (2020, p. 11).

In reply, it should first be noted that the above objection by Page and Linford assumes the static theory of time according to which later events already exist and hence (according to them) might be able to cause earlier events. According to the static theory, our ordinary experience of time flow and the present is regarded as illusory, and in the absence of these one might have difficulty defining 'earlier than', 'beginning', and 'first moment'.[33] The static theory of time is controversial (for objections to the static theory and a defence of the dynamic theory of time, see, for example, Craig 2000a, b). In any case, I shall explain below that my Modus Tollens argument would work even if the static theory of time is true and that the KCA can still be defended, even though time as it is used in mathematical physics is a quantity without direction.

To begin, it should be noted that the static theory does not exclude earlier than/later than relations; on the contrary, it is the presence of those relations which makes the B-series a temporal series, rather than a McTaggart C-series.[34] It is true that on a B-theory there is difficulty providing an explanation for why one direction is earlier rather than later. Nevertheless, even if Page's speculation that earlier events actually depends on later events is true, then what we call earlier (and beginning) would in fact be later in my argument. Therefore, if an infinite causal regress and a causal loop is impossible (see Chap. 5), there would still be a First Cause.

Second, it has been explained previously in Chap. 2 that the laws of nature and equations of physics merely provide an incomplete description of physical reality without ruling out causation and causal properties (and causal direction) which operate at a more fundamental level as the ground of the regularities described by these laws and equations. Moreover, it has also been noted in Sect. 3.1 that, even if the static theory of time is true, there is still something unique about time which makes it different from space. For example, within the spacetime block, the durations next to the duration I exists is occupied by my parents and that, if they had not existed, I would not exist. (Whereas this is not so with spatial order: if my parents do not stand on my left or right, I would still exist.) Likewise, we know that hydrogen is causally necessary for the formation of water, but water is not causally necessary for formation of hydrogen. In short, there is still a certain dependence and an ordering of things/events which indicate the dependence, and I shall call these 'causal dependence' and 'causal order'. (A causal sceptic would acknowledge that we observe correlations between things/events, but doubt that there is causation. In reply, the assumption of causation is justified in light of the critique of the causal sceptic's position which others [e.g. Weaver 2019] have offered [e.g. 'it is improbable that correlations exist without causation'] and which are summarized in Chap. 2. In any case, my argument can easily be translated in terms of correlation, by substituting the term 'causal order/cause' with 'correlations', 'causally ordered' with 'correlated', and 'uncaused' with 'uncorrelated': (1) If x begins uncorrelated, then y would also begin uncorrelated. (2) It is not the case that y begins uncorrelated. (3) Therefore, it is not the case that x begins uncorrelated.[35])

Third, as explained in Chap. 2, if something X has a temporal extension, the extension is finite, and it has temporal edges (that is, it does not have a static closed loop or a changeless phase that avoids an edge), then it has a beginning. The upshot is that a relevant defence of my argument for the Causal Principle 'whatever begins to exist has a cause' (in the context of defending the KCA) is independent of the temporal direction of causality. If physical reality as a whole is finite in temporal extent in whatever dimensions and having temporal edges (i.e. without a static closed loop or a changeless phase that avoids an edge), it would imply that physical reality has a beginning of existence.

Now the argument in Sect. 5.6 of this book refutes a static closed loop by demonstrating that it is viciously circular, while the argument in Chap. 6 demonstrates that if there is a changeless phase that avoids an edge, this (initially) changeless phase of reality must have libertarian freedom and thus would be a Creator rather than an impersonal physical reality which is constantly changing. Given that physical reality does not have a changeless phase that avoids an edge, and given that a 'change' is an event that has 'edges' at the state of having gained or having lost a property within a finite duration of time, physical reality as a whole would be finite in temporal extent if there is no infinite number of changes, and the latter is demonstrated by the arguments in Chap. 5. Given these arguments, physical reality as a whole would be finite in temporal extent in whatever dimensions. We then ask whether such a physical reality with temporal 'edges' (i.e. with a beginning) exists without causally necessary conditions.

Now if the number of earlier and later events in the scenario described by Linford and Page are finite and the series exists without a beginningless First Cause as a causally necessary condition, then the series of events would violate reflexivity; that is, the earlier events would be dependent on later events, which are dependent on earlier events.

Moreover, the two forms of Modus Tollens argument which I have defended above would still apply. That is,

I. If our spacetime block is finite in temporal extent in whatever dimensions and having 'edges' and uncaused, then there would not be any cause which would make it the case that only our spacetime block rather than other possible finite spacetime blocks as well as many other actual things/events with finite temporal extension within our block and having 'edges' exists without cause.
II. The properties of our spacetime block and the properties of those other possible/actual spacetime blocks/things/events which differentiate between them would be had by them in the concrete world only when they had already existed in the concrete world.
III. The circumstances are compatible with those other spacetime blocks/things/events existing.

As argued previously, (I), (II), and (III) jointly imply that there would be no essential difference between them where existing uncaused is concerned, and therefore those other spacetime blocks/things/events would also be uncaused. But this is contrary to fact given the causal order we observe. Therefore, it is not the case that our spacetime block is finite in temporal extent in whatever dimensions and having 'edges' and uncaused.

It might be objected that, given that *x* ('our spacetime block within which events are causally ordered/correlated') is logically possible, if *y* (increasing in strength of electric field) is part of a causally ordered/correlated *x* and *x* begins uncaused, *y* would naturally have causal relations/correlations with other parts of *x*.[36]

In reply, this objection begs the question by assuming that the events within *x* (our spacetime block) would still be causally ordered/correlated if *x* begins uncaused/uncorrelated. Whereas I am not begging the question by arguing that the events within *x* would not be causally ordered/correlated if *x* begins uncaused/uncorrelated, because the premises of my argument (i.e. I, II, and III) do not assume this conclusion, but jointly imply this conclusion. To elaborate, the following four claims should be distinguished:

1. It is possible that *x* ('a spacetime block within which events are causally ordered/correlated').
2. It is possible that *x* ('a chaotic spacetime block in which *y* begins uncaused/uncorrelated').
3. It is possible that *x* (a chaotic spacetime block in which *y* is uncaused/uncorrelated) begins uncaused.
4. It is the case that *x* ('a spacetime block within which events are causally ordered/correlated') begins uncaused.

While 1 is true and I grant that 2 and 3 are true for the sake of the argument, (I), (II), and (III) jointly imply that 4 is false, since (I), (II), and (III) jointly imply that there would be no difference between '*x*' and 'events within *x*' (such as '*y*') where beginning uncaused is concerned. In other words, our spacetime block would have been very different (i.e. the events within our spacetime block would have been causally disordered/

uncorrelated) if our spacetime block begins uncaused. Hence, it is not the case that our spacetime blocks begins uncaused.

## 3.12 Epistemological Objections

It might be objected that, perhaps unknown to us, something did begin to exist uncaused faraway a long time ago in another universe with very different metaphysical principles/laws of nature.

However, 'beginning long ago in a place far away' by itself would not be relevant consideration, because 'begins to exist uncaused' implies that the time and location at which it begins (whether long ago or far away) would not be a cause for its beginning rather than other things/events (say, a rapid increasing in strength of electric fields under certain circumstances) around us where beginning to exist uncaused is concerned. The metaphysical principles/laws of nature of that unknown universe cannot be a cause for why things begin to exist uncaused in that world but not in our universe, since beginning to exist uncaused implies that the laws or principles are not the cause.

One might object that to argue that because we do not see things beginning to exist uncaused around us therefore it didn't happen long ago is to commit the black swan fallacy. In reply, the black swan fallacy is a fallacy concerning inductive reasoning ('because a person X has observed many white swans and no black swans, therefore X does not think that black swans exist'—this is fallacious because X has not observed all swans). Whereas my argument for the Causal Principle is deductive, not inductive. My argument is not simply saying 'because we do not see things beginning to exist uncaused around us therefore it didn't happen long ago'. Rather, my argument is saying, 'if something (say, the universe) begins to exist uncaused long ago, then (because of I, II, and III) other things would begin to exist uncaused around us, but they don't; therefore, it is not the case that something begins to exist uncaused long ago'.

It has been objected that we cannot know whether the Causal Principle applies to the universe itself because we cannot observe the universe as a whole and confirm that the universe has a cause.[37] One might also raise the concern that the word *cause* gets its meaning from our use of language involving our experiences, but it's not clear whether it applies to the situation at the beginning of the universe, which is far beyond our experiences. As Nagel (2004) notes in his review of Rundle's (2004) book *Why There Is Something Rather Than Nothing*, 'the most difficult philosophical question posed by Rundle's critique is whether such efforts to use words to indicate something that transcends the conditions of their ordinary application make sense.' Drees (2016, p. 199) raises the concern that 'at the boundaries of physical cosmology our notions of time and causality break down'. It therefore is doubtful whether a 'Big Bang' as the limiting event of standard cosmology provides a stable model for 'the first event'. He concludes that when the conceptuality of space and time changes, possible answers to the question 'Why is there something rather than nothing?' need not to be thought of in temporal or causal terms.

The assumption of the first objection is that we can only know by direct observation, but this assumption is false. There are many events which we have not directly observed, but which we can know did happened by inferring from what we observe. Moreover, there are ways of knowing which are not dependent on observation. For example, I know that the statement 'there cannot be shapeless squares' is true, and I can know this without having to observe the entire universe to make sure that there are no shapeless squares anywhere. I just need to understand that the existence of a shape (e.g. square) implies that it is not shapeless. As explained in Chap. 1, the laws of logic are necessarily true because a violation of the laws of logic would be non-existent. The laws of logic would hold even in conditions far beyond our experiences, such as at the origin of universes and of time, in the microphysical world and in Kant's 'noumenal world'—there cannot be shapeless squares in such conditions too.

The fact that the laws of logic are necessarily true implies that the conclusion of a deductively valid argument from true premises must be true, and I have already explained that the Modus Tollens argument for the Causal Principle is deductively valid, and that its premises are true. The argument implies that the Causal Principle not only applies within the universe, but to everything without restriction, including the universe itself, the microphysical world, and the 'noumenal world'.

Hence, with regard to Nagel's and Drees' concern, if 'beginning' does not apply to the universe at its earlier stages, then following the laws of logic (which as explained above cannot be violated even in conditions far beyond our experiences) what that implies is that the universe would be 'beginningless'. I discuss this possibility in Chaps. 5 and 6. If 'beginning' does apply but 'cause' does not apply to the beginning of the universe, then following the laws of logic (which as explained above cannot be violated even in conditions far beyond our experiences) what that implies is that the beginning of the universe would be 'uncaused'. If that is the case, (1.1) there would not be any cause which would make it the case that only universe rather than some other things (e.g. the beginning of a rapid increasing in strength of electric fields under certain circumstances around me) begins to exist uncaused. Moreover, (1.2) the properties of the universe and the properties of those other things which differentiate between them would be had by them only when they had already begun to exist. Additionally, (1.3) the circumstances around us are compatible with the beginning of existence of those other events. As explained previously, (1.1), (1.2), and (1.3) imply that there would be no essential difference between them where beginning to exist uncaused is concerned. Thus, if the universe begins to exist uncaused, it cannot be the case that only the universe begins to exist uncaused. In that case, the beginning of those other things would also be uncaused. But this is contrary to my experience. I (thankfully!) do not experience such events happening without causes such as (say) having to switch on the electric field generator. Therefore, it is not the case that the universe begins to exist uncaused.

## 3.13 Conclusion

In Chaps. 2 and 3, I have defended the Causal Principle 'whatever begins to exist has a cause' against objections and developed an argument which demonstrates that the principle is true. Contrary to the claims of some scientists and philosophers, fundamental physics does not exclude efficient causation, and quantum physics has not shown that the Causal Principle is violated given that (1) quantum events do not begin to exist without causal antecedents, (2) our current understanding of physics is limited, and (3) there are viable deterministic interpretations of quantum phenomena. On the other hand, (1) an inductive argument, (2) an argument from the concept of non-being, and (3) a Modus Tollens argument have been offered in the literature in support of the Causal Principle. I developed the Modus Tollens argument in response to objections. The argument states that, if *x* begins uncaused, then *y* which begins to exist would also begin uncaused, which is not the case; therefore, the antecedent is not the case. Against Oppy's claim that only *x* (the initial state of reality: ISOR) begins uncaused, I have shown that one cannot simply claim that this is a brute fact, for if there is no relevant difference between (say) ISOR and *y*, this would imply that they are the same where beginning to exist uncaused is concerned. One also cannot appeal to abstract objects to provide the necessary metaphysical grounding for Oppy's claim, because any such grounding would have to be concrete in order to ground the difference between *x* and *y* in the concrete world. I go on and demonstrate that there isn't any concrete grounding because (I) there would not be any causally antecedent condition which would make it the case that *x* rather than *y* begins uncaused, (II) the properties of *x* and the properties of *y* which differentiate between them would be had by them only when they had already begun to exist, and (III) the circumstance is compatible with the beginning of *y*. In particular, against the appeal to current spatial considerations, pre-existent things or laws of nature preventing things/events from beginning uncaused around us, I have shown that such considerations would not prevent events such as electric fields increasing in strength and energy-conserving changes from beginning uncaused, in which case our experiences would have been very different from what they are. I have also defended a second form of Modus Tollens argument which shows that, if our spacetime block begins uncaused, then some other possible spacetime blocks would also begin uncaused initially and collide with ours, causing massive disruption, which is not the case; therefore, the antecedent is not the case. Although I have defended two forms of Modus Tollens argument in this chapter, it should be noted that any one of the two forms of Modus Tollens argument would be sufficient to refute Oppy's claim. Against the objection that the first form of argument only shows that things could begin to exist uncaused now, not that they would, I reply that I am referring to actual events, and arguing that there would be no difference between them where beginning to exist uncaused is concerned if one of them begins uncaused. I also explained that my argument is consistent with libertarian freedom. I conclude that the Causal Principle would hold even in conditions that are far beyond our experiences, such as at the origin of the universe, the microphysical world, and in Kant's 'noumenal world', for if it does not, our experiences would be very different from what they are.[38]

In conclusion, the Modus Tollens argument demonstrates the absurd consequences which would follow if something $x$ begins without any causally antecedent condition which makes it the case that $x$ (rather than other things) begins to exist. The conclusion of this argument implies that whatever $x$ begins to exist *depends* on the cause which *makes it the case* that $x$ (rather than other things) begins to exist.

Finally, in addition to the Modus Tollens argument defended above, which is already sufficient for establishing the conclusion 'if our universe has a beginning, it has a cause', another independent argument can be offered for this conclusion (as explained in Sect. 3.1, any one of these arguments is sufficient; thus, my case for the KCA does not depend on the Modus Tollens argument or the following argument). I have explained previously in Sect. 3.8.4 that it is unreasonable to think that our universe with its billions of stars and highly ordered laws of physics (which indicates that our universe is a huge, interconnected, highly ordered structure; see Chap. 4) fundamentally began uncaused; that it just happened to be like that without anything determining it to be like that. Such a conclusion is worse than magic—which at least has a magician—and it is worse than chance, which is highly unlikely (see Chap. 4) but at least has prior conditions which ground the probabilities. Most scientists recognize the implausibility of simply saying that the universe is 'fine-tuned' by chance; thus, they have proposed the multiverse hypothesis in an attempt to address this implausibility. In Chap. 4, I shall show that this attempt fails, but the point here is that saying that the universe began uncaused and was 'fine-tuned' is even more implausible than saying that the universe began by chance and was 'fine-tuned'. Thus, it is more reasonable to conclude that, if our universe had a beginning, the highly ordered systems of our fine-tuned universe, together with the highly ordered laws of physics, fundamentally came from something—a Cause, which is arguably highly powerful and intelligent. I shall present the arguments for this conclusion in detail in the following chapters. In particular, I shall first discuss the evidences for the fine-tuning and order of the universe in Chap. 4, and I shall demonstrate that the universe has an ultimate beginning in Chaps. 5 and 6.

## Notes

1. I thank Wes Morriston for helpful discussion.
2. Vilenkin states that quantum mechanics does not rule out the possibility that a lump of matter can turn into a tiger but only indicate that the probability of such events would be very low (https://www.youtube.com/watch?v=pGKe6YzHiME; 48:40). Such purported events would not be uncaused because there are causally necessary conditions such as the lump. Such events having a very low probability of happening is disanalogous to uncaused events happening all the time, which is what my argument predict would happen if something (e.g. our universe) begin uncaused.
3. Rodriguez-Pereyra (2019) notes that the acceptance of properties is compatible with being a nominalist. He explains, 'Nominalism has nothing against properties, numbers, propositions, possible worlds, etc., *as such*. What Nominalism finds uncongenial in entities like properties, numbers, possible worlds and propositions is that they are supposed to be universals or abstract objects … What is required of nominalists who accept the existence of numbers, properties, possible worlds and propositions is that they think of them as particulars or concrete objects.' Trope-theorists may regard S as a trope.
4. As he did during our debate: https://www.youtube.com/watch?v=a8NrTv-Durc&t=129s.
5. In our debate, Oppy did not grant (I), (II), and (III)—to grant it would be suicidal for his case! However, not granting it does not imply he has rebutted the reasons which I gave which show that (I), (II), and (III) are in fact implied by Oppy's theory.
6. I thank Zhang Jiji for helpful discussions on this point and in what follows.
7. Pearce (2021a, b) argues that grounding necessitarianism rests on questionable assumptions and even full metaphysical grounding is compatible with indeterminism.

8. I thank Don Page for discussing this objection.
9. For grounding indeterminism, see Pearce (2021a, b).
10. I thank Zhang Jiji for suggesting this line of argument.
11. Abstract objects may make a difference in the abstract world. For example, 'there is a possible world W in which Abel kills Cain (and not vice versa as described in the Book of Genesis). Although possible worlds, with their objects, are usually regarded as abstract objects, it may, nevertheless, be said that in W, Abel causes (or partly causes) the death of Cain' (Erasmus 2018, p. 95).
12. I thank Lee Wang Yen for helpful discussions on this point.
13. I thank Lee Wang Yen for raising this objection.
14. Indeed, Dumsday himself does not use it this way.
15. I thank Joe Schmid for suggesting this.
16. https://hughjidiette.wordpress.com/2020/12/09/graham-oppy-vs-andrew-loke-on-the-kalam-my-critique-of-lokes-argument.
17. I thank Alex Malpass for asking this question.
18. I thank Zhang Jiji for raising this objection.
19. I thank Wes Morriston for putting the matter this way.
20. I thank Zhang Jiji for helpful discussion.
21. I thank Alex Malpass for this formulation.
22. I thank Hugh Jidiette for raising this question.
23. I thank Andrew Brenner for raising this objection.
24. I thank Andres Luco for raising this objection.
25. Wolff (2013, pp. 900–901) notes that Fine is not expressing a linguistic truth but 'real definitions': the idea that what really makes a thing that thing is not up to our linguistic or conceptual choices.
26. Dumsday (2019, p. 124) notes that 'if substantivalism is true then in principle intrinsic properties can be attributed to space. For instance, space can rightly be spoken of as having a structure of its own, perhaps even as having causal powers. Space can also be a subject of change, insofar as those intrinsic properties may be alterable; think for instance of the idea in contemporary cosmology that space has been expanding continually since the big bang.'
27. I thank Nick Morris for raising this question.
28. I thank Hugh Jidiete for helpful discussion.
29. I thank Fox for raising this objection.
30. I thank Hugh Jidiette for raising this objection.
31. I thank Dion Jones for raising this objection.

32. Collins (2009, p. 270) notes that 'the laws of physics are not strictly speaking time-reversal invariant—since time-reversal symmetry is broken in weak interactions, notably the decay of neutral kaons'.
33. I thank Tim Maness for mentioning this point at the 2020 AAR conference.
34. I thank William Lane Craig for this point.
35. I thank Don Page for helpful discussion.
36. I thank Don Page for helpful discussion concerning this objection.
37. https://www.youtube.com/watch?v=oD06eEbrzjs.
38. For replies to other objections against the Modus Tollens argument and the Causal Principle, see also Loke (2017a, chapter 5).

## Bibliography

Almeida, Michael. 2018. *Cosmological Arguments*. Cambridge: Cambridge University Press.

Bigelow, John, Brian Ellis, and Caroline Lierse. 1992. The World as One of Kind. *British Journal for the Philosophy of Science* 43: 371–388.

Bird, Alexander. 2007. *Nature's Metaphysics*. Oxford: Oxford University Press.

Bliss, Ricki, and Kelly Trogdon 2014. Metaphysical Grounding. *The Stanford Encyclopedia of Philosophy* (Winter 2014 Edition), ed. Edward N. Zalta. http://plato.stanford.edu/archives/win2014/entries/grounding/. Accessed 21 January 2017.

Carrier, Richard. 2018. *The Problem with Nothing: Why The Indefensibility of Ex Nihilo Nihil Goes Wrong for Theists*. https://www.richardcarrier.info/archives/14486.

Chalmers, A. 1999. Making Sense of Laws of Physics. In *Causation and Laws of Nature*, ed. H. Sankey. Dordrecht: Kluwer.

Collins, Robin. 2009. The Teleological Argument: An Exploration of the Fine-tuning of the Universe. In *Blackwell Companion to Natural Theology*. Oxford: Wiley-Blackwell.

———. 2018. The Argument from Physical Constants. In *Two Dozen (Or So) Arguments for God*, ed. Jerry Walls and Trent Dougherty. Oxford: Oxford University Press.

Craig, William Lane. 2000a. *The Tensed Theory of Time: A Critical Examination*. Synthese Library 293. Dordrecht, the Netherlands: Kluwer Academic Publishers.

———. 2000b. *The Tenseless Theory of Time: A Critical Examination*. Synthese Library 294. Dordrecht, the Netherlands: Kluwer Academic Publishers.

Craig, William Lane, and James Sinclair. 2009. The Kalām Cosmological Argument. In *The Blackwell Companion to Natural Theology*, ed. William Lane Craig and J.P. Moreland. Chichester: Wiley-Blackwell.

Drees, Willem. 2016. The Divine as Ground of Existence and of Transcendental Values: An Exploration. In *Alternative Concepts of God: Essays on the Metaphysics of the Divine*, ed. Andrei Buckareff and Yujin Nagasawa. Oxford: Oxford University Press.

Dumsday, Travis. 2019. *Dispositionalism and the Metaphysics of Science*. Cambridge: Cambridge University Press.

Edwards, Jonathan. 1830. *The Works of Jonathan Edwards*. New York: G. & C. & H. Carvill.

Erasmus, Jacobus. 2018. *The Kalām Cosmological Argument: A reassessment*. Cham: Springer Nature.

Feser, Edward. 2013. *Aristotle on Method and Metaphysics*. London: Palgrave.

Fine, Kit. 2002. The Varieties of Necessity. In *Conceivability and Possibility*, ed. Tamar Gendler and John Hawthorne. Oxford: Clarendon.

Gould, Paul, ed. 2014. *Beyond the Control of God? Six Views on the Problem of God and Abstract Objects*. London: Bloomsbury.

Lange, Marc. 2016. *Because Without Cause: Non-Causal Explanations in Science and Mathematics*. Oxford: Oxford University Press.

Loke, Andrew. 2017a. *God and Ultimate Origins: A Novel Cosmological Argument. Palgrave Frontiers in Philosophy of Religion Series*. Cham, Switzerland: Springer Nature.

Lewis, Peter. 2016. *Quantum Ontology*. Oxford: Oxford University Press.

Lewis, Geraint, and Luke Barnes. 2016. *A Fortunate Universe: Life in a Finely Tuned Cosmos*. Cambridge: Cambridge University Press.

Linford, Dan. 2020. The *Kalām* Cosmological Argument Meets the Mentaculus. *The British Journal for the Philosophy of Science*. https://doi.org/10.1093/bjps/axaa005.

Linnemann, N. 2020. On Metaphysically Necessary Laws from Physics. *European Journal for Philosophy of Science* 10: 23.

Loke, Andrew. 2012. Is an Uncaused Beginning of the Universe Possible? A Response to Recent Naturalistic Metaphysical Theorizing. *Philosophia Christi* 14: 373–393.

———. 2017. *God and Ultimate Origins: A Novel Cosmological Argument. Palgrave Frontiers in Philosophy of Religion Series*. Cham, Switzerland: Springer Nature.

Mackenzie, Ruari, et al. 2017. Evidence Against a Supervoid Causing the CMB Cold Spot. *Monthly Notices of the Royal Astronomical Society* 470 (2): 2328–2338.

Mawson, T. 2011. Explaining the Fine Tuning of the Universe to Us and the Fine Tuning of Us to the Universe. *Royal Institute of Philosophy Supplement* 68 (2011): 25–50.

Miksa, Rad. 2020. Deny the Kalam's Causal Principle, Embrace Absurdity. *Philosophia Christi* 22: 239–255.

Nagel, Thomas. 2004. Review of Bede Rundle's Why there is Something. Rather than Nothing, Times Literary Supplement, May 7.

Oderberg, David. 2002. Traversal of the Infinite, the 'Big Bang,' and the *Kalām* Cosmological Argument. *Philosophia Christi* 4: 303–334.

Oppy, Graham. 1991. Craig, Mackie, and the Kalām Cosmological Argument. *Religious Studies* 27: 189–197.

———. 2010. Uncaused Beginnings. *Faith and Philosophy* 27: 61–71.

———. 2013a. *The Best Argument Against God*. London: Palgrave Macmillan.

———. 2013b. Ultimate Naturalistic Causal Explanations. In *The Puzzle of Existence*, ed. Tyron Goldschmidt. New York: Routledge.

———. 2015. Uncaused Beginnings Revisited. *Faith and Philosophy* 32: 205–210.

Pearce, Kenneth. 2021a. Monistic Metaphysical Rationalism Requires Grounding Indeterminism. Presentation at Global Philosophy of Religion Conference, Birmingham University.

———. 2021b. Foundational Grounding and Creaturely Freedom. *Mind*, fzab024. https://doi.org/10.1093/mind/fzab024.

Prior, Arthur. 1968. Limited Indeterminism. In *Papers on Time and Tense*. Oxford: Clarendon Press.

Rasmussen, Joshua. 2018. Review of God and Ultimate Origins. *European Journal for Philosophy of Religion* 10: 189–194.

Rodriguez-Pereyra, Gonzalo. 2019. Nominalism in Metaphysics. *The Stanford Encyclopedia of Philosophy* (Summer 2019 Edition), ed. Edward N. Zalta. https://plato.stanford.edu/archives/sum2019/entries/nominalism-metaphysics/.

Rundle, Bede. 2004. *Why There Is Something Rather Than Nothing*. Oxford: Oxford University Press.

Siegel, Ethan. 2018. Why Haven't We Bumped into Another Universe Yet. https://medium.com/starts-with-a-bang/ask-ethan-why-havent-we-bumped-into-another-universe-yet-fa15b45b0ce9.

Stoeger, William. 2001. Epistemological and Ontological Issues Arising from Quantum Theory. In *Quantum Mechanics: Scientific Perspectives on Divine Action*, ed. Robert Russell et al. Berkeley: Center for Theology and the Natural Sciences.

Tahko, Tuomas E., and E. Jonathan Lowe. 2020. Ontological Dependence. *The Stanford Encyclopedia of Philosophy* (Fall 2020 Edition), ed. Edward N. Zalta. https://plato.stanford.edu/archives/fall2020/entries/dependence-ontological/.

Weaver, C. 2019. *Fundamental Causation: Physics, Metaphysics, and the Deep Structure of the World*. London: Routledge.

Wilson, Alastair. 2020. *The Nature of Contingency: Quantum Physics as Modal Realism*. Oxford: Oxford University Press.

Wolff, J. 2013. Are Conservation Laws Metaphysically Necessary? *Philosophy of Science* 80 (5): 898–906.

# 4

# Fine-Tuning and Order of our Universe

## 4.1 Introduction

Various properties of the physical world have been suggested as indicative of the work of 'a designer with the intellectual properties (knowledge, purpose, understanding, foresight, wisdom, intention) necessary to design the things exhibiting the special properties in question' (Ratzsch and Koperski 2019). These properties include the 'fine-tuning' of the inorganic realm for supporting life, orderliness, uniformity, contrivance, adjustment of means to ends, particularly exquisite complexity, particular types of functionality, delicacy, integration of natural laws, improbability, the intelligibility of nature, the directionality of evolutionary processes, aesthetic characteristics (beauty, elegance, and the like), and apparent purpose and value (including the aptness of our world for the existence of moral value and practice) (Ratzsch and Koperski 2019). In this book, I shall focus on two such properties: fine-tuning and orderliness, although it should be noted that the other properties require explanation as well and I shall discuss them occasionally in what follows.



A. Loke, *The Teleological and Kalam Cosmological Arguments Revisited*, Palgrave Frontiers in Philosophy of Religion, https://doi.org/10.1007/978-3-030-94403-2_4

## 4.2 Fine-Tuning and Orderliness

### 4.2.1 Fine-Tuning

Concerning the 'fine-tuning' of the universe, Robin Collins explains,

> The fundamental structure of the universe is 'balanced on a razor's edge' for the existence of life .... This precise setting of the structure of the universe for life is called the 'fine-tuning of the cosmos'. This fine-tuning falls into three major categories: that of the laws of nature, that of the constants of physics, and that of the initial conditions of the universe. (Collins 2009, p. 202)

It has been objected that there could be other forms of life which do not require a fine-tuned universe (Stenger 2013). In reply, what the calculations have shown is that universes with different laws, constants, and boundary conditions would most likely give rise to much less structure and complexity, which would be incompatible with any kind of life, not merely life-as-we-know-it (Lewis and Barnes 2016, pp. 255–274). This is illustrated by the following two examples of fine-tuning:

First, the cosmological constant characterizes the energy density of the vacuum which is responsible for the acceleration of the universe's expansion. On theoretical grounds, one would expect it to be larger than its actual value by an immense number of magnitudes (between $10^{50}$ and $10^{123}$), but only values a few order of magnitude larger than the actual value are compatible with the formation of galaxies (Friederich 2018). Lewis and Barnes (2016, p. 164) remark:

> The (effective) cosmological constant is clearly fine-tuned. It's just about the best fine-tuning case around. There is no simpler way to make a universe lifeless than to make it devoid of any structure whatsoever. Make the cosmological constant just a few orders of magnitude larger and the universe will be a thin, uniform hydrogen and helium soup, a diffuse gas where the occasional particle collision is all that ever happens. Particles spend their lives alone, drifting through emptying space, not seeing another

particle for trillions of years and even then, just glancing off and returning to the void.

The fine-tuning of the cosmological constant has recently been challenged by physicist Fred Adams (2019), who argues that the life-permitting variation of the constants is wider in some respects than previously thought. Nevertheless, he acknowledges that 'Even if the parameters of physics and cosmology can deviate from their values in our universe by orders of magnitude, "unnaturally small" ratios are still required: For example, the cosmological constant can vary over a wide range, but must be small compared to the Planck scale' (pp. 141–2). In other words, the range is still not wide in an absolute sense, and 'fine-tuning' (at the level of the ratios) is still required.

Second, concerning entropy, the initial state of the Big Bang must be extremely highly ordered (i.e. low entropy) with a very high amount of usable energy. The probability that a universe chosen at random would possess the necessary degree of order that ours does (and so possesses a second law of thermodynamics according to which the universe is progressing from a state of order to states of increasing disorder) is 1 in $10^{10(123)}$. If the universe were less ordered than this, the matter in it would have collapsed through friction into black holes (which represent extreme states of disorder and incompatible with any form of life), rather than form stars (Holder 2004, pp. 38–39, citing Penrose 1989, pp. 339–345, who notes that $10^{10(123)}$ is a number so large that the noughts cannot be written down in full even if each of the $10^{80}$ protons of our universe were to be used to write down one nought).

Stenger (2013) objects that calculations of improbabilities often fail to consider the consequences of varying more than one parameter at a time. In reply, studies of the complete parameter space of (segments of) the Standard Model indicate that the life-permitting range in multidimensional parameter space is likely very small (Barnes 2012, Sect. 4.2). Without fine-tuning, the universe would have become a 'rubble' after the Big Bang, in which case not only 'life as we know it' would not exist, any organized matter with ability to reproduce would not exist. Against the supposition that proponents of fine-tuning erroneously presuppose that only carbon-based life is possible, Hawthorne and Isaacs (2018, p. 147) note that 'it would be very hard to have physical life in any form if an inhospitable cosmological constant led to a universe that expanded so rapidly that particles did not interact with one another or to a universe that collapsed back in on itself only moments after its generation'. Likewise, Rasmussen and Leon (2018, pp. 103-4) observe:

> A universe with nothing but empty space has no ingredients for life ... a million motionless particles will never produce an amoeba ... a universe with only particles that constantly repel each other will produce an endless scatter, with no complex unities, anywhere, ever ... a universe with things that only attract each other will only form a blob, forever.

### 4.2.2 Orderliness

#### 4.2.2.1 Introduction

Concerning the Teleological Argument from orderliness, an example is *The Fifth Way* of Aquinas' famous five proofs for the existence of God (*Summa Theologica*, Part I, Question 2, Article 3), in which Aquinas argues from 'things which lack knowledge, such as natural bodies ... acting always, or nearly always, in the same way'.[1]

It is an irresistible fact that the natural world appears to exhibit certain regular patterns of behaviour. When one gazes into the night sky, one cannot help but wonder why the stars and planets move according to a certain order. Likewise, the alternation of seasons, the formation of clouds and rain, the sustenance of life on earth, and so on are also in accordance with a certain order. This order is characterized by law-like regularities which are of a mathematical nature and are predictably the same everywhere in the universe.[2]

Cambridge physicist John Polkinghorne argues that those who work in fundamental physics encounter a world in which large-scale structures and small-scale processes are alike characterized by a wonderful order that is expressible in concise and elegant mathematical terms, citing Paul Dirac's well-known belief that the laws of nature should be expressed in beautiful equations (Polkinghorne 1998, p. 2).

Polkinghorne explains that mathematical beauty involves such qualities as economy and elegance, and that extensive consequences are found to flow from seemingly simple initial definitions, as when the endless baroque complexities of the Mandelbrot set are seen to derive from a specification that can be written down in a few lines. Polkinghorne writes,

> 300 years of enquiry have shown that it is just such mathematically beautiful theories that prove to have the long-term fertility of explanation that convinces us that they are indeed describing aspects of the way things are. In other words, some of the most beautiful patterns that the mathematicians can think about in their studies are found actually to be present in the structure of the physical world around us. (Polkinghorne 2011, pp. 72-3)

Nevertheless, McGrath (2018, pp. 118-119) observes that 'the concept of "beauty" is subjective and contested, leading some to make the "eminently rational decision" to pursue "indicators of truth in disregard of beauty." Properties of a theory that have at some point been considered to be aesthetically attractive have at other times been considered neutral or displeasing.'

Regardless of whether 'beauty' is present or not, the mathematical describability of the order is indisputable. With regard to this order, Oxford physicist Roger Penrose confesses that 'it remains a deep puzzle why mathematical laws should apply to the world with such phenomenal precision ... Moreover, it is not just the precision but also the subtle sophistication and mathematical beauty of these successful theories that is profoundly mysterious' (Penrose 2004, pp. 20–21).

After surveying the discoveries of the laws of nature in over 1000 pages of his magisterial book *The Road to Reality*, Penrose writes: 'The most important single insight that has emerged from our journey, of more than two and one-half millennia, is that there is a deep unity between certain areas of mathematics and the workings of the physical world' (ibid., pp. 1033–1034). Citing the highly esteemed mathematical physicist Eugene Wigner's (1960) lecture on the effectiveness of mathematics in the physical sciences, Penrose comments: 'Not just the extraordinary precision, but also the subtlety and sophistication that we find in the mathematical laws operative at the foundations of physics seem to me to be much more than the mere expression of an underlying 'order' in the workings of the world' (ibid., p. 1046n.34).

#### 4.2.2.2 Objection: Human Creation

Some might think that, because we have invented the mathematics to characterize the way our world operates, it is not surprising that the universe operates according to mathematical patterns. Carrier (2003) claims that '*any* universe composed of conserved and discrete objects arranged into patterns in a multidimensional space will *always* be describable by mathematics. We invented mathematics just for that purpose: to describe such things.' On this view, some sort of mathematical order or another has to apply to the universe, and one might claim that we just happen to live in the one we observe. Likewise, Livio notes that some have objected that mathematics is a human creation developed to characterize the operation of our world and to solve the problems our world presents. Nature, if it is explicable at all, has to be explicable in some form of language or model, and that mathematics is just that. Given this, it is hardly surprising that the universe operates according to mathematical patterns. Others have objected that mathematics may not explain every situation, and that to some extent scientists have cherry-picked what problems to work on based on those problems being amenable to mathematical treatment (Livio 2009, Chap. 9).

Wenmackers (2016) argues that our knowledge and use of mathematics may have arisen by the evolutionary process. For example, protomathematical capacities might have been useful in earlier evolutionary stages of our species; for example, being able to estimate and to compare the number of fruits hanging from different trees contributes to efficient foraging patterns. These capacities are therefore naturally selected and developed into our current power to think abstractly and to act with foresight. She concludes that the fact that our mathematical reasoning can be applied successfully is precisely why the traits that enable us to achieve this were selected in our biological evolution (p. 9).

Nevertheless, the above objections do not explain how physical entities could be of such a nature that allows a large number of phenomena to be mathematically describable and explicable in such a way that requires highly advanced intellect to work it out. The mathematics involved in describing our universe is not merely a few simple equations like 2+2=4, but highly sophisticated ones, and (contra Carrier) not any universe would be like *this*. Rather, the universe would have to be highly ordered, as implied by its describability by advanced mathematics.

The above point holds even if (as some have suggested) mathematics basically just describes conditionals of some sort or other, for the conditionals would not be as simple as 'if you were to have 2 things, and add to them 2 more things, then you would have 4 things'. Rather, it would be something like, 'if you were to have m, and add m to another m to another m … .9000000000000000 times, you will get a value for e, which is related to time x power, which is related to … etc.' This conditional implies a huge, interconnected, highly ordered structure. A highly ordered structure is far less likely to be explainable by chance compared to a simply ordered one (see Sect. 4.4), and in order for a highly ordered structure to arise from simple laws, a high degree of order must already be in place in order for this 'arising' to happen (see Sects. 4.5 and 4.6). My argument does not require an appeal to 'why God would particularly care about advanced mathematics' (see Sect. 7.6); rather, it is based on exclusion (Sect. 7.5).

The multitude of mathematical equations with numerous variables reflect a highly ordered arrangement of the distinct objects which composed the physical world described by these equations. The patterns of order in multidimensional space and natural laws with systemic applicability reflect a huge interconnected structure with multiple parts. It would be unreasonable to explain away such a structure by saying that some sort of order or another has to apply to the components, and we just happen to discover the one we observe. Physicist Michael Heller remarks that the mathematical equations in physics can be treated by physicists as expressing a kind of software of the universe (Heller 2013, p. 594), and one would think that there cannot be a software without a software programmer. To establish the conclusion of design however requires ruling out other alternative hypotheses, which I undertake in the rest of this book. The point here is simply that, while the objections by Carrier et al. may explain the applicability of simple calculations, they do not explain the high degree of ordering of the physical world that is presupposed by the applicability of high-level mathematics.

A Kantian might explain mathematical discovery by arguing that mathematics is the conceptual framework through which we experience the phenomenal universe, while claiming that we know nothing about the noumenal universe. Nevertheless, in order for such highly sophisticated mathematics to successfully characterize the way our world operates, the objective world, that is, the universe-in-itself, must have a high degree of order. As Einstein observes, 'even if man proposes the axioms of the theory, the success of such a project presupposes a high degree of ordering of the objective world, and this could not be expected *a priori*' (Goldman 1997, p. 24).

Einstein's argument is not based on the mere presence of order within our universe; it is the high degree of rationality and intelligibility of the order which the argument is based on. The particles of the universe are related to one other and many particles behave similarly, and the question that needs to be answered is, 'Why are their relations and behaviour so rational, intelligible, highly ordered and forming such a huge interconnected structure, instead of being crude, simple and having an almost featureless order?'

Even at the quantum level, where things are often regarded as messy and counter-intuitive, various mathematical equations such as Schrodinger's still hold, and this, as well as the widespread effectiveness of mathematics at the macro level, demands explanation.

Moreover, if ordering is an inevitable selection effect created by our act of perception as the Kantian asserts, why do we still find some things disordered or yet unintelligible and not see everything as a teleological structure (Barrow and Tipler 1986, p. 91, citing Janet, Trendelenburg and Herbart)? Holder (2004, p. 4) notes:

> Kant's position regarding the human imposition of order also does not seem to square with how scientists see the world. For example, quantum theory seems to be forced on us by the reality of the external world, which exhibits such strange and startling phenomena at the micro-level, rather than being a human creation imposed on the world.

In other words, contrary to the Kantian, the counter-intuitive nature of quantum physics indicates that the mathematical equations that are used to describe it are not merely the creation of our own minds, and those seeking an understanding as completely as possible must therefore ask what it could be that links together the reason within (mathematical thinking) and the reason without (the structure of the physical world) in this remarkable way (Polkinghorne 2011, pp. 72–73).

Additionally, as demonstrated in Chap. 1, the laws of logic is not merely our way of thinking but reflect the way mind-independent reality is (e.g. there cannot be a shapeless square in the mind-independent reality), and therefore these laws can be used to formulate an argument by exclusion for a Designer (see below).

#### 4.2.2.3 Platonic Objection

It has been suggested that the reason why the 'laws of physics' are so well explained by mathematical descriptions is related to the postulation that the nature of the space of mathematical reality is Platonic (Penrose 2004, p. 1029).

However, the postulated existence of a Platonic world with abstract mathematical objects still does not explain why the Platonic world could be mapped onto the physical world via the power of human mental activity, and how mindless physical entities could have this orderly behaviour (Frederick 2013). Philosopher Roger Trigg (1993, pp. 186-187) observes that mathematical theories can exist but still not be about anything. And Stephen Hawking (1988, p. 126) had asked: 'even if there is only one possible unified theory, it is just a set of rules and equations. What is it that breathes fire into the equations and makes a universe for them to describe?'

Cosmologist Max Tegmark (2008) has replied with a radical proposal that our physical universe is equivalent to an abstract mathematical structure.[3] This looks like Pythagoreanism reborn, whereby physical objects are somehow reduced to abstract mathematical structures (Dumsday 2019, p. 35). This proposal is related to a view known as Ontic Structural

Realism (OSR). Proponents of OSR claim support from our best current theories in physics; for example, they claim that

> traditional conceptions of individuality break down at the quantum level, such that the notion of particles as fundamental individual 'objects' with intrinsic identities should be abandoned in favour of fundamental relations; that the metaphysics of quantum field theory is best interpreted along structuralist lines, insofar as symmetries are best seen there as ontologically prior to fields; that the metaphysics of quantum gravity is best interpreted along structuralist lines, since particles are best seen there as deriving their identities from their structural context; that structuralism provides for a superior account of the metaphysics of spacetime; and that structuralism allows for a novel way of defusing the traditional debate over whether matter at the fundamental level is continuous or discrete, and, relatedly, provides a plausible way of reconceiving wave-particle duality. (Dumsday 2019, pp. 27-29)

However, not all proponents of OSR defend *eliminativist* OSR, which claims that, at the fundamental level, relations exist but objects either do not exist ('There are no things. Structure is all there is', Ladyman et al. 2007, p. 131)[4] or they exist but are nothing over and above their place/function in the relational structure which is ontological prior over them and the bearers of any property. There are more moderate versions of OSR which claim that objects and relations are symmetrically dependent with no ontological priority obtaining between them, and that there are objects which have 'an intrinsic identity defined partly in terms of the possession of intrinsic properties and partly in terms of their place/function in the structure. As such, their identity is not wholly reducible to their structural role, yet they cannot exist independently of the structure' (ibid., p. 30). These more moderate versions of OSR are compatible with the arguments I defend in this book.

On the one hand, there is no conclusive argument that compels the acceptance of the eliminativist version of OSR over the moderate version because, as Ladyman et al. (2007, p. 9) themselves observe, 'science, usually and perhaps always, underdetermines the metaphysical answers we are seeking.' They admit:

> Of course, all the considerations from physics to which we have appealed do not logically compel us to abandon the idea of a world of distinct ontologically subsistent individuals with intrinsic properties. As we noted, the identity and individuality of quantum particles could be grounded in each having a primitive thisness, and the same could be true of spacetime points. (p. 154)

On the other hand, the more moderate versions 'allow for the option of explaining the concretization of structure by reference to the concretization of its component objects, since on these versions of OSR the latter have at least some intrinsic identity conditions of their own, which could perhaps include whatever it is that provides for concretization' (ibid., p. 36). Thus, objects are not 'purely speculative philosophical toys' (cf. Ladyman et al. 2007, p. 154) but explain concretization.

Moreover, the *eliminativist* OSR of Tegmark collapses the distinction between abstract and the concrete physical; this is metaphysically dubious, since unlike physical entities, abstract entities do not have causal powers. Hence, Tegmark's proposal that our universe is an abstract mathematical structure still does not explain how the entities in our universe could causally interact in the orderly way noted above. Ladyman et al., who affirm *eliminativist* OSR, say, 'What makes the structure physical and not mathematical? That is a question that we refuse to answer. In our view, there is nothing more to be said about this that doesn't amount to empty words' (Tegmark 2008, p. 158), claiming that standard methods of distinguishing the concrete from the abstract by appealing to causal efficacy are unworkable for fundamental physics (pp. 159–161). However, as I have argued in Chaps. 2 and 3, causation is necessary for and compatible with fundamental physics, and what grounds one event (change) following another (i.e. what grounds their relation) are causal properties, and the Modus Tollens argument for Causal Principle demonstrates that events do not begin uncaused. While a naturalist Platonist might be able to explain the permanence of mathematical truths by appealing to timeless abstract objects, abstract objects by themselves cannot explain why physical entities follow complicated mathematical truths, since abstract objects have no causal power to make physical entities behave in such a way.

Why then are the events in our physical universe like this? Why is it the case that the sequence of events can be described by mathematical equations which indicate a high degree of ordering? How could unthinking mindless physical entities and forces have such an orderly behaviour? As Danny Frederick asks, 'What is to stop some bits of matter moving in ways which are inconsistent with natural laws; or the same piece of matter moving at one time in a way which accords with natural laws but at another time in a way which is inconsistent with them?' (Frederick 2013, p. 271).

Frederick argues that, while natural laws may be regarded as ceteris paribus rather than exceptionless laws (i.e. they may be default regularities that hold in the absence of outside interference), and while natural laws should be understood as descriptions of what is happening rather than rules for natural objects to follow, nevertheless the question still remains as to how the events in the universe could happen in such a manner describable by natural laws. He notes that statements of natural law are modal descriptions rather than mere descriptions: unlike mere descriptions, modal descriptions describe the limits to what can happen and can be used for prediction. He also observes that it would not help to point out that microphysics shows that the fundamental laws of nature are statistical, for one could then ask how the changes of unthinking physical entities could so arrange themselves over time as to exhibit a probability distribution (ibid.).

The pressing question, therefore, remains: The universe does not have to be like this, but why is it like this? Throughout history, a number of eminent scientists have come to the conclusion that the most plausible explanation is that the universe is the work of a Supreme Intelligent Mind who imposed a rational order onto the mindless physical entities. For example, Einstein writes:

> Certain it is that a conviction, akin to religious feeling, of the rationality or intelligibility of the world lies behind all scientific work of a higher order … This firm belief, a belief bound up with deep feeling, in a superior mind that reveals itself in the world of experience, represents my conception of God.[5]

Paul Dirac, one of the pioneering geniuses of quantum theory and a deeply avowed atheist in his younger days, came to acknowledge the plausibility of a Designer after years of research in physics when he says:

> It seems to be one of the fundamental features of nature that fundamental physical laws are described in terms of a mathematical theory of great beauty and power, *needing quite a high standard of mathematics for one to understand it* … One could perhaps describe the situation by saying that God is a mathematician of a very high order, and He used very advanced mathematics in constructing the universe. (Dirac 1963)

### 4.2.3 Summary

To sum up the views of the scientists cited above, the following features of our universe have been noted:

1. Fine-tuning
2. The existence of orderly patterns of events which can be described by advanced mathematics (see also the discussion of laws of nature in Chap. 2)

In what follows, we shall examine which hypothesis best explains both of these features. It may be that some hypothesis or combinations of hypotheses can explain (1) but not (2), or (2) but not (1), and therefore fail because what needs to be explained are both of these features taken together.

## 4.3 A Logically Exhaustive List of Categories of Possibilities

In his writings, Richard Dawkins has repeatedly warned of the danger of jumping to the conclusion of design. He cites as example the argument from the apparent design of living organisms, which he thinks is a God-of-the-gaps argument (i.e. an argument based on gaps in our existing knowledge). He argues that in the past it was thought that the improbability of dragonfly's wing or an eagle's eye originating by chance implied that these were designed, and that this conclusion resulted from a failure to see the possibility of the alternative explanation of Darwinian evolution. He argues,

> After Darwin, we all should feel, deep in our bones, suspicious of the very idea of design. The illusion of design is a trap that has caught us before, and Darwin should have immunized us by raising our consciousness … A full understanding of natural selection encourages us to move boldly into other fields. It arouses our suspicion, in those other fields, of the kind of false alternatives that once, in pre-Darwinian days, beguiled biology. Who, before Darwin, could have guessed that something so apparently designed as a dragonfly's wing or an eagle's eye was really the end product of a long sequence of non-random but purely natural causes? (Dawkins 2006, pp. 139, 141)

Dawkins raises an important point. Nevertheless, one should also be careful not to make the fallacious argument that, because many things once thought to be divinely designed actually do have natural explanations, therefore all things have natural explanations. The correct way to proceed is to assess, on a case-by-case basis, which explanation is the best for each case. To assess the case concerning the mathematical describable order of physical entities and to address Dawkins' concerns, I shall demonstrate that a logically exhaustive list of categories of alternative hypotheses can be devised, and that various objections can be given to rule out each of these categories.

The failure to consider alternative hypotheses is evident in William Dembski's widely discussed book *The Design Inference*, in which Dembski attempts to demonstrate that regularity, chance, and design are logically exhaustive and competing modes of explanation. He writes:

> Whenever explaining an event, we must choose from three competing modes of explanation. These are regularity, chance, and design. To attribute an event to a regularity is to say that the event will (almost) always happen. To attribute an event to chance is to say that probabilities characterize the occurrence of the event, but are also compatible with some other event

happening. To attribute an event to design is to say that it cannot reasonably be referred to either regularity or chance. Defining design as the set-theoretic complement of the disjunction regularity-or-chance guarantees that the three modes of explanation are mutually exclusive and exhaustive. (Dembski 2006, p. 36)

However, Dembski glosses over the possibility that regularity, chance, and design can be combined in various ways, and his subsequent use of his three competing modes of explanation for explaining biological structures has been criticized for ignoring various evolutionary pathways.[6] Such a pathway has been proposed for cosmology as well (see the discussion of Smolin's proposal below), and regardless of the merits of this proposal, it is important that this theoretical possibility be considered. Moreover, Dembski fails to consider the option that the event may be 'Uncaused', as has been postulated by Hawking for the Big Bang (see Chap. 6). Incomplete considerations of alternative explanations such as Dembski's serve as a warning that we should be more rigorous in our assessment of alternative explanations with regard to the Teleological Argument. Consider also Monton's claim that 'when people observe features of the universe, they sometimes infer that the feature occurred as a result of design, and they sometimes infer that the feature occurred some other way—by chance, necessity, coincidence, unguided natural processes, or what have you' (Monton 2010, p. 208). The qualifying phrase 'what have you' is too slack and does not address the sort of concerns raised by Dawkins.

Various forms of design arguments have been suggested in the literature, for example, significance testing (If E has a low probability and is specified, it is due to intelligent design), inductive sampling, analogical, Bayesian, likelihoodist, and abductive (IBE) (Sober 2019). The problem of unconsidered alternative explanations besets all of them. For example, concerning Bayesianism and Inference to the Best Explanation (IBE), which are widely used by contemporary philosophers, Ratzsch and Koperski (2019) observe,

substantive comparison can only involve known alternatives, which at any point represent a vanishingly small fraction of the possible alternatives.

> Choosing the best of the known may be the best we can do, but many would insist that without some further suppressed and significant assumptions, being the best (as humans see it) of the (humanly known) restricted group does not warrant ascription of truth, or anything like it.[7]

In response to Craig's argument that an infinite mind can explain the connections between abstract, physical, and mental which Penrose admits are mysteries, Penrose replies he does not see why an infinite mind is the only solution because there could be other possibilities which we still do not know of and cannot verify. On the other hand, appealing to God can be used to solve any problem, so it is not helpful.[8]

Now I am not claiming that the Teleological Argument must be able to eliminate all the other alternative explanations in order to be of any value. To require the elimination of all the possible alternatives may be too demanding a requirement for reasonable belief, since such a criterion is not fulfilled even by all rational inferences in the natural sciences or in everyday life (Bird 2005, pp. 26–28). Nevertheless, the concerns noted in the preceding paragraphs indicate that it would be desirable if the argument can be made more rigorous such that all the possible alternatives can indeed be eliminated.

The above concerns can be addressed by devising a logical exhaustive list of possible explanations and an exclusion of all the alternative categories of explanations such that the conclusion of design follows logically rather than being merely appealed to solve a problem. Concerning the Teleological argument defended here, the logically exhaustive list of categories of possibilities is demonstrated by the rigorous use of the Law of Excluded Middle, and is as follows:

1. The fine-tuning and order of the universe is either fundamentally Uncaused, or it is fundamentally due to either 1.1, 1.2, or 1.3[9]:

1.1. random cause(s) ('Chance').

1.2. non-random cause(s), in which case either.

1.2.1. it is fundamentally due to non-intelligent, non-random cause(s) ('Regularity'), or

1.2.2. it is fundamentally due to intelligent, non-random cause(s) ('Design').

1.3. a combination of random and non-random causes, in which case either.

1.3.1. it is fundamentally due to a combination of non-intelligent, non-random cause(s) + random cause(s) ('Combinations of Regularity and Chance'), or

1.3.2. it is fundamentally due to a combination of intelligent, non-random cause(s) + random cause(s) +/- non-intelligent, non-random cause(s) (e.g. Evolutionary Creationism: involves a Designer).

2. The fine-tuning and order of the universe is not[10] fundamentally due to Chance, Regularity, and Combinations of Regularity and Chance, and it is not fundamentally Uncaused.

3. Therefore, the fine-tuning and order of the universe is fundamentally due to Design.

It should be noted that my argument by exclusion does not require 'perfect' elimination ('rule out') understood as demonstrating that other possible hypotheses have zero probability. It only requires showing that their probability is so low that they can be eliminated as reasonable alternatives to Design even if we assign them very generous probability estimates (see Sect. 7.5), *and this is how the 'not' in the above syllogism should be understood.*

From the above syllogism, it can be seen that all possible hypotheses belong to the following categories: (i) Chance, (ii) Regularity, (iii) Combination of Regularity and Chance (e.g. natural selection + random variation, as in the case of naturalistic evolution), (iv) Uncaused, and (v) Design (the Designer may or may not have used processes such as evolution).

Although each of these categories has been discussed before in the literature, a logical demonstration that these are the only possible categories of hypotheses has not been published before, despite the huge amount of literature on the Teleological Argument over the centuries, hence the unique contribution of this book. It should be noted that such a list can be used for other types of Teleological Argument with respect to other cases of apparent design as well, by simply replacing 'the existence of mathematically describable order and fine-tuning of the universe' with other features of apparent design in question. Because of its utility, this list contributes to the discussion of the Teleological Argument in general.

One might raise the worry that new, previously unconsidered hypotheses could all be lumped together in the catch-all basket, and that 'without knowing the details of what specific unconsidered hypotheses might look like, there is simply no plausible way to anticipate the apparent likelihood of a novel new hypothesis' (Ratzsch and Koperski 2019). In reply, I shall show that there is an *essential feature* of each of the categories alternative to design which renders it unworkable as an ultimate explanation for the fine-tuning and order of the universe. As noted earlier, these alternative categories are (i) Chance, (ii) Regularity, (iii) Combination of Regularity and Chance (e.g. natural selection + random variation, as in the case of naturalistic evolution), and (iv) Uncaused. Because the terms chance, random, and the related notion of probability have multiple meanings, I shall first clarify my usage of these terms before evaluating the alternative categories in turn.

Broadly speaking, there are two main concepts of probability: (1) an epistemic notion and (2) a non-epistemic notion, better known as physical probability (Eagle 2019).

(1) The epistemic notion of probability can be further subdivided into objective and subjective interpretations (Holder 2004, p. 74):

(1.1) Objective interpretation of epistemic notion of probability (this includes 1.1.1. *classical* and 1.1.2. *logical/evidential* probability). This refers to objective evidential support relations (e.g. 'in light of the relevant seismological and geological data, California will *probably* experience a major earthquake this decade') (Hájek 2019). It measures the extent to which the evidence is entailed by the hypothesis (Holder 2004, p. 74, citing Swinburne).

(1.1.1.) The classical interpretation 'assigns probabilities in the absence of any evidence, or in the presence of symmetrically balanced evidence. The guiding idea is that in such circumstances, probability is shared equally among all the possible outcomes, so that the classical probability of an event is simply the fraction of the total number of possibilities in which the event occurs … for example, the classical probability of a fair die landing with an even number showing up is 3/6' (Hájek 2019).

A related notion is the Principle of Indifference, which Collins (2009, p. 234) states as follows:

> When we have no reason to prefer any one value of a variable p over another in some range R, we should assign equal epistemic probabilities to equal ranges of p that are in R, given that p constitutes a 'natural variable.' A variable is defined as 'natural' if it occurs within the simplest formulation of the relevant area of physics.

Applying the principle to the argument from Fine-tuning, Collins (2009, p. 234) writes:

> Since the constants of physics used in the fine-tuning argument typically occur within the simplest formulation of the relevant physics, the constants themselves are natural variables. Thus, the restricted Principle of Indifference entails that we should assign epistemic probability in proportion to the width of the range of the constant we are considering.

The epistemic probability is argued to be very small, because for a fine-tuned constant C, Wr/WR << 1, where Wr is the width of the life-permitting range of C, and WR is the width of the set of values for which we can make determinations of whether the values are life-permitting or not (Collins 2009, pp. 244, 252). Likewise, Lewis and Barnes (2016, pp. 286-7) reason that, if all we knew was that a certain universe obeyed the laws of nature, without specifying the values of the constants of nature and initial conditions, the probability that that universe would contain life forms is extremely small.

Following Hume, it might be objected that our universe is the only universe of which anyone had experience, invalidating it as the basis of an inductive inference. However, while this universe is the only one we experienced, we can still think about how it could have been different. Ratzsch and Koperski (2019) observe:

> If we let ***C*** stand for a fine-tuned parameter with possible values in the range [0, x], and if we assume that nature is not biased toward one value of ***C*** rather than another such that each unit subinterval in this range should be assigned equal probability, then fine-tuning is surprising insofar as the

life-permitting range of ***C*** is tiny compared to the full interval, which corresponds to a very small probability.

Critics accuse the Principle of Indifference of extracting information from ignorance, and argue that in a state of ignorance, it is better to assign imprecise probabilities or to eschew the assignment of probabilities altogether (Hájek 2019).

In reply, concerning the problem of assigning prior probability of the constants and initial conditions of a given theory (e.g. the probability of a constant having a value in a certain small range, without any knowledge about our universe), Lewis and Barnes (2016) note that 'we cannot calculate the posterior at all without some estimate of the prior probability' (p. 287). However, this is not a big problem because 'if our data are very good, then our conclusions won't depend much on the prior probability' (p. 288). In fine-tuning cases, 'the speed and severity with which disaster strikes as one tiptoes through parameter space show that the probability of a life-permitting universe, given the laws but not the constants, will be very small for any honest (and non-fine-tuned!) prior probability' (ibid.).

In other words, if there are some factors which we are ignorant of which entail that the probability is not small (a concern raised in Hossenfelder 2019), those factors would need to be 'fine-tuned'.

(1.1.2) Logical theories of probability allow for the possibilities to be assigned unequal probabilities depending on the evidence (Hájek 2019). While the best beliefs to have are those that are logically probable on our rightly basic beliefs, to the extent to which an investigator's standards are close to the correct ones, he/she will use rightly basic beliefs and logical probability (Holder 2004, pp. 75–76).

(1.2.) Subjective interpretation of epistemic notion of probability (subjective probability). This refers to an agent's degree of confidence referring to a graded belief (e.g. 'I am not sure that it will rain in Canberra this week, but it *probably* will') (Hájek 2019).

(2) Non-epistemic notion of probability, also known as physical probability (this includes the *frequentist*, *propensity*, and *best-system* interpretations): this applies to various systems in the world, independently of what anyone thinks (Hájek 2019). The *frequentist* interpretation relates to the

outcome of many trials of an experiment, such as many tosses of a fair coin (Holder 2004, p. 73). Whereas the propensity interpretation refers to

> the extent to which one or more events cause another event. The outcome of my toss of a coin may be determined completely by the impulse I impart to it, the angle at which my thumb strikes it, the atmospheric conditions at the time, and so on; and so the coin may have a physical probability of 1 of landing heads on a particular toss. Indeed, if determinism were true all physical probabilities would be 0 or 1. Most physicists, however, believe that quantum theory is ontologically indeterminate and so the physical probability of a quantum event, such as the radioactive decay of an atom within a certain time, has a physical probability between 0 and 1. (Ibid.)

An example of the best system interpretation is '*the Mentaculus*', which attempts to provide a complete probability map of the universe (see Chap. 2).

Evaluation of different interpretations of probability:

> As noted above, there are different interpretations of probability which are suited for different contexts of discussions. Which of the above interpretations is suitable for discussing the probabilities of the hypotheses concerning the fine-tuning and order of the universe in the context of the argumentation of this book?

The non-epistemic notion of probability (physical probability) is not appropriate, because according to the standard view of physical possibility, 'alternative physical laws and constants trivially have physical probability zero, whereas the actual laws and constants have physical probability one' (Friederich 2018).[11]

The subjective epistemic notion is also not appropriate, because the arguments in this book do not concern the psychological state of any particular individual, but the state of the universe.

Therefore, an objective epistemic notion of probability is the only appropriate one for the purposes of this book. I will be using both the classical interpretation and the logical/evidential interpretation where appropriate. In particular, by arguing that there are essential properties of each of the alternative hypotheses to design which render it unlikely, I will be attempting to construct logical probabilities concerning that hypothesis and showing that the probability is low on the basis of evidence.

Broadly speaking, there are two main concepts of 'random':

(1) An epistemic notion: referring to those processes whose outcomes we cannot know in advance, that is, *unpredictable* (Eagle 2019).

(2) A non-epistemic notion: the non-epistemic notion may be subdivided as follows:

(2.1) A non-epistemic notion used to characterize the disorder and patternlessness of an entire collection of *outcomes* of a given repeated process. On Eagle (2019)'s conception,

> randomness indicates a lack of pattern or repetition … randomness is fundamentally a product notion, applying in the first instance to sequences of outcomes, while chance is a process notion, applying in the single case to the process or chance setup which produces a token outcome … randomness is indifferent to history, while chance is not. Chance is history-dependent.

On the basis of this conception, he argues that there are counterexamples to the Commonplace Thesis (CT) 'Something is random iff it happens by chance.' One interesting potential counterexample involves coin tossing. 'Some have maintained that coin tossing is a deterministic process, and as such entirely without chances, and yet which produces outcome sequences we have been taking as paradigm of random sequences' (ibid.). Eagle (2019) also argues it is possible for a chancy and indeterministic process to produce a non-random sequence of outcomes.

(2.2.) A non-epistemic notion used to characterize a *process*. Eagle (2019) notes that some philosophers deliberately use 'random' to mean 'chancy' and acknowledges that this process conception of randomness is perfectly legitimate, but complains that it makes the Commonplace Thesis a triviality and does not cover all cases of randomness.

Eagle notes that some have defined randomness as indeterminism, but this view

> makes it difficult to understand many of the uses of randomness in science … This view entails that random sampling, and random outcomes in chaotic dynamics, and random mating in population genetics, etc., are not in fact random if determinism is true, despite the plausibility of their being so. It does not apparently require fundamental indeterminism to have a randomized trial, and our confidence in the deliverances of such trials does not depend on our confidence that the trial design involved radioactive decay or some other fundamentally indeterministic process. Indeed, if Bohmians or Everettians are right (an open epistemic possibility), and quantum mechanics is deterministic, the view that randomness is indeterminism entails that nothing is actually random, not even the most intuitively compelling cases. (Ibid.)

Hence, Eagle concludes that the view that randomness is indeterminism should be rejected (ibid.).

The term 'chance' also has a variety of meanings:

(1) Epistemic notion:

(1.1) Synonymous with an epistemic notion of random, that is, unpredictable. 'Something that happens unpredictably without discernible human intention or observable cause, e.g. "Which cards you are dealt is simply a matter of chance"' (*Merriam-Webster Dictionary*, definition 1a)

(1.2) Synonymous with an epistemic notion of probability. 'The possibility of a particular outcome in an uncertain situation … the degree of likelihood of such an outcome e.g. a small chance of success' (*Merriam-Webster Dictionary*, definition 4)

(2) Non-epistemic notion: chance is often used synonymously with physical probability (Eagle 2019). It is also used for the juxtaposition of unrelated causal trajectories (e.g. car crashes, when two people meet by accident) (Ellis 2018).

Evaluation of different interpretations of 'random' and 'chance'.

As noted above, there are different interpretations of 'random' and 'chance' which are suited for different contexts of discussions. Which of the above interpretations is suitable for the use of these terms in my syllogism demonstrating the logically exhaustive list of categories of possibilities as explained above?

The epistemic notion is not appropriate: the syllogism is not referring to what we can predict, but what is the case. The definition of randomness as indeterminism is also inappropriate, because of the reasons Eagle explained (see above). Rather, by using the term 'random causes' in my syllogism and labelling this as the 'Chance hypothesis', I intend to represent a common usage in the scientific literature relevant to certain forms of hypotheses which have been postulated as possible explanations for 'fine-tuning', such as the inflationary cosmology and multiverse scenarios. For example, cosmologist Andreas Albrecht writes,

> One typically imagines some sort of chaotic primordial state, where the inflation field is more or less *randomly* tossed about, until by sheer *chance* it winds up in a very rare fluctuation that produces a potential-dominated state ... Inflation is best thought of as the 'dominant channel' from *random chaos* into a big bang-like state. (Albrecht 2004, pp. 384-5; italics mine)

The above description by Albrecht uses the terms 'random' and 'chance' as a non-epistemic notion to characterize something that brought about (i.e. caused) a fluctuation resulting in a bigbang-like state. In other words, 'random' and 'chance' is used in a non-epistemic sense to describe causes that bring about a variety of outcomes with varying degree of order and/or specificity. This definition of 'random' and 'chance' is compatible with determinism (and indeterminism); if determinism is true, the varying outcomes are determined to the varying conditions of the cause(s); if indeteminism is true, a cause in the exact same condition may produce different outcomes. To hypothesize that a causal process produced multiple universes such that one that is fine-tuned resulted by chance is analogous to saying that in a game a machine randomly tossed three fair dice multiple times such that this process resulted in the winning ordered combination of 'triple six' by chance.

By using the terms 'random' and 'chance' I am not attempting to discuss the Commonplace Thesis nor to cover all cases of randomness and chance, nor am I using the term 'chance' as 'physical probability' in my syllogism, as Eagle does in his article. Hence, my use of the term 'Chance hypothesis' to label 'random causes' is not susceptible to Eagle's objections to the process notion of randomness noted above.

In summary, I am using the term 'random' in 'random causes' (and labelling this as the Chance hypothesis) in a non-epistemic sense to describe causes that bring about a variety of outcomes with varying degree of order and/or specificity. This contrasts with the 'Regularity' hypothesis, whereby causes bring about outcomes that are not varied, and the 'Design' hypothesis, whereby causes have freedom to intentionally bring about outcomes which may be varied or not varied and for a purpose (cf. Dawes 2007, p. 73, who defines 'design' to mean 'the work of some intentional agent acting purposefully'). To evaluate whether each of the five hypotheses—(i) Chance, (ii) Regularity, (iii) Combination of Regularity and Chance, (iv) Uncaused, and (v) Design—is true on the basis of evidence, I will be using probability in an epistemic objective sense.

I shall now proceed to evaluate the various categories of hypotheses, starting with the Chance hypothesis.

## 4.4 Chance Hypothesis

### 4.4.1 The Argument from Selection Bias and Chaos

With regard to the mathematically describable order of our universe, Wenmackers (2016, p. 10) objects that it may just be due to our selection bias, for the majority of possible mathematical variations are not applicable to our world in any way (p. 10). Moreover, we can never be sure that the application of mathematics to the world is perfect, since empirical precision is always limited. Wenmackers note the objection that the fact that there is some part of mathematics at all that works well requires explanation, even if this does not constitute all or most of mathematics (pp. 10–11). Wenmackers replies that the alternative case in which no mathematics would describe anything in the universe and a world in which processes cannot be summarized or approximated in a meaningful way would not help us to have evolved in this world (Wenmackers 2016, pp. 10–14).

However, the question is, why our world should be such that allows for evolution? As Einstein argues,

> A priori, one should expect a chaotic world, which cannot be grasped by the mind in any way … Even if man proposes the axioms of the theory, the success of such a project presupposes a high degree of ordering of the objective world, and this could not be expected a priori. That is the 'miracle' which is being constantly reinforced as our knowledge expands. There lies the weakness of positivists and professional atheists. (Goldman 1997, p. 24)

Wenmackers (2016, p. 13) objects by claiming that

> random processes are very well-behaved: they consist of events that may be maximally unpredictable in isolation, but collectively they produce strong regularities. It is no longer a mystery to us how order emerges from chaos. In fact, we have entire fields of mathematics for that, called probability theory and statistics, which are closely related to branches of physics, such as statistical mechanics.

However, the randomness that she is referring to is epistemic ('may be maximally unpredictable'). In actuality, the so-called chaos has a high degree of underlying order which is described by the complex equations formulated by statisticians (Bishop 2017). Likewise, the so-called self-organization process (e.g. crystallization) which describes overall order arising from interactions between apparently disordered parts has a high degree of underlying order involving the interactions. The question posed by Einstein is, why should there be any high degree of ordering at all? (One might reply that the high degree of ordering is explained by another level of ordering; this possibility is discussed under the Regularity hypothesis in Sect. 4.5, and also under the Uncaused hypothesis in Chap. 7)

Steiner (1998, pp. 24-26) observes that, in order for mathematics to be applicable for predicting observations of physical entities, the properties of physical entities must remain reasonably stable over time. For example, there are four coins in my pocket, after removing two coins, I should have two coins left, but if the coins are unstable such that they disintegrate very quickly, I would not observe two coins when I check my pocket. 'The number of coins in my pocket … stay constant long enough for humans to count them … The coins in my pocket are usually the same whether or not I walk around the house, put candies in my pocket, too, and so forth' (p. 26). What explains this stability over time? Various properties of a particle, for example, could have changed so quickly that makes mathematical predictions impossible. While one might suggest that there could have been various constraints that prevent the existence of the alternative disordered schemes, the question remains as to why the constraints should exist in such a well-ordered way that resulted in the mathematically describable behaviour.

Genuine randomness is extremely improbable as a causal explanation for the order noted above in view of the fact that one could conceive of a potentially infinite[12] number of alternative ways in which the behaviour of mindless physical entities in the universe is disordered. A *particle, for example, could have moved in billions*[13] *of alternative directions at every moment, other than consistently in the direction describable by any form of mathematical equation.* As noted earlier, 'random causes' is supposed to describe causes that bring about a variety of outcomes with varying degrees of order and/or specificity, without favouring one rather than the other alternatives. Thus, following the Principle of Indifference, if the universe was fundamentally brought about by random causes, then each one of the billions of possible ways of the behaviour of mindless physical entities in the universe should be assigned equal probability. This means the probability of any one of them—including the probability that it moves *consistently in the direction describable by any form of mathematical equation*—is extremely low. Against the criticism that the Principle of Indifference extracts information from ignorance, it can be replied that, if there are some factors which we are ignorant of which entail that the probability for mathematically describable order is not small, those factors would need to be 'fine-tuned' (i.e. ordered by regularity, regularity and chance, design, or combinations of these; see below); it would not be purely random.

Finally, Wenmackers' argument from selection bias and chaos does not explain the fine-tuning of the universe (nor is it intended to).

### 4.4.2 Anthropic Principle

With regard to Fine-tuning, some scientists deny the conclusion of design by arguing that, if these conditions were not 'fine-tuned', we won't be here to observe them; since we are here, we should not be surprised about the fine-tuning.

However, this reply is too superficial. Philosopher John Leslie provides the analogy of a criminal who was dragged before a firing squad of 100 trained marksmen, all of whom missed when the command to fire was given and the criminal found himself alive. It would be ridiculous for the criminal to think that 'since I am still alive, I should not be surprised that all of them missed!' (Leslie 1982, p. 150). On the contrary, the observation that all the marksmen missed requires an explanation other than chance. Perhaps the 100 marksmen had conspired to spare him, or perhaps it was a miracle; in any case, it is unreasonable to attribute his survival to chance.

Sober (2019, p. 73) claims that the fine-tuning case and the firing squad case differ by arguing that, in fine-tuning, the sequence is as follows: $t_1$: constants are set, $t_2$: you are alive, $t_3$ you observe you are alive; while in the firing squad case, $t_1$: firing squad decides, $t_2$: you are alive (just before they fire), $t_3$: you observe you are alive. Sober claims that, in the case of fine-tuning, if you are alive at $t_2$, the constants must be right at $t_1$, $t_2$, and $t_3$; thus, the probability of your observing at $t_3$ that the constants are right is the same regardless of whether it was God or chance that set the values of the physical constants at $t_1$. However, in the case of the firing squad, if you are alive at $t_2$, that leaves open what the firing squad decided at $t_1$ what it will do just after $t_2$; thus, your observing at $t_3$ that you are alive provides evidence about the squad's decision at $t_1$. Thus, the fact that you are alive at $t_2$ induces an Observation Selection Effect in the fine-tuning case but not in the firing squad case. Nevertheless, this still does not explain why are the constants right at $t_1$. As argued previously, why the constants are right at $\mathrm{t}_1$ still requires a reasonable explanation other than chance.

### 4.4.3 Improbable Event Happens

A sceptic might object that even though the apparent probability of a fine-tuned and ordered universe occurring by chance is outrageously tiny, it still could have happened by chance. After all, improbable events happen all the time. For example, the probability of someone winning a lottery involving thousands of participants is outrageously tiny, but still it happened. The probability of clouds, snowflakes, and so on taking the particular beautiful forms that they do is outrageously tiny and these forms may appear to be designed, but we know that they are the result of natural forces.

In response, the cases cited above are disanalogous to the case concerning order and fine-tuning. In a lottery all the participants are equally qualified to win. Likewise, among the millions and millions of possible forms which clouds, snowflakes, and so on can take, a large proportion of them are 'suitably qualified' to appear beautiful or take a certain recognizable pattern or another—this is called pareidolia: a common psychological phenomenon. By contrast, it is not the case that all the values or a large proportion of values among the billions of possible values[14] which (say) those physical constants can take would have 'qualified' to allow for life after the Big Bang. On the contrary, the proportion of possible values which would allow for life is extremely small; as explained above, the overwhelming majority of possible values would not allow for any form of life at all—indeed, they would be devoid of structure and pattern. (Lewis and Barnes 2016, p. 164: 'Particles spend their lives alone, drifting through emptying space, not seeing another particle for trillions of years and even then, just glancing off and returning to the void.') As explained earlier, an explosion such as the Big Bang would most likely have resulted in disorder and debris, rather than a universe which expands for billions of years and which allows life to originate and survive. Similar to the scenario of the 100 marksmen who missed the criminal, survival in such circumstances requires an explanation other than chance. Likewise, it is not the case that each possible behaviour of particles among the billions of possible behaviours would have resulted in a consistently mathematically describable order. On the contrary, as explained earlier, the proportion of such possible behaviours is extremely small.

The above observations illustrate the fact that we are not just talking about improbable events, but an event which is improbable and has a *specificity*, that is, a universe that is highly ordered and which has the capacity for allowing the production of functional objects, in particular embodied intelligent life. The idea of specificity can be illustrated by the analogy of an archer who shoots arrows at a wall. After the event,

> she could make herself appear to be a skilled archer by simply painting bull's-eyes around whatever places on the wall an arrow falls. But the pattern thus created would not be a specification; it would be a fabrication. If the bull's-eye already exists, on the other hand, and she sets out to hit it and succeeds, it represents a specification. (Dawes 2007, p. 71, citing Dembski)

The idea of painting a bull's-eye around wherever the arrow falls is analogous to whoever is the winner in the lottery case. In this case, any place on the wall has equal chance of being the bull's eye of an arrow shot randomly, just as any participant in the lottery has equal chance of being the winner. By contrast, it is not the case that any of the possible values of those physical constants allows for life; on the contrary, the vast majority of possible values do not allow for life, and the range of possible values that allow for life is extremely small; a small deviation from the existing values would result in a lifeless universe (thus, the values are highly specified in this sense). To fall within such a small range which (unlike the rest of the range of possible values) allows for life would be analogous to falling within a small region of the wall which (unlike the rest of the wall) has been marked out as the bull's eye before the arrow is shot.

Moreover, the features of 'being highly ordered and allowing for the production of functional objects such as embodied intelligent life' are 'special' because:

(1) Functionality is often associated with design (Ratzsch and Koperski 2019; although as noted at the end of this section I do not claim that this type of specified complexity by itself is a reliable criterion for detecting design). To illustrate, if one were to discover in the midst of a jungle a structure which has the capacity for allowing the production of

motorcars, one would reasonably conclude that it was designed. The reason is because it is unreasonable to think that the components of this structure were fundamentally brought together and assembled by Chance, Regularity, or Combinations of Regularity and Chance, or that the structure began to exist Uncaused, and (as shown above) the only remaining explanation is Design. It is true that there are also other arrangements of the components of the structure which are very unlikely. Nevertheless, the overwhelming proportion of the possible arrangements of the components (e.g. wiring not attached to assembly line, door panels not fitting the vehicle frame, etc.) would not allow for the production of anything functional. Therefore, the arrangement of the components which allow for the production of motorcar is 'special' and warrants an explanation. Likewise, as implied by the discussion in Sect. 4.1, the overwhelming proportion of possible universes would not allow for the production of functional objects such as living cells. Thus, the fact that our universe allows for the production of living cells warrants an explanation.

It might be objected that, unlike the structure (factory?) which allows for the production of motorcar, our universe does not seem to be organized towards producing life; indeed, most parts of our universe are inhospitable to life, and hence are not specified or functional in the same sense as the components of the structure. On the other hand, Carroll objects that our universe is too fine-tuned for life. He writes,

> If the reason why certain characteristics of the universe seem fine-tuned is because life needs to exist, we would expect them to be sufficiently tuned to allow for life, but there's no reason for them to be much more tuned than that. The entropy of the universe, for example [seems] much more tuned than is necessary for life to exist .... [F]rom purely anthropic considerations, there is no reason at all for God to have made it that small. (Carroll 2016, p. 311)

I shall discuss the objection concerning inhospitality towards life in greater detail in Sect. 7.3. At this point I would like to highlight the fact that, while it is true that our universe is not fully analogous with the factory-like structure, there is nevertheless a point of analogy, namely, just as the overwhelming proportion of the possible arrangements of the components would not allow for the production of anything functional, the overwhelming proportion of possible universes would not allow for the existence of functional objects such as living cells. The relevant sense of specificity is that in both cases the extremely narrow range of possibilities that allow for the existence of functionality is somehow actualized.

Contrary to Carroll, this relevant sense of specificity does not require the fine-tuning to be *solely* for the existence of life, rather than (say) for the existence of life *and* other features such as (for example) certain aesthetic features of our universe. Hence, Carroll's objection is based on a mistaken assumption. Barnes (2019) replies that

> low entropy initial conditions over the observable universe (as opposed to merely in our Solar System, for example) are necessary for our beautiful night sky, from what we see with our naked eye to our biggest telescopes. On a clear night, far away from city lights, try staring deeply into the Milky Way for a while and see if you're compelled to shout, 'not worth it!'

(2) Embodied intelligent living things can have plenty of meaningful physical interactions with one another and can be aware of God and can 'communicate and establish a deep relation of love with God, if God exists at all … Intelligent life can actualize moral values in the world' (Chan and Chan 2020, p. 8).[15] Thus, if a good God exists, 'God would have good reason to create intelligent lives (as well as a universe in which intelligent lives can emerge and flourish' (ibid.).

Sinhababu (2016) offers an objection to the fine-tuning argument for God's existence by suggesting the metaphysical possibility of alternative psychophysical laws that permit a wider range of physical entities to have minds, such that 'Whenever two electrons were a prime number of centimeters apart, they could have the mental states involved in heartfelt communication about their histories. Every subsequent time they were a whole number of meters apart, they could fondly remember each other' (p. 425). He argues that such psychophysical laws are possible if a non-physical God having a Mind is possible (pp. 426–427).

However, the point remains that, if the universe is not fine-tuned, the universe would be deprived of physical interactions with particles 'drifting through emptying space, not seeing another particle for trillions of years and even then, just glancing off and returning to the void' (Lewis and Barnes 2016, p. 164). While God can create alternative psychophysical laws or disembodied intelligent beings (e.g. angels), that still does not answer the question 'Why our physical universe is so special, that is, allowing for so many physical interactions and highly ordered?' In a similar vein, Hawthorne and Isaacs (2018, pp. 147-148) respond to the objection that there is no special expectation that God would make physical life rather than non-physical life by arguing that this objection does not actually make much of a difference to the fine-tuning argument, because the fact is that there is physical life which is more likely given theism than atheism.

Accepting the conclusion that specified events with extremely low probability happened as a result of chance is unreasonable. Are we seriously going to believe that the 100 marksmen missed by chance? Consider also the case of suspected plagiarism in which two essays submitted to a professor by two different students are word-for-word identical. It is very improbable that such 'specified' events happen by chance. While there are other arrangements of the words of the essays which are also very unlikely, the overwhelming proportion of the possible arrangements of the words would result in essays that are not identical, rather than two essays that are word-for-word identical. Hence, most professors would rightly insist on investigating for plagiarism.[16] Yet the improbability of a highly ordered and life-permitting universe is far greater than these examples! While we can imagine that specified events with extremely low probability (e.g. the case of suspected plagiarism) happened as a result of chance, we should regard such conclusions as belonging only to the imagination but not to reality.

It should be noted that, while my argument here makes use of 'specified complexity' to argue against the Chance hypothesis, I do not claim (as Dembski does) that specified complexity by itself is a reliable criterion for detecting design (Dembski 2002, p. 24). One of the main criticisms against Dembski's use of the idea of specified complexity is that critics object that counterexamples from evolutionary biology can be found. However, my book does not make this claim. Indeed, I think that specified complexity by itself is not a reliable criterion for detecting design because additional arguments need to be provided to rule out other alternatives to design (such as the evolutionary alternative; see below), and I provide such arguments in what follows. Thus, my book avoids the criticism against Dembski.

### 4.4.4 The Problem of Normalizing Probabilities

Against conceptual probability, it has been objected that, from a logical point of view, the full interval of the possible values of the fine-tuned parameter is from 0 to ∞, and since the range is infinite, there is no sense in which life-friendly universes are improbable; the probabilities are mathematically undefined (McGrew, McGrew, and Vestrup 2001).

Lewis and Barnes (2016, p. 286) reply that 'these kinds of "what to do with infinity" problems are often encountered in the physical sciences, especially in cosmology, and so these objections cannot succeed against fine-tuning without paralyzing probabilistic reasoning in all of physics'. Ratzsch and Koperski (2019) propose:

> One solution to this problem is to truncate the interval of possible values. Instead of allowing ***C*** to range from [0, ∞), one could form a finite interval [0, ***N***], where ***N*** is very large relative to the life-permitting range of ***C***. A probability distribution could then be defined over the truncated range … The argument for fine-tuning can thus be recast such that almost all values of ***C*** are outside of the life-permitting range. The fact that our universe is life-permitting is therefore in need of explanation.[17]

It should be noted that the fine-tuning argument concerns the concrete universe, not abstract logically possible worlds. Collins (2009, p. 249) argues that, where our concrete physical universe is concerned, the range of the possible values of the fine-tuned parameter is not infinite, noting that 'the so-called Plank scale is often assumed to be the cutoff for the applicability of the strong, weak, and electromagnetic forces' (see also the argument against concrete infinities in Loke (2012b; 2017a, chapter 2)). Therefore, 'the limits of our current theories are most likely finite but very large, since we know that our physics does work for an enormously wide range of energies. Accordingly, if the life-permitting range for a constant is very small in comparison, then … that there will be fine-tuning' (Collins 2009, p. 249.).

## 4.4.5 Multiple Universes

### 4.4.5.1 Introducing Various Types of Multiverse Hypothesis

Many scientists have suggested that perhaps there are many universes which have been formed, such that eventually one that is fine-tuned would be formed by chance. Collins (2009, p. 257) explains: 'Just as in a lottery in which all the tickets are sold, one is bound to be the winning number, so given a varied enough set of universes with regard to some life-permitting feature F, it is no longer surprising that there exists a universe somewhere that has F.' The multiverse hypothesis is often combined with the anthropic principle to suggest that, given a large variety of universes, 'it is neither surprising that there is at least one universe that is hospitable to life nor—since we could not have found ourselves in a life-hostile universe—that we find ourselves in a life-friendly one' (Friederich 2018). Some have used the concept of infinity to postulate a spatially infinite universe or an infinite number of universes, given which anything that is possible would happen. Somewhere in such an infinite universe/infinite number of universes, there would be regions exhibiting some degree of order, and since life cannot exist where there is no order, we will find ourselves in one of those regions with order.

There are different types of multiple universes theories: some postulate the simultaneous existence of many universes (spatial multiverse theories), others postulate one universe arising after another consecutively (temporal multiverse theories) (Gale 1990). Various philosophical postulations and scientific mechanisms have been proposed for various multiverse theories. For example, while most philosophers accept the use of the language of possible worlds as a way to talk about necessity and possibility (modal logic), philosopher David Lewis speculates that all possible worlds exist concretely (modal realism) (Lewis 1986). Hugh Everett's Many Worlds interpretation of quantum mechanics has also been used to postulate the existence of infinite branches of spacetime (parallel worlds) resulting from quantum splitting; this interpretation of quantum theory has been used by some cosmologists to explain the cosmic coincidences (Holder 2004, pp. 52–53). Many physicists have suggested that the process of inflation resulted in causally isolated spacetime regions ('island universes'), and that the process is 'eternal' in the sense that the formation of island universes never ends, resulting in the production of an infinite number of island universes (Vilenkin and Tegmark 2011, citing Guth 2000).

It should be noted that the postulation of a multiverse per se is not contrary to theism, for it is possible that God created a multiverse (call this the '*theistic* multiverse hypothesis'). Thus, proving the existence of more than one universes per se will not refute theism. However, the use of the postulation of multiverse by atheists to explain away God/Designer (i.e. claiming that the fine-tuning and order of our universe can be explained by the multiverse such that there is no need for a designer; call this the '*atheistic* multiverse hypothesis') is beset with several problems, which I shall explain below.

#### 4.4.5.2 Insufficient Evidence for the *Atheistic* Multiverse Hypothesis

On the one hand, there is insufficient reason or evidence for thinking that any of the atheist multiverse scenarios is true. Concerning Lewis' modal realist hypothesis, by speculating that all possible worlds exist concretely, Lewis is no longer talking about possible worlds as such; rather, he is speculating that the actual world is far more extensive than we thought. In other words, if we found out that his hypothesis is true, 'we would simply have learned that the actual world is richer than we thought—that it contains all of these island universes' (Pruss 2009, p. 36, attributing to Van Inwagen). However, there is no good evidence which shows that such concrete worlds really exist. As for Everett's interpretation, it is not proven as well; there are other possible alternative deterministic interpretations of quantum physics such as Bohm's pilot-wave model (see Chap. 2). On the other hand, Everett's interpretation (according to which every possibility is actual) is beset with the so-called measure problem (see below).[18]

While some evidence for inflationary cosmology (which is claimed to have brought about multiverses) has been proposed, this has been disputed by other cosmologists, and the problem with testing multiverse hypothesis remains (Friederich 2018). It should be noted that the so-called Eternal Inflation Model explained by Vilenkin and Tegmark (2011) does not mean eternal in the past without a beginning; rather, it is postulated to be eternal in the future in the sense that it has no end. In fact, Vilenkin (2015) himself argues for an ultimate beginning of the universe, thus accepting premise 2 of Craig's formulation of the Kalām Cosmological Argument, namely, 'The Universe began to exist.' Given that an actual infinite regress of events is impossible (see Chap. 5), it must still be finite in the past in the sense of having a first event.

Moreover, the claims that 'In an eternally inflating universe, anything that can happen will happen; in fact, it will happen an infinite number of times' and 'inevitably, an unlimited number of bubbles of all possible types will be formed in the course of eternal inflation' (Vilenkin and Tegmark 2011) are based on the assumption that the future is an already existing actual infinite rather than a potential infinite. However, the assumption that it is an actual infinite is unproven and falsified by Mawson's argument and by other arguments discussed in Chap. 5; thus, the future (if it is indeed infinite) should be regarded as a potential infinite.[19] Vilenkin and Tegmark (2011) state: 'that's how we test any scientific theory: we assume that it's true, work out the consequences, and discard the theory if the predictions fail to match the observations.' Mawson's argument explained below does just that: it shows how the prediction of 'anything that can happen will happen' fails to match the observations. Claiming that inflation can stretch continuous space indefinitely does not imply that an actual infinite is actually reached. As Ellis et al. (2004, p. 927) note, 'Future infinite time also is never realized; rather, the situation is that whatever time we reach, there is always more time available' (see Chap. 5). Indeed, more recently, Tegmark himself has advocated the rejection of the actual infinite because of the so-called measure problem (see Sect. 4.4.5.3 below).

Some purported evidence of multiple universes (e.g. claims of universes collisions leaving behind 'scars' on the CMB; this has been disputed by other scientists, as noted in Chap. 2), even if confirmed, only implies that there is more than one universe but does not imply that there is an infinite number or a large number of them. It should be noted that, in order for the multiverse hypothesis to explain the fine-tuning and order of our universe, a huge number of varied universes would be required, but there is no conclusive evidence that such a huge number of varied universes exist. The evidence for inflation does not by itself imply the evidence for an actual infinite number of universes, as illustrated by cosmologist George Ellis' (2007, Sect. 2.8) acceptance of the former but rejection of the latter (see below).

#### 4.4.5.3 Arguments against the *Atheistic* Multiverse Hypothesis

On the other hand, there are powerful scientific and philosophical objections against the *atheistic* multiverse hypothesis.

First, currently popular 'multiverse' scenarios which suggest the formation of baby universes that eventually become causally independent of the mother universe are contrary to the Generalized Second Law of Thermodynamics (Curiel 2019, citing Wall 2013a, 2013b).

Second, Ellis (2007, Sect. 9.3.2) observes that 'the concept of infinity is used with gay abandon in some multiverse discussions, without any concern either for the philosophical problems associated with this statement' (Ellis 2007, Sect. 8.1). Recall the discussion on multiverse mentioned earlier whereby some have postulated an actual infinite number (or a very large number) of universes to explain the fine-tuning of the universe. Following philosopher Tim Mawson, one can object that, on such a hypothesis in which every possibility (or very large number of possibilities) is actual, the probability of any universe in which we can more or less continually and consistently understand through induction is infinitely (or extremely) small. The reason is because at every moment there would be (roughly speaking) an infinite (or very large) number of ways in which things 'go wrong' with respect to our beliefs arrived at by induction and only one way in which things 'go right'.[20] Yet the mathematically describable order of our universe indicates that our universe is one in which we can more or less continually and consistently understand through induction.

One might reply by arguing that the probability of such a universe is indeed infinitely (or extremely) small, but because an ordered universe is necessary for the survival of life, we would still find ourselves in such a universe due to the anthropic principle. However, the survival of life would only require us to live in an ordered universe up to this present moment. There are an infinite number of ways the next moment might go wrong. But as I am typing this, the next moment has arrived and this has gone right in spite of its infinitesimal small probability if there were an infinite number of universes. Thus, it is far more likely that there isn't an infinite/large number of universes. As Holder (2004, p. 126) notes regarding the problem concerning the persistence of order in this universe,

> presumably in an infinite ensemble of possible universes, many will be identical to ours up to, say, the present moment or midnight on 31 October 2008, and then dissolve into chaos … imagine a monkey sitting at a typewriter for untold aeons. The animal is vastly more likely to produce 'To be or not to be' at some stage and then sink into chaos than to produce the whole of Hamlet. Similarly, random selection of universes from a vast ensemble is far more likely to produce a solar system embedded in chaos, or a finely-tuned epoch followed by chaos, than a universe with the order, and persistence of that order, which our universe actually possesses.

Indeed, more recently, cosmologist Max Tegmark (who had earlier advocated an actual infinite eternal universe scenario, as noted in Chap. 4) has advocated the rejection of the infinite because of the so-called measure problem, which he calls 'the greatest crisis facing modern physics'. The problem is that, if inflationary cosmology were to result in an actual infinite number of universes, then 'whatever experiment one makes … there will be infinitely many copies of you … obtaining each physically possible outcome … So, strictly speaking, we physicists can no longer predict anything at all!' (Tegmark 2015). However, we do live in a universe in which physicists can predict many events. Therefore, the antecedent is false.

Third, the *atheistic* multiverse scenario faces the Boltzmann Brain problem. Collins explains,

> This is the problem that, under naturalistic views of the mind, it is enormously more likely—on the order of $10^{10(123)}$ times more likely—for observers to exist in the smallest bubble of order required for observers, than in a universe that is ordered throughout. (The order being referred to here is measured by entropy—the lower the entropy, the higher the order.) Yet, we do not exist in a bubble of low entropy, but in a universe with low entropy throughout. (Collins 2018, pp. 90-91)

Craig (2012) notes that 'appeal to an observer self-selection effect accomplishes nothing because … most observable worlds will be Boltzmann Brain worlds'.

In other words,

1. If atheist multiverse scenario is true, it is overwhelmingly probable that we would observe that we are isolated brains surrounded by thermal equilibrium. (Prediction)
2. We do not observe that we are isolated brains surrounded by thermal equilibrium.
3. Therefore, it is overwhelmingly probable that the atheist multiverse scenario is false. (Adapted from Lewis and Barnes 2016, pp. 317–318)

Lewis and Barnes (2016, p. 322) note: 'The multiverse has a tightrope to walk. Too few varied universes, and it will probably fail to make a life-permitting one at all. Too many non-fine-tuned universes, on the other hand, could result in a universe filled with Boltzmann Brains.' For the multiverse to walk this tightrope, it would need to be fine-tuned (ibid.). In other words, those life-permitting multiverse scenarios which are supposedly able to avoid the Boltzmann Brain problem would themselves require fine-tuning, and therefore they are not (by themselves) the ultimate solution to the fine-tuning problem.

Fourth, even if there are many universes, the process which led to their formation (whether involving string theory or not; see Sect. 4.5) would itself require fine-tuning in order to stably generate so many different kinds of universes (and ensure that they do not face other problems such as colliding and destroying one another), such that eventually one that is 'fine-tuned' (and describable by highly sophisticated mathematical equations) is generated by chance. As Collins (2018, p. 90) notes, 'anything that produces such a multiverse itself appears to require significant fine-tuning.'

As an illustration, consider the 'famous fine-tuning problem of inflation'. Lewis and Barnes (2016, pp. 172-173, citing Neil Turok) explain that in order for any form of life to exist in our universe, the universe must have a very specific amount of lumpiness: a *Q* value between one part in 1,000,000 and one part in 10,000. However, 'inflation can produce practically any value of Q, from zero to very large values. If Q is greater than one, the universe comes pre-loaded with black holes; this really is not a good idea. The properties of the inflation must be fine-tuned to produce the right value of Q, so again we replace one fine-tuning with another.' As Holder (2004, p. 136) observes, 'the fine-tuning required by inflationary models is a serious drawback since inflation was meant to explain fine-tuning!'

Finally, even if there are many universes, there must still be a divine First Cause, as shown by the arguments presented in Chaps. 5 and 6.

## 4.5 Regularity

It has been suggested that there could be fundamental general principles in nature which determined the laws and constants of physics of our universe (Einstein 1949, p. 63).

For example, Bird (2007, p. 212) suggests: 'If the law of gravitation is not fundamental but is derived from deeper laws (as physicists indeed believe) then it could well turn out that the value of G is constrained in a way that we do not yet understand. In which case it might be, for all we know, that the value of G is necessary.'

There are two problems with this kind of suggestion.

First, it does not solve the fine-tuning problem because the fundamental principles or laws do not uniquely determine a fine-tuned universe. 'Physics is blind to what life needs. And yet, here we are' (Lewis and Barnes 2016, p. 181). For example, according to our present understanding of string theory (the most promising candidate 'theory of everything'), string theory does not predict the state of our universe but allows for a vast landscape of possible universes (Hawking 2003). Susskind notes:

> The two concepts—Landscape and megaverse [i.e. multiverse]—should not be confused. The Landscape is not a real place. Think of it as a list of all the possible designs of hypothetical universes. Each valley represents one such design .... The megaverse, by contrast, is quite real. The pocket universes that fill it are actual existing places, not hypothetical possibilities. (2005, p. 381)

Thus, the string theory does not uniquely determine the laws and constants (Friederich 2018), nor does it determine the initial conditions such as the initial low-entropy condition.

The landscape, which is a large set of possibilities, 'can't of itself solve the fine-tuning problem; in fact, it's part of the problem. As an illustration, the large number of possible lottery tickets is precisely what makes winning unlikely' (Lewis and Barnes 2016, p. 305).

Second, the 'Regularity' hypothesis only pushes the question one step back: how could such mindless non-intelligent, non-random causes have this orderly behaviour, and how could such mindless causes generate a universe with such a high degree of mathematically describable order? As Frederick observes:

> It is obviously useless to point out that some laws can be explained in terms of other laws, for example, that we may explain why matter accords with Einstein's quantitative law of gravitation (a modification of Newton's inverse-square law) by invoking the law that a body will pursue the easiest course through undulating space-time. That just puts the puzzle back a step. How can it be that everybody always pursues the easiest course? The explanation of some laws in terms of others leaves unanswered the question of how mindless matter, or forces, can behave in a way which accords with a law. (Frederick 2013, p. 271)

(The hypothesis that this question can be pushed back ad infinitum because there is an infinite regress of non-intelligent, non-random causes is considered under the 'Uncaused' hypothesis, which is refuted by the arguments presented in Chaps. 5–7.)

## 4.6 Combination of Regularity and Chance

Consider (iii) 'Combination of Regularity and Chance'. Plato (*the Laws*, Chap. 10) mentioned those who denied the gods' existence had argued that the order we perceive in the universe is merely the product of the interaction of chance and regularity. A modern-day proponent would be Stenger (2000), who argues that the laws of physics do not need fine-tuning because they are based on a combination of symmetry and the random breaking of it. However, Stenger fails to explain 'why would randomly broken symmetry give rise to precisely the right set of laws required for life instead of the vast range of other possibilities?' (Collins 2013, pp. 37–38). This indicates that a fine-tuning of the breaking would be required.

Cosmologist Lee Smolin (1997) has proposed a naturalistic evolutionary scenario for universes. He suggests that the singularities inside black holes are the sources of new baby universe phases that resemble their parents. As each black-hole singularity individually produces a different universe phase and, in each case, there would be a slight readjustment to the fundamental physical constants, there could be some form of 'natural selection' of universes, where the fundamental constants slowly evolve to obtain 'fitter' universes in which there are proliferation of black holes and thus produce many 'children'. With further generations, universes with black holes and stars (including those which help support life) would come to dominate the population of universes within the multiverse. Smolin argues that there is some indication that the fundamental physical constants of our universe are indeed such as to favour a proliferation of black holes.

Other physicists such as Roger Penrose have criticized Smolin's proposal for the speculative nature of the idea that the fundamental physical constants are readjusted as new baby universes are formed from black-hole singularities. Penrose also criticizes Smolin for the geometrical implausibility of the idea that highly irregular singularities can magically convert themselves into (or glue themselves to) the extraordinarily smooth and uniform Big Bang that each new universe would need if it is to acquire a respectable Second Law of the kind that we are familiar with

(Penrose 2004, pp. 761–762). Moreover, 'it's probably easier just to create black holes directly in a lumpy Big Bang or by fluctuations in an inflating universe rather than go to all the bother of creating stars' (Lewis and Barnes 2016, p. 355). Given this, the proliferation of universes with stars that support life would not be likely.

Additionally, based on the discussion in the foregoing sections of this chapter, it can be seen that, for an evolution of universes (or other kinds of 'Combination of Regularity and Chance') to happen, a high degree of order (such that particles do not move in billions of alternative direction at each moment, etc.) and fine-tuning (in order to avoid the Boltzmann Brain problem, etc.) must already be in place. The existence of such an order and fine-tuning remains unexplained by the 'Combination of Regularity and Chance' hypothesis. (As argued in Sect. 4.5, multiverse theories do not provide a reasonable explanation for this initial order and fine-tuning as well.)

One might object that Darwin's work shows that the existence of order is not necessarily proof of deliberate creation, and that what applies to biology may well apply at other levels.

In reply, on the one hand, Darwin's work only applies to a certain kind of order, namely, 'intermediate order'. This is the order which, once certain ordered regularities (e.g. natural selection) are in place, certain complex systems may develop via a process over time. Indeed, as Kojonen (2021) argues, given the possibility that a Designer could work through secondary causes such as setting up these regularities and the initial conditions and using these to bring about different living organisms, and given that Darwinian explanations are actually compatible with the biological design argument in this sense, Darwinian evolution has not refuted the biological design argument at all (I argue that evolution is compatible with Christian theism in Loke 2022). Kojonen also notes, 'In the case of complex phenomena, it is often the case that there is not just a single "best explanation," but rather different facets of the phenomena are explained by different explanations. Getting the full explanation may require combining, rather than just contrasting explanations' (ibid., p. 88). In other words, in the case of biology, there may well be evidence of both evolution and design (at the deeper level of what makes evolution possible) that warrants the combination of both explanations.[21]

On the other hand, the argument offered here concerns 'order at a more fundamental level'. That is, it concerns the regularities which are required to be in place in order for 'Combination of Regularity and Chance' to be possible. This kind of order cannot in principle be explained by evolutionary theory, since the theory presupposes the existence of this kind of order.[22] As explained in the discussion on the Regularity hypothesis above (see Sect. 4.5), the postulation of this order leaves unanswered the question of how mindless matter can behave in a way which accords with this order. (The objector might reply by hypothesizing that this order is uncaused; he/she might suggest that the combination of chance and regularity could cause design-like complexity, starting from simpler uncaused elements.[23] In reply, my arguments in Chaps. 6 and 7 against the Uncaused hypothesis would rule out such a hypothesis.)

## 4.7 Conclusion

I have formulated an original deductive argument which demonstrates that the following are the only possible categories of hypotheses concerning 'fine-tuning' and 'the existence of orderly patterns of events which can be described by advanced mathematics': (i) Chance, (ii) Regularity, (iii) Combinations of Regularity and Chance, (iv) Uncaused, and (v) Design. I have shown that there is an essential feature of (i) Chance, (ii) Regularity, and (iii) Combinations of Regularity and Chance which renders them unworkable as the ultimate explanation for the fine-tuning and order. The only remaining hypotheses are Uncaused and Design. One key issue is whether physical reality has a beginning, for if it does, then given the Causal Principle established in Chaps. 2 and 3 it is not uncaused. To address the key issue, I shall first discuss whether an actual infinite regress of events is possible and whether there is a First Cause in the next chapter.

## Notes

1. Aquinas also argues that natural bodies 'act for an end' and 'obtain the best result'. My argument does not require this aspect of his argument concerning final causes and value. Evans (2018, pp. 112-113) comments that Aquinas is directing our attention to two features of the natural world, orderliness and value. It is the feature of orderliness which I shall focus on in this book (I shall leave the discussion on value to another occasion): We observe natural bodies 'acting always, or nearly always, in the same way'.
2. Ellis (2007, section.3.3 n.41) notes: 'The effective laws may vary from place to place because for example the vacuum state varies; but the fundamental laws that underlie this behaviour are themselves taken to be invariant.'
3. Tegmark argues that his hypothesis is motivated by the assumption that there exists an external physical reality completely independent of us humans (External Reality hypothesis), which he thinks implies that a 'theory of everything' has no baggage; that is, it must be well-defined also according to non-human sentient entities (say, aliens or future supercomputers) that lack the common understanding of concepts that we humans have evolved, for example, 'particle', 'observation', or indeed any other English words. But his argument begs the question against the possibility that entities denoted by words such as 'particle' truly exist in a way that cannot be reduced to mathematics, and regardless of whether aliens are able to understand them or not.
4. In answer to the objection that relations are impossible without relata, they argue that the relata are other relations (pp. 154–155).
5. Einstein (1960, p. 262). While some have regarded Einstein as a pantheist on account of his statement 'I believe in Spinoza's God who reveals himself in the orderly harmony of what exists, not in a God who concerns himself with the fates and actions of human being' (*The New York Times*, April 25, 1929), Einstein himself clearly denied being a pantheist or atheist ('I am not an atheist, *and I don't think I can call myself a pantheist*', cited in Jammer 1999, p. 48, italics mine; Cf. Stanley (2009, pp. 192-193), who neglected the phrase in italics. Einstein's citation of Spinoza should perhaps be understood as follows: like Spinoza, he does not believe that there is a God *who is concerned with human affairs*. Likewise, his statement 'The word god is for me nothing more than the

expression and product of human weaknesses, the Bible a collection of honorable, but still primitive legends which are nevertheless pretty childish. No interpretation no matter how subtle can (for me) change this' in his letter to the philosopher Eric Gutkind on 3 January 1954 can be understood in its context as an opposition to the God of religion rather than God as a Designer of the cosmos. Given what is said above, one should perhaps say that Einstein was a Deist.

6. Dembski does consider the combination of chance and regularity in his other book *No Free Lunch* (Dembski 2002), but his so-called explanatory filter in *The Design Inference* does not do so; I cite the latter merely to warn that we should be more rigorous in our assessment of alternative explanations with regard to the Teleological Argument.
7. Against Swinburne, who makes extensive use of an appeal to simplicity in support of theistic arguments, McGrath notes that 'problems with the use of the criterion of simplicity remain. It is difficult to define and operationalize the notion, and to provide it with an independent epistemic foundation. Furthermore, it is by no means clear that simplicity is a sign of truth, or even an indicator of the potential long-term success of a theory. Perhaps unsurprisingly, many philosophers of science now tend to see simplicity therefore as a desirable quality for theories, while recognizing that many theories deemed to be valid or successful are not simple' (McGrath 2018, p. 117).
8. https://www.youtube.com/watch?v=9wLtCqm72-Y
9. 1.1, 1.2, and 1.3 involve causes which may or may not be uncaused. I shall argue that there is an uncaused Designer in Chap. 7.
10. See the paragraph below the syllogism for how 'not' is to be understood.
11. Friederich (2018) goes on to note: 'If the laws and constants that physics has so far determined turned out to be merely effective laws and constants fixed by some random process in the early universe which might be governed by more fundamental physical laws, it would start to make sense to apply the concept of physical probability to those effective laws and constants…However, the fine-tuning considerations…do not seem to be based on speculations about any such process, so they do not seem to implicitly rely on the notion of physical probability in that sense.' To assume that random process is the case would be begging the question against the other hypotheses.
12. Not an actual infinite; see Section 4.4.4.
13. Not an actual infinite; see Section 4.4.4.

14. Collins (2009, p. 247) warns against thinking in terms of possible universes which are actual infinite in number and susceptible to the criticisms in McGrew, McGrew, and Vestrup (2001) (I argue in Chap. 5 that actual infinities can only exist in the abstract but not in the concrete and possible universes are abstract). Rather, one should think in terms of possible values within a concrete range; see Section 4.4.4.
15. Goff (2019, p. 114) overemphasized this point by stating that 'unless life/intelligent life is objectively of great value, the fine-tuning needs no explanation'. I don't think 'value' is a necessary condition. As argued above, functional complexity by itself would require an explanation. To use the analogy mentioned above, even if the structure is not valuable and what it produces is not valuable, the discovery of such a structure would still require an explanation.
16. https://www.youtube.com/watch?v=yto4jXOOen8
17. Ratzsch and Koperski (2019) note: 'A more rigorous solution employs measure theory. Measure is sometimes used in physics as a surrogate for probability. For example, there are many more irrational numbers than rational ones. In measure theoretic terms, ***almost all*** real numbers are irrational, where 'almost all' means all but a set of zero measure. In physics, a property found for almost all of the solutions to an equation requires no explanation; it's what one should expect. It's not unusual, for instance, for a pin balancing on its tip to fall over. Falling over is to be expected. In contrast, if a property that has zero measure in the relevant space were actually observed to be the case, like the pin continuing to balance on its tip, that would demand a special explanation. Assuming one's model for the system is correct, nature appears to be strongly biased against such behavior.'
18. While some have claimed that our improbable existence itself is evidence for multiverse, others have pointed out that the reasoning for this claim commits the inverse gambler's fallacy (Goff 2021).
19. The future cannot be potential infinite if static theory of time is true.
20. Modified from Mawson (2011). Mawson does not speak of very large numbers, but the 'maximal multiverse hypothesis', which postulates every possible universe being actual.
21. It is beyond the scope of this book to discuss the biological design argument in greater detail; see Kojonen (2021) for a well-balanced discussion. For the purposes of the argument in this paragraph, it suffices to note that evolution has not eliminated design.

22. For the arguments for this in the biological realm, see Glass (2012); Kojonen (2021).
23. I thank an anonymous referee for suggesting this objection.

## Bibliography


Adams, Fred. 2019. The Degree of Fine-Tuning in Our Universe – And Others. *Physics Reports* 807: 1–111. arXiv:1902.03928v1 [astro-ph.CO] (Page Number Follows the arXiv Version).

Albrecht, A. 2004. Cosmic Inflation and the Arrow of Time. In *Science and Ultimate Reality*, ed. J. Barrow, P. Davies, and C. Harper. New York: Cambridge University Press.

Barnes, Luke. 2012. The Fine-Tuning of the Universe for Intelligent Life. *Publications of the Astronomical Society of Australia* 29: 529–564. https://doi.org/10.1071/AS12015.

———. 2019. A Reasonable Little Question: A Formulation of the Fine-Tuning Argument. *Ergo* 6 (42).

Barrow, J., and F. Tipler. 1986. *The Anthropic Cosmological Principle*. Oxford: Oxford University Press.

Bird, Alexander. 2005. Abductive Knowledge and Holmesian Inference. In *Oxford Studies in Epistemology*, ed. Tamar Gendler and John Hawthorne. Oxford: Oxford University Press.

———. 2007. *Nature's Metaphysics*. Oxford: Oxford University Press.

Bishop, R. 2017. Chaos. In *The Stanford Encyclopedia of Philosophy*, ed. E. Zalta. https://plato.stanford.edu/archives/spr2017/entries/chaos/.

Carrier, Richard. 2003. Fundamental Flaws in Mark Steiner's Challenge to Naturalism in The Applicability of Mathematics as a Philosophical Problem. The Secular Web. https://infidels.org/library/modern/richard_carrier/steiner.html.

Carroll, Sean. 2016. *The Big Picture*. London: Oneworld.

Chan, K., and M. Chan. 2020. A Discussion of Klass Landsman's Criticisms of the Fine-Tuning Argument. *Theology and Science*.https://doi.org/10.1080/14746700.2020.1755544.

Collins, Robin. 2009. The Teleological Argument: An Exploration of the Fine-tuning of the Universe. In *Blackwell Companion to Natural Theology*. Oxford: Wiley-Blackwell.

———. 2013. The Fine-Tuning Evidence is Convincing. In *Debating Christian Theism*, ed. J. P. Moreland, Chad V. Meister, Khaldoun Sweis. Oxford: Oxford University Press.

———. 2018. The Argument from Physical Constants. In *Two Dozen (Or So) Arguments for God*, ed. Jerry Walls and Trent Dougherty. Oxford: Oxford University Press.

Craig, William Lane. 2012. #285 Invasion of the Boltzmann Brains. *Reasonable Faith with William Lane Craig*. https://www.reasonablefaith.org/question-answer/fine-tuned-universe/invasion-of-the-boltzmann-brains.

Curiel, Erik. 2019. Singularities and Black Holes. In *The Stanford Encyclopedia of Philosophy*, ed. Edward Zalta. https://plato.stanford.edu/archives/spr2019/entries/spacetime-singularities/.

Dawes, Gregory. 2007. What Is Wrong with Intelligent Design. *International Journal for Philosophy of Religion* 61: 69–81.

Dawkins, Richard. 2006. *The God Delusion*. London: Bantam Press.

Dembski, William. 2002. *No Free Lunch: Why Specified Complexity Cannot Be Purchased Without Intelligence*. Lanham: Rowman & Littlefield.

———. 2006. *The Design Inference: Eliminating Chance Through Small Probabilities*. Cambridge: Cambridge University Press.

Dirac, Paul. 1963. The Evolution of the Physicist's Picture of Nature. *Scientific American* 208: 45–53.

Dumsday, Travis. 2019. *Dispositionalism and the Metaphysics of Science*. Cambridge: Cambridge University Press.

Eagle, Antony. 2019. Chance Versus Randomness. *The Stanford Encyclopedia of Philosophy*, ed. Edward Zalta. https://plato.stanford.edu/archives/spr2019/entries/chance-randomness/.

Einstein, Albert. 1949. Autobiographical Notes. In *Albert Einstein: Philosopher-Scientist*, ed. P. Schilpp. Peru, IL: Open Court Press, p. 63.

———. 1960. *Ideas and Opinions*, Trans. Sonja Bargmann. New York: Crown publishers.

Ellis, George. 2007. Issues in the Philosophy of Cosmology. In *Philosophy of Physics*, ed. J. Butterfield and J. Earman. Amsterdam: Elsevier.

———. 2018. Necessity, Purpose and Chance. In *God's Providence and Randomness in Nature: Scientific and Theological Perspectives*, ed. Robert Russell and Joshua Moritz. West Conshohocken: Templeton Press.

Ellis, George, U. Kirchner, and W. Stoeger. 2004. Multiverses and Physical Cosmology. *Monthly Notices of the Royal Astronomical Society* 347: 921–936.

Evans, C. Stephen. 2018. The Naïve Teleological Argument. In *Two Dozen (Or So) Arguments for God*, ed. Jerry Walls and Trent Dougherty. Oxford: Oxford University Press.

Frederick, Danny. 2013. A Puzzle About Natural Laws and the Existence of God. *International Journal for the Philosophy of Religion* 73: 269–283.

Friederich, Simon. 2018. Fine-Tuning. In *The Stanford Encyclopedia of Philosophy*, ed. Edward Zalta. https://plato.stanford.edu/archives/spr2018/entries/fine-tuning/.

Gale, G. 1990. Cosmological Fecundity: Theories of Multiple Universes. In *Physical Cosmology and Philosophy*, ed. J. Leslie. New York: Macmillan.

Glass, David. 2012. Can Evidence for Design Be Explained Away? In *Probability in the Philosophy of Religion*, ed. V. Harrison and J. Chandler. Oxford: Oxford University Press.

Goff, Philip. 2019. Did The Universe Design Itself? *International Journal for Philosophy of Religion* 85: 99–122.

———. 2021. Our Improbable Existence Is No Evidence for a Multiverse. https://www.scientificamerican.com/article/our-improbable-existence-is-no-evidence-for-a-multiverse/.

Goldman, R. 1997. *Einstein's God*. Northvale, NJ: Joyce Aronson Inc.

Guth, Alan. 2000. Inflation and Eternal Inflation. *Physics Reports* 333: 555–574.

Hájek, Alan. 2019. Interpretations of Probability. In *The Stanford Encyclopedia of Philosophy*, ed. Edward Zalta. https://plato.stanford.edu/archives/fall2019/entries/probability-interpret/.

Hawking, Stephen. 1988. *A Brief History of Time*. London: Bantam.

———. 2003. *Cosmology from the Top Down*. http://arxiv.org/abs/astro-ph/0305562.

Hawthorne, J., and Y. Isaacs. 2018. Fine-tuning Fine-Tuning. In *Knowledge, Belief, and God*, ed. M. Benton, J. Hawthorne, and D. Rabinowitz. Oxford: Oxford University Press.

Heller, M. 2013. Deep Questions on the Nature of Mathematics. *Notices of the American Mathematical Society* 60: 592–594.

Holder, Rodney. 2004. *God, the Multiverse, and Everything*. Aldershot: Ashgate.

Hossenfelder, Sabine. 2019. Screams for Explanation: Finetuning and Naturalness in the Foundations of Physics. *Synthese*. https://doi.org/10.1007/s11229-019-02377-5.

Jammer, Max. 1999. *Einstein and Religion*. Princeton: Princeton University Press.

Kojonen, Rope. 2021. *The Compatibility of Evolution and Design*. Palgrave Frontiers in Philosophy of Religion Series. Cham: Springer Nature.

Ladyman, James, Don Ross, David Spurrett, and John Collier. 2007. *Every Thing Must Go: Metaphysics Naturalized.* Oxford: Oxford University Press.

Leslie, John. 1982. Anthropic Principle, World Ensemble, Design. *American Philosophical Quarterly* 19 (2): 141–151.

Lewis, David. 1986. *On the Plurality of Worlds*. New York: Basil Blackwell.

Lewis, Geraint, and Luke Barnes. 2016. *A Fortunate Universe: Life in a Finely Tuned Cosmos*. Cambridge: Cambridge University Press.

Livio, Mario. 2009. *Is God a Mathematician?* New York: Simon & Schuster.

Loke, Andrew. 2022. *The Origin of Humanity and Evolution: Science and Scripture in Conversation*. London: T & T Clark.

Mawson, T. 2011. Explaining the Fine Tuning of the Universe to Us and the Fine Tuning of Us to the Universe. *Royal Institute of Philosophy Supplement* 68 (2011): 25–50.

McGrath, Alister. 2018. *The Territories of Human Reason: Science and Theology in an Age of Multiple Rationalities*. Oxford: Oxford University Press.

McGrew, Timothy, Lydia McGrew and Eric Vestrup. 2001. Probabilities and the fine-tuning argument: A sceptical view. *Mind* 110 (440): 1027–1038.

Monton, Bradley. 2010. Design Inferences in an Infinite Universe. In *Oxford Studies in Philosophy of Religion, Volume II*. Oxford: Oxford University Press.

Penrose, Roger. 1989. *The Emperor's New Mind: Concerning Computers, Minds and the Laws of Physics*. Oxford: Oxford University Press.

———. 2004. *The Road to Reality: A Complete Guide to the Laws of the Universe*. London: Random House.

Polkinghorne, John. 1998. *Believing in God in the Age of Science*. New Haven: Yale University Press.

———. 2011. *Science and Religion in Quest of Truth*. New Haven: Yale University Press.

Pruss, Alexander. 2009. The Leibnizian Cosmological Argument. In *The Blackwell Companion to Natural Theology*, ed. William Lane Craig and J.P. Moreland. Chichester: Wiley-Blackwell.

Rasmussen, Joshua, and Felipe Leon. 2018. *Is God the Best Explanation of Things: A Dialogue*. Cham: Springer Nature.

Ratzsch, Del, and Jeffrey Koperski. 2019. Teleological Arguments for God's Existence. In *The Stanford Encyclopedia of Philosophy*, ed. Edward N. Zalta. https://plato.stanford.edu/archives/sum2019/entries/teleological-arguments/.

Sinhababu, N. 2016. Divine Fine-Tuning vs Electrons in Love. *American Philosophical Quarterly* 53: 423–432.

Smolin, Lee. 1997. *The Life of the Cosmos*. New York: Oxford University Press.

Sober, Elliott. 2019. *The Design Argument*. Cambridge: Cambridge University Press.

Stanley, Matthew. 2009. Myth 21: That Einstein Believed in a Personal God. In *Galileo Goes to Jail: and Other Myths About Science and Religion*, ed. Ronald Numbers. Cambridge, MA: Harvard University Press.

Steiner, Mark. 1998. *The Applicability of Mathematics as a Philosophical Problem*. Cambridge, MA: Harvard University Press.

Stenger, Victor. 2000. Natural Explanations for the Anthropic Coincidences. *Philo* 3 (2): 50–67.

———. 2013. The Universe Shows No Evidence of Design. In *Debating Christian Theism*, ed. J.P. Moreland, Chad V. Meister, and Khaldoun Sweis. Oxford: Oxford University Press.

Tegmark, Max. 2008. The Mathematical Universe. *Foundations of Physics* 38: 101–150.

———. 2015. Infinity Is a Beautiful Concept – And It's Ruining Physics. https://www.discovermagazine.com/the-sciences/infinity-is-a-beautiful-concept-and-its-ruining-physics.

Trigg, Roger. 1993. *Rationality and Science: Can Science Explain Everything?* Oxford: Blackwell.

Vilenkin, Alexander. 2015. The Beginning of the Universe. *Inference: International Review of Science* 1 (4) https://inference-review.com/article/the-beginning-of-the-universe. Accessed 22 April 2015.

Vilenkin, Alexander, and Max Tegmark. 2011. The Case for Parallel Universes. *Scientific American*. https://www.scientificamerican.com/article/multiverse-the-case-for-parallel-universe/.

Wall, Aron. 2013a. The Generalized Second Law Implies a Quantum Singularity Theorem. *Classical and Quantum Gravity* 30 (16): 165003. Preprint: arXiv:1010.5513v4 [gr-qc].

———. 2013b. Corrigendum: The Generalized Second Law Implies a Quantum Singularity Theorem. *Classical and Quantum Gravity* 30 (19): 199501.

Wenmackers, Sylvia. 2016. Children of the Cosmos. In *Trick or Truth?: The Mysterious Connection Between Physics and Mathematics*, ed. Anthony Aguirre, Brendan Foster, and Zeeya Merali. Springer Nature: Cham.

Wigner, Eugene. 1960. The Unreasonable Effectiveness of Mathematics in the Natural Sciences. *Communications on Pure and Applied Mathematics* 13 (1): 1–14.

# 5

# Arguments for a First Cause

## 5.1 Introduction

An important issue in philosophy of religion debates concerns whether there is a First Cause, and if so, what is the First Cause. I shall address the first question in this chapter and the second question in the next. I shall begin by considering some of the scientific cosmological models relevant to addressing the first question, before presenting philosophical arguments against an infinite regress of causes.

## 5.2 Scientific Issues

According to the so-called Standard Version of the Big Bang (also known as the Friedmann–Lemaître–Robertson–Walker [FLRW] model), matter-energy began to exist at the initial cosmological singularity of the Big Bang. As cosmologists Barrow and Tipler explain, at the initial cosmological singularity, 'space and time came into existence; literally nothing existed before the singularity, so, if the universe originated at such a



A. Loke, *The Teleological and Kalam Cosmological Arguments Revisited*, Palgrave Frontiers in Philosophy of Religion, https://doi.org/10.1007/978-3-030-94403-2_5

singularity, we would truly have a creation ex nihilo' (Barrow and Tipler 1986, p. 442).

Over the years a number of scientists have proposed alternative models of the Big Bang, and these cosmological models may be classified under the following types:

Type (1): Originates from a finite past ex nihilo: for example, Vilenkin's (1982) 'Creation from Nothing' model
Type (2): Originates from Closed Timelike Curves (CTCs) where the universe 'creates itself', for example, Gott and Li (1998)
Type (3): Originates from a timeless initial state, for example, the Hartle–Hawking no-boundary proposal (1983)
Type (4): Originates from an actual infinite regress
Type (5): Involves a reversal of time

Type (1) has been discussed in Chaps. 2 and 3, while Type (3) will be discussed in Chap. 6. I shall discuss Types (2), (4), and (5) in this chapter, beginning with Type (4).

The following are examples of Type (4) cosmologies:

- Eternal Inflation model (Linde 1994; Aguirre 2007)
- Baum–Frampton (2007) phantom bounce cosmologies
- Veneziano and Gasperini's (2003) 'pre-Big Bang theory' based on analogues of the dualities of string theory
- 'Ekyroptic universe' initiated by a collision between pre-existing 'branes' in a higher dimensional spacetime (Steinhardt and Turok 2005)
- Conformal cyclic cosmology (CCC) model (Penrose 2010; see below)
- Static Quantum Multiverse model (Nomura 2012)

There are other proposals, such as the Loop Quantum Gravity model (Bojowald et al. 2004) and Poplawski's (2010) Black Hole model (which proposes that our universe might have originated from a black hole that lies within another universe), which are not committed to whether there is an actual infinite temporal regress (e.g. whether that universe was born from a black hole in another universe, which was born from a black hole in another universe, and so on) or a finite past.

To begin with our evaluation of Type (4) proposals, it should first be noted that none of these proposals are proven given that we do not currently have a well-established theory of quantum gravity, without which these proposals are, in the words of Ellis (2007, section 2.7), 'strongly speculative, none being based solidly in well-founded and tested physics'. Ellis (2007, section 9.3.2) also argues that it is not possible for science to prove that the universe is past infinite; 'observations cannot do so, and the physics required to guarantee this would happen … is untestable.' It has been explained in Chap. 4 that, while (1) there are evidences for inflationary Big Bang cosmology and that multiverse is possible, this does not imply (2) an infinite inflation and infinite multiverse in the past, as illustrated by Ellis' acceptance of (1) but rejection of (2).

The conformal cyclic cosmology (CCC) model (Penrose) proposes that the universe cycles from one aeon to the next, with each 'aeon' involves a big bang followed by an infinite future expansion that eventually results in the big bang of the next aeon. Penrose claims that anomalous regions have been found in the CMB temperature maps which results from the Hawking radiation from supermassive black holes in a cosmic aeon prior to our own (An et al. 2020). Other scientists are unconvinced by this purported evidence, objecting that there is no statistically significant evidence for the presence of such Hawking points in the CMB (Jow and Scott 2020). Moreover, even if there are such points, there could be alternative explanations other than cycles of aeons. In any case, such points do not prove that there are aeons which are infinite in the past or that there are an infinite cycles of prior aeons.

On the other hand, cosmological models which attempt to avoid a beginning face various difficulties related to the Second Law of Thermodynamics, the Borde–Guth–Vilenkin (BGV) theorem, acausal fine-tuning, and/or having an unstable or a metastable state with a finite lifetime (Craig and Sinclair 2009, pp. 179–182; Bussey 2013; Wall 2013a, 2013b), as well as philosophical arguments against possibility of an infinite regress (Ellis 2007, see section 5.3; it is a pity that a number of cosmologists have continued to ignore these objections; see, for example, Susskind 2012). For example, Vilenkin (2015) argues that the BGV theorem contradicts eternal past inflation. One might try to escape the BGV theorem by (1) postulating an earlier quantum era in which the theorem does not apply. However, Wall's work on the Generalized Second Law of Thermodynamics (GSLT) indicates that the GSLT would still apply on the quantum era and implies a beginning. One might also try to escape the BGV theorem by (2) postulating a reversal to the arrow of time (Aguirre 2007). To elaborate, some forms of bounce cosmologies which postulate that the universe was born from an entropy-reducing phase in a previous universe and the entropy reverses at the boundary condition (Linford 2020) have been proposed to avoid some of these problems. While it has been objected that such models which attempt to avoid a beginning by postulating a reversal of the arrow of time nevertheless have a type of 'thermodynamic beginning' which still requires an explanation (Wall 2014), Carroll replies:

> A thermodynamic beginning is not a beginning—it happens in the middle. It's a moment in the history of the universe from which entropy is higher in one direction of time and the other direction of time. There is no room in such a conception for God to have brought the universe into existence at any one moment. (Craig and Carroll 2015)

Nevertheless, there is a deeper problem which Carroll neglected (this neglect may be related to Carroll's dismissal of the Causal Principle, which I have responded to in Chap. 2), and that is the problem of causal dependence (Fig. 5.1):

B (Boundary condition at which there is entropy reversal)

Increasing entropy <-------|-------> increasing entropy

$\ldots x_2 x_1 y_1 y_2 \ldots$

U1 U2

**Fig. 5.1** Bounce Cosmology with entropy reversal

According to these types of bounce cosmologies, event $x_1$ is the first event of universe 1 (U1), while event $y_1$ is the first event of universe 2 (U2). Events $x_1$, $x_2$, ... and events $y_1$, $y_2$, ... have a beginning (i.e. these events are finite temporally and have 'edges'; see Chaps. 2 and 3). Given the Causal Principle that whatever begins to exist has a cause (see Chap. 3), if the beginning of $x_2$ is causally dependent on the beginning of $x_1$, what brought about $x_1$ at the boundary condition? Likewise, given that the beginning of $y_2$ is causally dependent on the beginning of $y_1$, what brought about $y_1$ at the boundary condition?

Linford suggests that the universes to either side of the interface might be interpreted as the simultaneous causes of each other. He writes:

> Supposing (as Craig argues) that there are no instants ... Instead, there exists an interval of time whose boundary is the interface between the two universes. Consequently ... for every existent temporal interval, U1 and U2 co-exist. There is no need to introduce an independent cause for the interface, and we can interpret U1 and U2 as the simultaneous causes of each other. (p. 24)

However, the view that U1 and U2—or more precisely, $x_1$ (the first event of U1) and $y_1$ (the first event of U2)—are simultaneous causes of each other violate the irreflexivity of causation and amounts to a vicious circularity: on this view the beginning of existence of U1 is dependent on the beginning of existence of U2, which depends on the beginning of existence U1 in order to begin to exist (see the critique of causal loop and Gott and Li's (1998) proposal below).

One might claim that causal dependency stops at the boundary condition (or moment of minimal entropy), and it is this condition that all later events depend upon in either time direction.[1] This implies that the boundary condition is the First Cause that brought about $x_1$ and $y_1$. However, the boundary condition is something that is temporally finite and has 'edges', and as argued in Chap. 2, such a thing would require a cause, and thus cannot be the first cause. Moreover, as I shall explain in Chap. 6, the independently existing First Cause must be something with (1) the capacity to initiate the first event, and also (2) the capacity to prevent itself from initiating the first event. These two capacities describe libertarian freedom. Therefore, the First Cause has libertarian freedom and hence is a Creator; it cannot be an impersonal boundary condition which Carrol describes. It would be of no use for Carroll to argue that such bounce cosmologies are mathematically possible, for it has already be explained in Chap. 1 that, even if a cosmological model is mathematically possible, it cannot be a correct model of the cosmos if it is metaphysically impossible.

Against Craig and Sinclair (2009, 2012), who have argued the low-entropy interface between universes should be understood as the beginning of two universes and this beginning requires an efficient cause beyond either of the two universes, Linford (2020) attempts to present a dilemma for them.

Linford argues that, on the one hand, if the direction of time is reducible, then efficient causation would most likely reducible as well given the Mentaculus (see Chap. 2), and thus even if the interface should be interpreted as an absolute beginning for two universes, these universes would probably not require a cause (p. 3). I have already argued in Chap. 2 why this horn of the dilemma is false.

In any case, let us consider the second horn of the dilemma. Linford argues that, on the other hand, if the direction of time is not reducible (as Craig in fact argued),[2] then 'we are left without reason to think that the direction of time aligns with the entropy gradient … then the direction of time need not point away from the interface in two directions. So, Craig and Sinclair's interpretation of the interface as an absolute beginning is unjustified' (pp. 16–17), and 'there's no longer reason to suppose that the direction of efficient causation would align with the entropy gradient' (pp. 16–17). In that case, events in a cosmological epoch of higher entropy could be the causes of events in an epoch of lower entropy. Linford writes:

> the fact that geodesics can be extended through the interface provides some reason to think that the interface should be interpreted as a transition from one universe to another and not as an absolute beginning for two universes. Again, we can imagine God—or the fictional agent occupying the view-from-nowhere—watching as metaphysical time passes, the world unfurling through ages of entropy decrease, until the entropy begins to increase once more. (pp. 26–7; cf. Halper 2021, p. 161, who argues that on Loop Quantum Cosmology [LQC], one has 'two classical space-times joined by a quantum bridge giving an hour-glass structure but with no reversal of the arrow of time'.)

However, if the direction of time does not point away from the interface in different directions but in the same direction, this would imply that the universe was born from an entropy-reducing phase in a previous universe and would violate the Generalized Second Law of Thermodynamics (see Wall 2013a, 2013b).

Linford (2020, pp. 18–19) claims that 'we already know that the second law of thermodynamics is a statistical regularity that admits of exceptions' without engaging with Wall's work on the general second law. In the entry on 'Singularities and Black Holes' in *The Stanford Encyclopedia of Philosophy*, Curiel (2019) notes that the recent important work by Wall (2013a, 2013b) indicates that the Generalized Second Law 'seems to admit of proof in ways much more mathematically rigorous than does the ordinary Second Law … Indeed, the Generalized Second Law is the only known physical law that unites the fields of general relativity, quantum mechanics, and thermodynamics. As such, it seems currently to be the most promising window we have into the most fundamental structures of the physical world.'

Halper (2021, p. 161) objects by arguing that Wall's results may not hold in full quantum gravity, concluding that 'this really underlies the view from cosmology today. Without a well-verified theory of quantum gravity, we cannot meaningfully describe the origin of our expanding universe and so we are in no position to say that cosmology implies a beginning to the universe.' Halper fails to note that the same problem (i.e. the lack of a well-verified theory of quantum gravity) also besets the cosmological models (which he cited) which affirms a beginningless physical reality. Wall's work is a good counterargument to those cosmologists who (despite the lack of a well-verified theory of quantum gravity) have argued that a beginningless model of physical reality is physically possible. Moreover, Halper also fails to note Wall's (2013a) statement that, since the (fine-grained) Generalized Second Law of Horizon Thermodynamics is 'widely believed to hold as a consequence of the statistical mechanical properties of quantum gravitational degrees of freedom, *it is a good candidate for a physical law likely to hold even in a full theory of quantum gravity*' (p. 2; italics mine). Wall also shows that the Generalized Second Law implies a 'quantum singularity theorem', which indicates that, even when quantum effects are taken into account, spacetime will still be geodesically incomplete inside black holes and to the past in cosmological models (like the currently most well supported ones, which start with a Big Bang singularity) (Wall 2013a, 2013b).

Moreover, if (as Linford suggests) the direction of time point in the same direction, then given that an infinite regress of events is impossible (as indicated by the philosophical arguments below), which implies that an infinite regress of transitions is impossible, there must still be a first event which requires a First Cause given the Causal Principle defended in Chap. 2. It should be noted that, while on such models time goes towards infinity in both entropy directions, this does not imply that an actual infinity can be obtained in both directions; rather, this can be understood as potential infinite[3] which is always finite at any time.

It might be thought that the law of conservation of energy which states that the energy of a closed system must remain constant implies that 'energy cannot be created' and that the universe has no beginning and end. However, what 'energy cannot be created' means is just that the total energy of an isolated system is constant. The fact that this law does not imply that the universe has no beginning can be seen by considering the Zero Energy universe Theory. This theory postulates that, 'in the case of a universe that is approximately uniform in space, one can show that the negative gravitational energy exactly cancels the positive energy represented by the matter. So the total energy of the universe is zero' (Hawking 1988, p. 129).

When this is applied to the beginning of the universe, it indicates the possibility that the Big Bang has no material cause. Hawking states that 'When the Big Bang produced a massive amount of positive energy, it simultaneously produced the same amount of negative energy. In this way, the positive and the negative add up to zero, always. It's another law of nature' (Hawking 2018, p. 32). Cosmologist Sean Carroll suggests that this theory can be used to explain why the Big Bang is consistent with the first law of thermodynamics, which states that the total energy of an isolated system is constant. For if the positive energy and negative energy of our universe balanced up to zero, then there is no violation of the first law of thermodynamics if our universe began to exist from zero energy.[4] As explained in Chap. 2, while this response implies that our universe has no material cause, it does not imply that our universe has no efficient cause; indeed, it does not imply that there is no requirement for a Creator to make the total energy of zero to be the way they are such that galaxies and gravity are formed while the total energy remain constant at zero. Given the deductive (Modus Tollens) argument for the Causal Principle 'whatever begins to exist has a cause', where 'cause' is either an efficient cause or a material cause (see Chap. 3), one should still ask what is the efficient cause which made the positive and negative energy to be the way they are (see Chap. 6). One can argue that God created the universe without using pre-existing material (i.e. ex nihilo), and this would not be a violation of the first law of thermodynamics because the total amount of energy of our universe after creation is still zero.[5] In any case, the determination of the equation of the law of conservation of energy assumes a closed physical system; hence, it does not rule out a supernatural cause creating it supernaturally, unless we beg the question by assuming that a closed physical system is all there is.[6]

Finally, the limitations of science with regard to the realism–anti-realism debate in physics and cosmology should be noted. Moreland and Craig (2003, p. 334) observes that, in the history of science, many theories (e.g. aether theory) have explained phenomena, generated fruitful research and accurate predictions, yet were later abandoned as false. Hawking recognizes this limitation when, in speaking of the reach of science, he affirms anti-realism by stating that science consists only of models, but when he speaks of scientific results, he argues with inappropriate confidence for a self-contained universe, thus contradicting himself (Giberson and Artigas 2007, p. 118).

I do not wish to overstate the limitations of science by claiming that all models should be interpreted in anti-realist terms. Rather, we should decide on a case-by-case basis and do so if there are good philosophical reasons to think so. The importance of philosophical considerations have already been demonstrated in Chap. 1, in particular, it has been demonstrated that metaphysical considerations are more fundamental than mathematical considerations. Thus, 'if we have good philosophical reasons for believing that the spacetime universe had a beginning a finite time ago, then if a "successful" scientific model runs counter to this belief, it may be best to interpret the model in antirealist terms' (Moreland and Craig 2003, p. 344). The five arguments against an infinite regress which

I discuss below are based on similar metaphysical considerations which are derived from understanding the nature of the world.

## 5.3 Introducing the Philosophical Arguments Against an Infinite Regress of Causes and Events

There are at least five arguments which have been offered to rule out an actual infinite regress of causes and events:

1. The argument from the impossibility of concrete actual infinities
2. The argument from the impossibility of traversing an actual infinite
3. The argument from the viciousness of dependence regress
4. The argument from the Grim Reaper paradox (Pruss 2018; Koons 2014)
5. The argument from Methuselah's diary paradox (Waters 2013)

Each of these arguments is independent of the other, and any one of these arguments would be sufficient to demonstrate that an actual infinite regress is not the case. Therefore, it is not enough for the objector of finite regress to rebut one of these arguments; rather, the objector would need to rebut all five of them (and perhaps others). It is important to emphasize this point, because it has often been wrongly assumed (e.g. by Zarepour 2020, p. 17) that if one rebuts (say) the argument against concrete actual infinities (aka the Hilbert Hotel Argument), then one has rebutted the KCA. (Zarepour fails to consider that, even if a concrete actual infinite [e.g. an actual infinite number of stars] is possible, this does not imply that an actual infinite can be traversed [e.g. finish counting an actual infinite number of stars one after another], nor does it imply that an infinite causal regress is not vicious, etc.) The argument against concrete actual infinities has been subjected to numerous objections in recent literature; I don't think the objections are compelling and I have replied to many of them elsewhere and developed a new version of the argument (Loke 2012, 2014b, 2016b, 2016c, 2017b, 2021a). However, to reply to the rest of them here would take up too much space, and is in any case unnecessary for the main argument of this book, since (as I have explained earlier) the KCA does not depend on this argument. Hence, I shall reserve my reply for future publications.

Arguments 4 and 5 have been defended at length by others elsewhere (Pruss 2018; Koons 2014; Waters 2013). A modified and easy-to-follow version of argument 4 can be briefly stated as follows[7]:

A piece of paper is passed down from the past to the present, from person to person. If it is blank, someone would write his unique name on it; if it is not blank, it is simply passed on. Suppose the past is infinite. When the paper reaches you, what is written? Something must have been written by then, yet nothing could have been written because any name that might have been written would have been a different name, that is, the name of the person before. In other words,

1. If the past is infinite, then the paper being passed down from the past to the present would have a name written on it.
2. If the past is infinite, then the paper being passed down from the past to the present could not have any name written on it.
3. Anything that entails a contradiction cannot exist.
4. Therefore, the past cannot be infinite. (From 1, 2, and 3)

In addition, Pruss (2018) has argued that an actual infinite causal chain results in a number of other paradoxes, and the simplest and most elegant solution is to accept causal finitism, which implies an uncaused First Cause (see also the Gong Peal paradox defended in Luna and Erasmus 2020).

In what follows, I shall briefly discuss the argument from the impossibility of traversing an actual infinite and the argument from the viciousness of dependence regress.

## 5.4 Argument Against Traversing an Actual Infinite

Think about a series of events (whether microscopic or macroscopic). Suppose event$_1$ begins at time $t_1$, event$_1$ causes event$_2$ at $t_2$, event$_2$ causes event$_3$ at $t_3$, and so on. The number of events can increase with time, but there can never be an actual infinite number of events at any time, for no matter how many events there are at any time, the number of events is still finite: If there are 1000 events at $t_{1000}$, 1000 events is still a finite number; if there are 100,000 events at $t_{100,000}$, 100,000 events is still a finite number, and so on. This illustrates that an actual infinite is greater than the number which can be traversed one after another at any time, because finite (one) + finite (another) = finite.

Since it is impossible to traverse an actual infinite number of events *from* event$_1$, it is likewise impossible to traverse an actual infinite number of earlier events *to* event$_1$, given that the number of events required to be traversed in both cases is the same. Thus, the number of events earlier than event$_1$ (and likewise, the number of earlier causes and durations) cannot be an actual infinite. Therefore, there must be a first event. As cosmologists Ellis et al. observe:

> a realized past infinity in time is not considered possible from this standpoint—because it involves an infinite set of completed events or moments. There is no way of constructing such a realized set, or actualising it. (Ellis et al. 2004, p. 927)

To illustrate, the set of earlier years is formed by a one-by-one process, for example, one year (e.g. 2018) followed by one year (2019). 'One' year is a finite number. Finite + finite = finite. Finite + finite cannot form an actual infinite. Hence, it is impossible to form or complete an actual infinite number of years. Thus, the number of earlier years cannot be actual infinite.

As an example of a series of events, one can think of a person marking a stroke each year. An argument against an actual infinite series of earlier events can be formulated as follows:

1. If the series of earlier (yearly) events is an actual infinite, then a person marking a stroke each year would have experienced[8] the accumulation of an actual infinite series of strokes by a one-by-one process.
2. It is not possible for a person marking a stroke each year to experience the accumulation of an actual infinite series of strokes by a one-by-one process.
3. Therefore, it is not possible that the series of earlier (yearly) events is an actual infinite.

For premise 1, consider the series of years BCE. If the series is an actual infinite, then the series of strokes would be actual infinite, and if the series is finite, then the series of strokes would be finite. The phrase 'experienced the accumulation of an actual infinite series of strokes by a one-by-one process' means that the person would experience the existence of a series of strokes which is not given together all at once, but added one at a time (i.e. she would have added one stroke at 1 BCE, added one stroke at 2 BCE … etc.) and forming the series as they do so. The series of strokes is supposed to have been made up by each stroke; none of the strokes existed beginninglessly. Each of them was added at some finite point earlier in time, one by one. At each point in time only a finite number is added to the series, which is formed as a result.

Now, if we suppose that there is no beginning to the process, then there is *no specific point* in time at which the person would have the experience of the accumulation of an actual infinite number of strokes, for in that case *at every year (not any specific year)* there is supposed to be a totality of actual infinite. For example, the number would be actual infinite at 1 BCE (since she would have added one stroke at 2 BCE, one stroke at 3 BCE … etc.); likewise, it would be actual infinite at 2 BCE (since she would have added one stroke at 3 BCE, one stroke at 4 BCE … etc.). In other words, there would be no point in time at which there is a transition from having a finite number to having an actual infinite number of strokes. Rather, at each year BCE a stroke would have been added to a series of strokes that was already (supposedly) actually infinite. Nevertheless, we still need to ask how is *that* series of strokes constituted in the first place. Obviously, it is constituted by a one-by-one process, that is, one (finite) stroke being added (e.g. at 3 BCE) followed by one (finite) stroke being added (e.g. at 2 BCE). But can such a series be infinite?

Premise 2 is based on the fact that a one-by-one process cannot constitute an actual infinite series, because finite + finite = finite.

Objectors to the argument against traversing an actual infinite often claim that the argument begs the question by assuming a starting point, for if there is a beginningless series with an actual infinite number of earlier events, then an actually infinite sequence has already been traversed (Morriston 2013, pp. 26–27).

In reply, the argument against traversing an actual infinite is based on the nature of a one-by-one process, that is, finite + finite = finite, it is not based on starting at a point and therefore does not beg the question. To elaborate, without begging the question by assuming a starting point or by presupposing whether the number of events earlier than any time $t_p$ is infinite or not, think of a series of events in the midst of being constituted by a one-by-one process (Fig. 5.2):

There is one event P produced at time $t_p$ . (Note that I stated 'there is one event P produced at time $t_p$'; I did not state or assume that a total of one event has been produced by $t_p$, which is false if there are events earlier than $t_p$.) There is event P followed by event Q produced at $t_p$ and $t_q$ and together they constitute two events, there are events P, Q, and R produced at $t_p$, $t_q$, and $t_r$, respectively, and together they constitute three events, and so on. The series of events is constituted by each event. The series is constituted by a finite number (e.g. 'one') of event/s and a finite number ('another') of event/s, and together they constitute a finite number of events, not-possibly an actual infinite number of events.

The above conclusion is not based on presupposing 'a particular time as a starting point' or that the number of earlier events is not actual infinite; thus, it is not question begging. Rather, the conclusion is based on the nature of the one-by-one sequential process (Finite + Finite = Finite), and how any concrete series of events is constituted. For example, P could

time ------tn---------to---------*t*p--------- *t*q --------- *t*r----------*t*s----------

event ----->N -----> O ------> P--------> Q -------> R---------->S-------->....

**Fig. 5.2** A series of events being constituted

be (say) marking a stroke on a piece of paper at a certain year and Q could be adding a stroke the next year in the process of forming a series of strokes. Consider the following demonstration by mathematical induction which, contrary to its name, is a deductive proof which shows that the process would result in a natural number and hence finite number of strokes: A set S is an inductive set if for every element *x* of S, the successor of *x* is also in S. Let *n* be a natural number. Since the set of natural numbers is an inductive set, $n + 1$ is a natural number for all natural *n*. If $n + m$ is a natural number, then $n + (m + 1) = (n + m) + 1$ is also natural number since the set of natural numbers is an inductive set. Therefore, $n + m$ is a natural number (which would be a finite number) for all *n* and *m* by mathematical induction.[9]

A sceptic might object that the above argument only implies that the series of events that happened between time $t_p$ and any time later than $t_p$ is finite; it does not imply that the series of events prior to $t_p$ is finite, and therefore does not exclude a beginningless series of events with an actual infinite number of earlier events.

This objection however fails to note that the above argument is intended to show how any series of events is constituted in the first place, and that 'any series of events' would include the series of events prior to $t_p$ which, following the above argument, would be finite as well. In other words, the above argument is intended to illustrate how *any* series of events constituted by a finite number ('one') of events being added followed by another finite number ('another') of events being added would result in only a finite series of events. Now the series of earlier events *is* such a series of events. Thus, the number of earlier events must be finite.

Concerning the phrase 'the accumulation of a series of strokes by a one-by-one process', a sceptic might object by citing Morriston, who notes that, if the series of years is a beginningless series in which every year BCE was preceded by another, then 'each year BCE would have been 'added to' a running total of years that was already infinite' (Morriston 2021, p. 8n.5). In other words, while it is impossible to start with finitude and constitute an actual infinite, in the case of a beginningless series of events one does not start with finitude; rather, an actual infinite is already constituted at each moment. At each moment a finite number (one) is added to an actual infinite (and not to a finite number). Thus, if the number of earlier years is an actual infinite, then a person marking a stroke each year would experience a finite number adding to an actual infinite number of strokes at each year. Finite + infinite + = infinite

In reply, it should be noted that the point I am making is different from Morriston's. His point is that, if the series of earlier events is actually infinite, then there already is an infinite series, to which a finite element is added. In other words, Morriston is thinking in terms of adding a finite element to *the series of events* which already exists. This does not answer the more fundamental question of how the series is constituted by its individual elements in the first place (and one must be careful not to beg the question by assuming that a beginningless series can exist).

Whereas I am thinking of the more fundamental question of how *any series of events* is constituted by its elements. I am thinking in terms of what is being added (i.e. one finite element followed by one finite element) to constitute the series (that is the meaning of the phrase 'the accumulation of a series of strokes by a one-by-one process'). Finite + finite = finite thus refers to the (finite) elements that constitute the series; it is more foundational than the series which they metaphysically ground. The series of events is constituted by one (finite) element being added followed by one (finite) element being added. The essential feature of this one-by-one sequential process is that the series of strokes is supposed to have been made up by each stroke; none of the strokes existed beginninglessly. Each of them was added at some finite point earlier in time, one by one. This point is illustrated by Fig. 5.2 and the mathematical induction proof, which is about how each element which are added into the series at each year constitute the whole, without begging the question either way. That is, without assuming whether the total number of elements in the series of earlier events is already infinite or not, the fact remains that the series is constituted by its elements, and each element is a finite quantity (i.e. 'one' event is a finite quantity), and together they constitute a finite quantity because finite + finite = finite; therefore, the series of earlier events cannot be actually infinite. This conclusion is arrived at without first assuming that the total number of elements in the series is not infinite, and hence the argument does not beg the question. In other words, the problem with postulating a beginningless series of events is that, even though it supposedly does not start with finitude but already has an infinitude, nevertheless the infinitude is still supposed to have been constituted by a one-by-one sequential process, which, as shown above, is impossible. It is true that, if the number of earlier years is an actual infinite, then a person marking a stroke each year would experience a finite number adding to an actual infinite number of strokes at each year. However, the problem is that the actual infinite number of strokes which supposedly already exist at each year is supposed to have been experienced by the person to have been constituted by a one-by-one process, which as shown above is impossible.

In summary, the objector claims that there is a distinction between constituting a series of later events from a particular event (a beginning) and constituting a series of earlier events to a particular event, and while it is impossible to constitute an actual infinite in the former case, it has not been shown to be impossible in the latter case. Contrary to this claim, I have argued that the number of elements required to be constituted by a one-by-one process is the same for both cases, and this implies the same impossibility for both cases because finite + finite = finite. Thus, the objection fails.

Against the explanation that 'any finite quantity plus another finite quantity is always a finite quantity', Malpass (2019) objects that the question is, '*how long we have been doing it for*. The lesson seems to be that if you only count for a finite amount of time, then you cannot construct an actual infinite by successive addition, but if you do it for an actually infinite amount of time, then you can.' In other words, if George has been counting −1 at $t - 1$, −2 at $t - 2$ … he would have counted an actual infinite and there would be no longer any more number of a negative infinite series to count.

However, such an objection ignores how the series of counting is constituted in the first place. It is constituted by one (finite) element being added followed by one (finite) element being added, which (as explained above) cannot constitute an infinite number. The set (an abstraction) of earlier events is grounded in the existence of its concrete members (each event), which constitute the set. Since the series of earlier events is constituted by individual (i.e. a finite number of) event being added followed by individual(finite) event being added, the series must be finite. Moreover, the supposition that there is already an actual infinite time in the past entails the possibility that finite + finite = infinite, but as explained previously the consequent is impossible; hence, the antecedent is impossible.

A sceptic might object that, if the process of one (finite) element being added followed by one (finite) element being added has been happening from an actual infinite past (i.e. if one repeats this process an infinite number of times), then an actual infinite number of elements would have been added to the series.

My reply is that an actual infinite number of elements is not supposed to have been added all at once, but one after another. To repeat a process actual infinite number of times, one needs to first proceed one time after another, but the problem is that the result of that process is always finite at any time, because 'one time' ('finite') after 'another' ('finite') implies finite + finite which is equal finite. Thus, the result of that process is always finite, because the number of each stroke is a finite number, and finite+ finite cannot be infinite. Therefore, one cannot have been adding from an actual infinite past since this entails the impossible consequence that finite + finite can be infinite. This reply does not beg the question because there is an independent proof that natural (finite) number + natural (finite) number = natural (finite) number using mathematical induction (see above).

In their defence of the argument against traversing an actual infinite, Craig and Sinclair (2009, p. 124) state that this argument would not work on a static theory of time because a series that was formed sequentially, one event occurring after another such that the collection grows with time, presupposes a dynamic theory of time. Whereas on the static theory of time, a series of events is not formed by addition of later events (which come-to-be) to earlier events; rather, the collection of events is a collection whose members all tenselessly coexist and are equally real. It would be like an infinitely long ruler with an infinite number of markings, with different features at each marking.

Craig himself has responded to this objection by providing a number of arguments against the static theory of time and defending the dynamic theory of time (Craig 2000a, 2000b). In what follows, I shall argue that, even if the static theory of time is true, it remains the case that a series is *experienced* to be forming over time by successive addition, and if the number of earlier events is infinite, the formation of an infinite series by successive addition would be experienced by someone living as long as time exists, but, as argued above, this is not possible.

A sceptic would object that on the static theory of time our experience of a series forming over time by successive addition is illusory, just as our experiences of the flow of time and the 'time arrow' is illusory; on a static theory there is no 'progression' and any specification of temporal points is arbitrary.[10]

Nevertheless, even if the static theory of time is true, it remains the case that our experience of 'a series forming over time by successive addition' exists, even if this experience is illusory. For example, we can obviously mark a stroke on a piece of paper at time $tp$, add another stroke at $tq$ and so on, hence forming a series of strokes. In other words, we can still experience a series forming over time by the addition of a finite number of element(s) (e.g. 'one event') followed by the addition of a finite number of element(s) (e.g. 'another event'). This is different from the infinite ruler case which does not involve the experiencing of adding one element after another over time. Likewise, we have experiences of progress and of specifying each year, for example, marking each year with (say) a stroke and adding another stroke the following year (finite + finite). In other words, even if there is no progress (given the static theory), yet it would seem to the person that there is progressive addition of strokes in the sense explained above. These experiences exist even if they are illusory. Illusory just means that our experience of the world is not true of the world; nevertheless, it remains true of the world that we have the experience. For example, it seems to me that I was in Hawaii last night, but in fact I wasn't at Hawaii; it was a dream. Nevertheless, it is true of the world that I had a dream in which it seems to me that I was at Hawaii. Now my seemings cannot involve a logical or mathematical impossibility. For example, it is impossible that I dream about a shapeless square or 2 + 2 = 5. Such impossibilities cannot exist anywhere, not even in our illusions or dreams. Hence, anything that entails the possibility of a person experience (whether as an illusion or not) such an impossibility must be false. Now if there is a beginningless series of events in which the number of earlier years is an actual infinite, this entails that a person experiencing the marking of each year with (say) a stroke and adding another stroke the following year (finite + finite) would experience the series constituted by what is added, and what is added is not added all at once, but one by one; that is, she would *experience*[11] finite + finite = infinite.[12] (This point remains valid regardless of whether the static theory of time is true.) The consequent is mathematically impossible. Hence, the antecedent is impossible.

Against the argument for the impossibility of an infinite regress, Russell (1969, p. 453) objects that there could be an actual infinite series of negative integers ending with minus one and having no first term. Likewise, Graham Oppy asks us to consider the series ..., $-n$, ..., −3, −2, −1. He writes: 'In this series, each member is obtained from the preceding member by the addition of a unit' (Oppy 2006a, p. 117).

In response, it should be noted that a negative or positive cardinal number series is a case of abstract actual infinite which exists timelessly rather than constituted by a one-after-another temporal process. Thus, it does not provide a counterexample to the claim that an actual infinite cannot be constituted by a one-after-another process in the concrete world. While each member of the abstract negative cardinal number series ..., $-n$, ..., −3, −2, −1 is 'obtained' from the preceding member by the addition of a unit, this 'obtaining' is in the form of timeless mathematical ordering relation. It is not the case that the abstract number −2 (say) is brought about in time by the addition of a unit to −3. Rather, the abstract numbers −2 and −3 have always existed timelessly, and this is unlike a causal series of concrete entities existing in time (note that static theory of time is not timeless, see Chap. 2). One can have an abstract actual infinite number of negative numbers each of which is timelessly separated from zero by a finite number of negative numbers. The existence of each of the number in the series is not causally dependent on any previous number, nor is it dependent on the actual infinite number which exists outside of the series. However, to constitute a series by a one-after-another causal process is a different matter. In contrast with a series of timeless numbers, a temporal series of causes and effects is constituted over time by adding one element after another, and each effect in time is causally dependent on a prior cause. The process proceeds one after another, constituting a finite number at any time. While the number of (say) cardinal numbers is an (abstract) actual infinite, the number of future events that has been constituted is clearly not an actual infinite because it is impossible to traverse an actual infinite as argued above, and I have argued that likewise the number of earlier events cannot be actual infinite as well.

In summary, a number series is a set the elements of which can be ordered one after another; it is not a series which is constituted over time by adding one element after another. Even if there could have been an actual infinite number of elements (e.g. abstract numbers) each of which is a finite quantity, this does not imply that there could have been an actual infinite series formed by successive addition of elements each of which is a finite quantity. On the contrary, a successive addition of elements each of which is a finite quantity cannot constitute an infinite series, because finite + finite = finite. That is why no infinite series can be formed by successive addition.

The point of the argument is that an actual infinite number of events cannot be constituted at any time. The objector suggests the hypothesis of a series of events that is infinitely long. However, in order to constitute an actual infinite number of events in the first place, the process has to proceed one event after another, and the problem is that the result of that process is always finite at any time. One does not constitute an actual infinite at any time, not at $t_{1000}$, $t_{100000}$, or $t_{1000000}$. As noted earlier, actual infinite stands outside of the series, timelessly and abstractly. As Copan and Craig (2017, p. 309) observe: 'Necessarily, given any finite number $n$, $n$+1 equals a finite number. Hence, $aleph_0$ has no immediate predecessor; it is not the terminus of the natural number series but stands, as it were, outside it and is the number of all the numbers in the series.' But here we are talking about what happens in a series of events in the concrete world, not timelessly and abstractly. The proponent of KCA is referring to what happens in the concrete world when he/she argues that an actual infinite number of events cannot have been constituted at any time $t$; therefore, the number of events earlier than $t$ is finite and hence has a first member (i.e. a first event) which (given the Causal Principle) has a First Cause. If one wants to talk about the timelessly abstract which has no causal powers, that would be irrelevant as an objection to the KCA and does not block the conclusion of the argument that there is a First Cause.

Morriston (2013, pp. 26–27) claims: 'From the fact that we cannot—beginning now—complete the task of enumerating all the events in a beginningless series, it does not follow that the present event cannot arrive or that a beginningless series of events that have already arrived is impossible. To suppose otherwise would be to confuse the items to be enumerated with the enumerating of them—it would be like arguing that there must be finitely many natural numbers because we can't finish counting them.'

In reply, constituting a series to the present from a beginningless past would require the number of events constituted to be actual infinite, but an actual infinite is too large to be constituted by a one-after-another process. The problem is not due to our ability to enumerate; rather, it is due to the nature of an actual infinite which is too large to be constituted by a one-after-another process. While there can be an infinite number of natural numbers in the abstract, to constitute an actual infinite in the concrete is a separate issue and the real issue here.

One might object that there could be an actual infinite number of points between (say) time $t_0$ and $t_1$ (just as there is an actual infinite of real numbers between 0 and 1) which is traversed in a manner similar to Zeno's supposed paradox of motion. In a similar vein, one might object that there could be an actual infinite number of events between $t_0$ and $t_1$. Where our universe is concerned, there might not be a first point of time at $t_0$ (Pitts 2008).

Three points can be made in response.

First, having a beginning does not require having a beginning point (Craig and Sinclair 2012, p. 99). It has been explained in Chap. 2 that something has a beginning if it has a temporal extension, the extension is finite, and it does not have a static closed loop or a changeless phase that avoids a boundary. Suppose $t_0$–$t_1$ is the first temporal interval. Even if there were an actual infinite number of points/events between $t_0$ and $t_1$ and there is no first point of time at $t_0$, the extension of $t_0$–$t_1$ is still finite because the points/events sum up to a duration that is finite in magnitude. Therefore (if it does not have a static closed loop or a changeless phase that avoids a boundary; see Sect. 5.6 and Chap. 6), the series of points/events would still have a beginning, and hence (given the argument in Chap. 3) would still require a cause. (To elaborate, suppose an increasing in strength of electric field occurred at time interval $t_8$–$t_9$. Even if there were an actual infinite number of events between $t_8$ and $t_9$ and there is no first point of time at $t_8$, the extension of $t_0$–$t_1$ is still finite and has temporal boundaries. We know from experience that such an increasing in strength of electric field is caused when [say] I switched on an electric field generator. Now, the Modus Tollens argument defended in Chap. 3 implies that, if our universe begins uncaused at time interval $t_0$–$t_1$, there would be no difference between that event and the increase in strength of electric field at $t_8$–$t_9$ where beginning to exist uncaused is concerned, and thus the latter would also begin uncaused, which is not the case; hence, the antecedent is not the case.)

Second, the argument for the impossibility of traversing an actual infinite has a crucial disanalogy with Zeno's paradox of motion. In the case of the argument, the events in a temporal series are actual. By contrast, in the case of Zeno's paradox, the interval traversed can be regarded as being potentially infinitely divisible and not actually infinitely divided. In other words, one can keep on dividing the interval by half without ever ending up with an actual infinite number of divisions. The claim that Achilles must pass through an infinite number of halfway points in order to cross the stadium begs the question by assuming that the whole interval is a composition of an infinite number of points which have been traversed (Craig and Sinclair 2009, p. 119). (One might object, 'but how could one distinguish between those that are actual events and those that are only potential events?'[13] In reply, this is a question about epistemology [How we distinguish?]. Whereas the arguments in this section concerns ontology: they demonstrate that, ontologically, there cannot be a traversal of an actual infinite in the concrete world. The epistemological question [How we distinguish?] is irrelevant as an objection to the argument which concerns ontology, and therefore fails to rebut the conclusion that there cannot be a traversal of an actual infinite in the concrete world. In other words, even if we are not able to distinguish between those that are actual events and those that are only potential events, it remains the case that—ontologically—there are only a finite number of events that are traversed.)

Third, I shall explain below that it is fallacious to think of time as a continuum of points.

Against this, one might object that, at the level of fundamental physics, events may be analogous to points. In other words, events may not be discrete entities in a causal chain; rather, causality could be continuous in nature.[14]

In reply, on the one hand, it has not been proven that spacetime is a continuum made up of actual infinite number of points. Cosmologist George Ellis notes that 'there is no experiment that can prove there is a physical continuum in time or space; all we can do is test spacetime structure on smaller and smaller scales, but we cannot approach the Planck scale.' A distinction should be made between mathematical models of the physical world and the physical world itself. Craig and Sinclair (2012, p. 100) explain that it has not been proven that space and time really are composed of an actual infinity of points rather than simply being modelled as such in general relativity. While infinities are useful mathematically, that does not imply that concrete infinities exist, just as the fact that imaginary numbers (e.g. $\sqrt{-1}$) are mathematically useful does not imply that that they correspond to concrete entities (they obviously don't!). Infinities, like imaginary numbers, can be regarded as useful abstract tools. Imaginary numbers work as a shorthand for mathematical operations involving real numbers. Likewise, infinities may work as approximations or generalizations. For example, the idea of infinities can be understood as approximations which have proven to be useful in addressing *problems* concerning pendulums, chemical decay, coagulation kinetics, diffusion, convection, economic equilibrium, and fluid and air flow. Tegmark explains:

> Consider, for example, the air in front of you. Keeping track of the positions and speeds of octillions of atoms would be hopelessly complicated. But if you ignore the fact that air is made of atoms and instead approximate it as a continuum—a smooth substance that has a density, pressure, and velocity at each point—you'll find that this idealized air obeys a beautifully simple equation explaining almost everything we care about: how to build airplanes, how we hear them with sound waves, how to make weather forecasts, and so forth. Yet despite all that convenience, air of course isn't truly continuous. I think it's the same way for space, time, and all the other building blocks of our physical world. (Tegmark 2015)

In some cases infinities can be understood abstractly as a limit concept, or as generalizations as in a polygon circle. Regarding infinitesimal calculus, Oxford philosopher A.W. Moore points out that mathematicians using calculus can uphold claims ostensibly about infinitesimals or about infinite additions, knowing that they are only making disguised generalizations about what are in fact finite quantities (Moore 2001, p. 73). Various scientists and philosophers have also argued that time and space could be a set of discrete entities of extended simples (i.e. a spatiotemporal entity that has no proper parts and does not have the shape and size of a point) (Van Bendegem 2011; Hagar 2014). Craig notes that 'Since the advent of quantum theory, philosophers, and physicists as well, have exhibited much greater openness to taking time and space to be discrete rather than dense. In fact, many think that the continuity of spacetime in general relativity is what needs to go if we are to have a unified physical theory of the world' (citing Butterfield and Isham 1999, section 3.2; Huggett and Wüthrich 2013). On the other hand, one might defend an alternative view that time might be continuous yet divide into a finite number of smallest parts of finite durations rather than divide into instants/points (see my reply to Puryear in Loke 2016a and Loke 2017a, chapter 2).

It has been explained above using the example from quadratic equation that, while what is mathematically impossible is metaphysically impossible, what is mathematically possible is not always metaphysically possible. Thus, the mathematical possibility of actual infinity does not imply that there are metaphysically possible in the concrete world. Thus, for example, while one can always decompose a real function in terms of infinitely many sinusoidal functions (Fourier series) with countably infinitely many coefficients (Pitts 2008), this can be regarded as true only for mathematical modelling (Chan 2019; for other examples, see Loke 2017a, chapter 2). Likewise, the infinite number of points within the intervals (0, 1) and (0, 2) and the one-to-one correspondences between them merely refer to abstractions; it does not imply that they can be realized as concrete entities. Likewise, while there is an *abstract* actual infinite number of points between (say) $t_0$ and $t_1$ just as there are an actual infinite number of real numbers between 0 and 1, this does not mean that they can exist in the concrete, nor does it imply that such a series can be constituted in the concrete.

While one might claim that it is possible to have a concrete point in between any two points,[15] it is logically invalid to infer from this to the conclusion that there is an actual infinite number of concrete points which can be traversed. This would be guilty of a fallacious modal operator shift, inferring from the true claim,

> 1. Possibly, there is some point at which *x* is divided.
>
> To the disputed claim,
>
> 2. There is some point at which *x* is possibly divided. (Craig and Sinclair 2009, p. 114)

It would be like arguing 'because a leaf could be any colour, therefore it can be every colour'. A leaf obviously cannot be of every colour at the same time because of metaphysical constraints. Likewise, there might be metaphysical constraints such as those which I have explained using the Christmas Present Scenario (Loke 2017a, Chapter 2) which prevent all the points from existing together concretely, even though it is possible that each of the points exists concretely.

On the other hand, Craig observes that the idea of spacetime being a continuum made up of actual infinite number of points results in the ancient Greek paradoxes of motion: Suppose time $t$ is the last point in time at which an object O is at rest, it would be impossible for O to begin to move. To reply that $t'$ is the first point in time at which O is in motion will not do, for $t$ is supposed to be the last point in time at which O is at rest, and one can always think of $t^*$ (where $t < t^* < t'$) for any $t$ or $t'$ (Craig and Sinclair 2012, p. 100; thus, for example, suppose time 0.00 sec is the last point in time at which an object O is at rest; to say that time 0.01 sec is the first point in time at which O is in motion will not do, for 0.00 sec is supposed to be the last point in time at which O is at rest, and therefore O would have already been in motion at (say) 0.005 sec, that is, before 0.01 sec. The solution to this paradox is to reject the theory that time is composed of an infinite number of points and to adopt the theory that time is composed of durations. In that case, one can suppose time $t$ is the last duration in time at which an object O is at rest, and that $t'$ is the first duration at which O is in motion, and that there is no duration $t^*$ in between $t$ and $t'$). Additionally, recent defences of the Grim Reaper Paradox provide convincing reason to think that spatiotemporal intervals are not composed of a dense infinity of points or instants (Craig 2018, p. 398). Moreover, I have shown above that there cannot be a traversal of an actual infinite of events because finite + finite = finite; therefore, this is a proof that there isn't an actual infinite number of point-events between (say) time $t_0$ and $t_1$ which is traversed.

Sorabji (2006, pp. 221–222) objects that a beginningless sequence does not face the same difficulties as an endless sequence because traversing the former would involve only one terminus (e.g. the present moment), whereas traversing the latter would involve two termini (e.g. the present moment and some future moment). 'And [having two termini] is what prevents the future series of traversed years from being more than finite' (Sorabji 2006, p. 222).

In reply, the cause of the impossibility of traversing an actual infinite in my explanation above is not due to the number of termini, but the fact that an actual infinite has greater number than the number which can be traversed one after another in time, because finite + finite = finite. Sorabji does not provide any solution to this difficulty.

One might object that the reason actual infinite is too large to be traversed by a one-after-another process is because one cannot arrive at the endpoint of that which has no end by beginning from a point (Leon 2011). Since there is no endpoint, one can always increase, and every point that is arrived at is always smaller than an actual infinite, but this is not the case if one does not begin from a point by arriving at the present from an actual infinite number of earlier events. However, whether there is an end or no end should not affect the number that can be traversed by a process. The reason is because, regardless of whether there is an end or not, the number traversed by a one-after-another process is finite, and this is due to the nature of the process in which finite + finite= finite.[16]

Pruss (2018, p. 152) objects that 'imagine someone who, according to our external time, now exists and will always exist. But she lives her life backwards. This person, thus, at this point can be said to have lived an infinite life. And if this is her moment of death, then she has completed that infinite life.' This objection begs the question by assuming that it is possible that someone now exists and will always exist and that the future is concrete actual infinity. It begs the question against the view that someone will exist forever only if the future is potential infinite in dynamic time. On the other hand, the argument against traversing an actual infinite which I explained above would rule out such a scenario, regardless of whether a person travels forwards or backwards in time.

One might object to KCA by suggesting that events should be understood as 'becoming'.[17] In reply, becoming (with no end) assumes a potential infinite, but events that are causally prior have already happened and therefore cannot be a potential infinite (Loke 2017a, chapter 2). Likewise, recursive function (a function that calls itself during its execution) involving an infinite loop is a potential infinite in its actual execution (perpetually increasing towards abstract actual infinity as a limit but never actually arriving at an actual infinite in time), and thus is inapplicable as well. (Note that a potential infinite is a series which increases towards actual infinity but does not arrive at actual infinity. The possibility of such a series is not the possibility of completing the counting an actual [infinite] series of past events. Rather, the *possibility* is the possibility of an *abstract* actual infinite *as a limit concept.* It is important to distinguish between abstract actual infinite and concrete actual infinite. The argument against traversing an actual infinite concerns the concrete; that is, a concrete actual infinite cannot be traversed. This is consistent with the existence of an abstract actual infinite as a limit concept.)

It has been objected by Dretske (1965) that, if starting from a point someone (e.g. George) does not stop counting, then George will count to infinity 'in the sense that he will count each and every one of the finite numbers'. Oppy (2006a, p. 61) likewise argues that 'one counts to infinity just in case, for each finite number N, one counts past N. But unless one stops counting, one will eventually reach any given finite N' (see also Malpass 2021).

However, the question that is relevant to the Kalām is not whether George will count an actual infinite (since 'will count' concerns future events rather than past events). Rather, the question is whether George can be at any particular time *t* and counts an infinite number successively by that time, and the answer is no: there is no time at which he could have counted an actual infinite number of elements by counting one element after another. Even if it is the case that George counts as long as time exists, actual infinity will always be greater than the numbers to which George has counted by time *t*. Thus, the fact is that there is no time at which an actual infinite *has been* counted (Loke 2014a).

Whereas to have counted as long as time exist if time is beginningless (e.g. George counts 0 at the year 2020, -1 last year [2019], -2 the year before that [2018] … etc.) would have required an actual infinite to have been counted at a particular time (say, the year 2020), which as explained previously is impossible. In other words, Dretske, Oppy, and Malpass are guilty of redefining 'traversing an actual infinite' without solving the problem in the context of debating the cosmological argument. (Malpass 2021's 'fills the future' argument is not relevant to the discussion concerning the past: if the number of durations earlier than [say] the year 2020 is an actual infinite, then we could indeed look back [say, in the year 2021] and say that the counting was completed in the year 2020, and the consequent [an infinite set of completed events] is what proponents of the Kalām are arguing against [i.e. via Modus Tollens] in the context of discussing whether the universe began to exist.)

The same problem besets Almeida's (2018, p. 52) objection that 'Mathematical induction is valid. If 1 has the property of being a number, and if for each finite *n*, if *n* is a number, then *n* + 1 is a number, then all of the positive integers are numbers. Likewise, if *n* is traversed, and if for each finite *n*, if *n* is traversed, then *n* + 1 is traversed, then all of the positive integers are traversed. Yes, the whole thing! No fallacy of composition is involved. The mistake is in believing that we would have to traverse something other than finite numbers in order to traverse an infinite series.' Almeida (2018, p. 53) acknowledges: 'it is also true that the clock never reaches the infinite time $t_{\aleph 0}$. But this is because no such time as $t_{\aleph 0}$ exists. If the clock ticks off the finite times in each Sn—as we have proven it does by mathematical induction – then the clock ticks off infinitely many times.'

In response, the fact remains that an actual infinite cannot be constituted (as Almeida acknowledges, 'the clock never reaches the infinite time $t_{\aleph 0}$'), whereas the constitution of a beginningless series of earlier events to the year 2020 would have required an actual infinite to be constituted, which as explained previously is impossible. In other words, Almeida is guilty of redefining 'traversing an actual infinite' as 'ticks off the finite times in each Sn' instead of 'reaches the infinite time $t_{\aleph 0}$', without solving the problem in the context of debating the cosmological argument. Now I do agree that mathematical induction is valid, but Almeida has misapplied it by wrongly defining the problem.

It might be asked whether an omnipotent God could traverse an actual infinite.[18] Many philosophers and theologians throughout history (e.g. Augustine, Anselm, Aquinas, Hoffman and Rosenkrantz) have explained that divine omnipotence does not require God to do what is metaphysically impossible (e.g. God cannot make a shapeless square because there cannot be any such thing for God to make; see further, Loke 2010, p. 526). Since it has been shown above that traversing an actual infinite is metaphysically impossible, there is no violation of divine omnipotence to say that God cannot traverse an actual infinite because there cannot be any such traversing for God to do.

## 5.5 The Argument from the Viciousness of Dependence Regress

The argument against a dependence regress had a long history. While Leibniz had argued against infinite regress by claiming that grounds can never be found in this way (see Cameron 2008), Hume (1779/1993) objected that if every item in a collection was causally explained by a preceding item, the whole collection would be explained (Hume–Edwards Principle). A Humean might therefore argue that, if there is an infinite chain of events with each event being grounded by the prior event that caused it, and the chain itself is grounded by another entity which is grounded by another, and so on, then there is no problem with an infinite regress (Cameron 2008, p. 11; Cameron eventually appeal to intuition and explanatory utility to argue against such a regress). One might reply that, if all we want is an account of why each thing exists, then an infinite regress is benign because each thing is explained by its cause; but if we want an account of why there are things at all, it is vicious because we have not explained where existence comes from (Bliss 2013,

p. 414; Cameron 2018). However, the objector might say that, in an infinite chain, there is always existence; thus, it is meaningless to ask where existence comes from. I shall reply to this objection in what follows by arguing that there is a problem of dependence and a need for the capacity to begin to exist which is not met by an infinite regress.

My argument can be formulated as follows:

1. A dependence regress (i.e. a regress in which each item depends on the prior one) is a vicious regress.
2. A causal regress is a dependence regress.
3. Therefore, a causal regress is a vicious regress.

While arguments based on dependence have often been used to argue against an infinite causal regress in the case of an essentially ordered series in the Thomistic Cosmological Argument, I have argued in Loke (2017a, chapter 3) that they can be used for an accidentally ordered temporal series as well. The Modus Tollens argument defended in Chap. 3 implies that whatever begins to exist would *depend* on causally necessary condition(s) (this also implies the transitivity of causal dependence understood as dependence on causally necessary condition(s)[19]). Moreover, since dependence is a kind of contingency (defined as a dependence on circumstances), my argument can also be understood as a kind of contingency argument which demonstrates that there is a necessary being which (as explained below and in Chap. 6) is an uncaused First Cause which is beginningless, initially changeless, and has libertarian freedom.

To illustrate the viciousness of a causal dependence regress, think about a series of train wagons in which each train wagon requires a preceding one to pull it if it is to begin to move. Before the last train wagon begins to move, the one before it has to begin to move, and before that train wagon begins to move, the one before it has to begin to move, and so on. No matter how many such train wagon there are, none of them would begin to move, because no prior wagon escapes from the problem of depending on a prior dependent member in order to begin movement (vicious regress). What is required is an engine, a First Puller which does not depend on another train wagon to pull it, and which has the independent capacity to bring about the beginning of movement.

Likewise, before I begin to exist, my parents has to begin to exist, and before they begin to exist, their parents have to begin to exist, and so on. No matter how many prior dependent causes there are, none of them would begin to exist, because no prior dependent cause escapes from the problem of depending on a prior dependent member in order to begin to exist (vicious regress). What is required is a First Cause which exists independently (i.e. not dependent on a prior entity). Since whatever begins to exist has a cause (as established in Chap. 3), this First Cause would be beginningless.

One might object by claiming that the above argument only works within a deterministic and reductionistic worldview, which neglects the fact that different branches of the natural sciences talk about their own causal explanations using different scientific theories (rather than uniformly as in the case of a series of wagons).[20]

In reply, first, it is not true that indeterminism has been proven; defensible deterministic interpretation of quantum physics exists (see Bricmont 2017).

Second and more importantly, in any case, the basic idea of causally necessary condition is compatible with both determinism and indeterminism, and it is compatible with different descriptions by different scientific theories in different branches of sciences. For example, quantum particles do not begin to exist from non-being. Rather, they emerge from the quantum vacuum with quantum field and which possesses 'zero-point energy (the energy remaining in a substance at the absolute zero of temperature (0 K), which gives rise to vacuum fluctuations)' (Daintith 2009). The fluctuation is therefore dependent on the quantum field and the energy in the quantum vacuum, which is a causally necessary condition. Likewise, hydrogen is a causally necessary condition for the formation of water.

Even though the emergence of quantum particles may be indeterministic while the formation of water is deterministic, and even though these two events are explained by different theories, nevertheless both theories (as well as other scientific theories) involve effects which are dependent on necessary conditions (which is what I mean by a cause). Even though the kinds of necessary conditions and the kinds of dependence might be different for each theory, they all involve necessary conditions and dependence nonetheless.

The train car analogy is only an analogy; the point of the analogy is that a regress of causally necessary conditions is a vicious dependence regress, and therefore requires an independently existing First Cause which does not require causally necessary condition. This point does not require the assumption that each effect is deterministically brought about by the causes, nor does it require the causes are describable by the same scientific theory. Even if each effect is not deterministically brought about and are not describable by the same scientific theory, it remains the case that the effects depend on causally necessary conditions and there cannot be an infinite regress of dependence; thus, there must still be an independent First Cause. Hence, the fact that different branches of the natural sciences talk about their own causal explanations does not affect my argument. On the contrary, it has been explained in Chap. 2 (e.g. using the observation that the causal term 'interaction' is fundamental to science) that causation itself is fundamental to science, and it has been explained in Chap. 3 that the Modus Tollens argument for the Causal Principle implies that whatever begins to exist would *depend* on causally necessary condition(s).

To speak of a causal chain is to speak of a chain of causally necessary conditions describable by various theories of science. The above examples show that we are able to say something in empirical/scientific terms about the great chain of causes, that is, the quantum vacuum as a causally necessary condition for quantum fluctuation and hydrogen as a causally necessary condition for the formation of water. The Modus Tollens argument in Chap. 3 shows that the causal principle is true and therefore each entity in the chain would have a causally necessary condition except the beginningless First Cause.

Consider another analogy (which I shall call the Debtors' Scenario). Suppose I have nothing and the only way for me to begin to have money is to get it from Justin, Suppose Justin has nothing and the only way for him to begin to have money is to get it from Alex. If everyone is like this, no one would ever begin to have money. As Schaffer (2016, p. 95) observes, 'One cannot be rich merely by having a limitless supply of debtors, each borrowing from the one before. There must actually be a source of money somewhere.'[21] Money would not simply emerged from the chain.[22] What is required is someone who does not need to get money from others and is able to have money, that is, a First Source of money.

This is analogous to the real world in which I have no existence before I begin to exist, and the only way for me to begin existing is to be brought about by prior causes (my parents). However, they also have no existence before they begin to exist, and the only way for them to begin existing is to be brought about by prior causes (my grandparents). If everyone is like this, no one would ever begin to exist just as no one would ever begin to have money in a series of debtors. What is required is something which does not need another thing to bring it about and is able to bring about other things, that is, a First Cause.

To elaborate, suppose entity/event $x$ has a beginning of existence and $x$ causally explains why $y$ begins to exist. $x$ can causally explain the beginning of $y$ only after $x$ begins to exist, but the problem is the prior entity $x$ cannot explain why $x$ itself begins to exist given the Causal Principle (established in Chap. 3) that whatever begins to exist has a cause. Thus, $x$ cannot explain why there are entities ($x$, $y$) which begin to exist; that is, $x$ has 0 capacity for explaining the beginning of existence of entities ($x$, $y$). If there is something $w$ which begins to exist prior to $x$, then $w$ can explain why ($x$, $y$) begin to exist, but the problem is that $w$ also cannot explain why itself begins to exist. Thus, $w$ cannot explain why there are entities ($w$, $x$, $y$) which begin to exist, and relies on there being a prior entity $v$ which also has 0 capacity for explaining the beginning of existence of entities ($v$, $w$, $x$, $y$) if $v$ itself begins to exist. If every prior entity has a beginning, then no prior entity escapes from the problem of having 0 capacity for explaining why there are entities which begin to exist; indeed, every prior entity has 0 capacity for explaining the beginning of existence of entities; thus, nothing could ever happen. What is required is an independently existing and beginningless First Cause which is not being sustained in existence, which does not need to depend on another thing to bring it about and which has the independent capacity to bring about other things by itself. Thus, given any change in reality whatsoever there must be such an uncaused First Cause. This First Cause does not begin to exist and hence does not face the problem of an infinite regress scenario in which all members of the series have beginnings of existence without anything having the capacity for explaining the beginning of existence of entities. Such a First Cause would answer the question 'Why is there something rather than nothing?'; that is, there would be no causal explanation for why the First Cause exists, since being beginninglessness and not being sustained in existence *implies* that it was not brought about.

One might ask whether my argument 'if every member of a causal series has no capacity to begin to exist without prior cause, then all the members would have no capacity to begin to exist without prior cause' commits the fallacy of composition. In reply, as Reichenbach (2021) observes, arguments of the part–whole type are not always guilty of this fallacy; it depends on the content of the argument. Sometimes the totality has the same quality as the parts because of the nature of the parts invoked. For example, if every member of a set of entities has 0 mass, then all the members (regardless of the number of members) would have 0 mass, because 0 + 0 + 0 … = 0. Likewise, if every member in a series of debtors has 0 capacity for explaining the beginning of existence of money, then all the members (regardless of the number of members) would have 0 capacity for explaining the beginning of existence of money, because 0 + 0 + 0 … = 0. Similarly, if every member of a causal series has 0 capacity for explaining why there are entities which begin to exist, then all the members (regardless of the number of members) would have 0 capacity for explaining why there are entities which begin to exist, because 0 + 0 + 0 … = 0.

One might object that, in an actual infinite regress of events, the actual infinite series itself does not begin to exist; thus, the whole series itself is exempt from the Causal Principle that beginning of existence has a cause, even though each part of the series (i.e. every event) has a cause.[23] In this case, the whole would have a property which the parts do not have. This is the point of disanalogy with the Debtors' Scenario: while the whole series of infinite number of debtors would not avoid the problem that the whole series has no money because each part has no money, the whole series of an actual infinite regress of events would have no beginning even though each part has a beginning.

In reply, it is trivially true that if an infinite regress causal series exist it would not have a cause. However, my argument is that such a beginningless series cannot exist in the first place because of the problem of vicious regress; that is, no prior dependent cause escapes from the problem of depending on a prior dependent member in order to begin to exist. Every member in such a causal series suffers from the problem of depending on prior member for beginning of existence, analogous to the Wagon Scenario, where every member suffers from the problem of depending on prior member for beginning of movement, without any source.

In other words, such a beginningless series does not get rid of the problem that all its members have beginnings and are dependent in the similar way that all the debtors are dependent, and this is the point of analogy with the Debtors' Scenario. It does not get rid of the problem that there is 0 capacity for explaining the beginning of existence of entities/events. Every member in such a causal series suffers from the problem of having 0 capacity for explaining the beginning of events, which is analogous to the Debtors' Scenario, where every member suffers from the problem of having 0 capacity for explaining the beginning of money in the chain. There needs to be a source somewhere. While on the infinite regress scenario the whole does not have beginning, it remains the case that all the members of the whole has a beginning (y has a beginning, x has a beginning, w has a beginning …), and my argument concerns how to explain the latter. One cannot avoid the latter by appealing to the former; one can only appeal to the former if one postulates a beginningless changeless (eventless) entity enduring through all durations, but such an entity would not bring about the events of our universe which is what requires explanation here. On the other hand, postulating an infinite number of changes which are infinitesimally closely ordered within a duration of time would not work as well (and in any case this postulation has been ruled out by the arguments in Sect. 5.4). The reason is because an infinite number of states each of which has 0 capacity for explaining why there are entities which begin to exist still does not get rid of the problem that there is 0 capacity for explaining the beginning of existence of entities within the series.

It might be objected that, while none of the entities in the series has capacity for causally explaining the beginning of existence of events, the series as a whole has such a capacity.

However, this would not work because every entity in the series has 0 capacity for explaining the beginning of existence of entities; therefore, collectively as a whole, 0 + 0 + 0 + 0 … = 0. (This is analogous to the Debtors' Scenario, where every member suffers from the problem of having 0 capacity for explaining the beginning of money in the chain; therefore, collectively as a whole there is no such capacity. This does not commit the fallacy of composition; see discussion above) This implies that, collectively, the earlier (or causally prior) entities would not have the capacity for explaining why there are later entities which begin to exist. This means that there is 0 capacity for explaining the beginning of entities within the infinite regress series, which implies that the beginning of my existence would not have occurred if the causal chain leading to my beginning of existence was such an infinite regress. But I had begun to exist. Therefore, there is no such infinite regress but rather there is a First Cause. The above argument is not based on our common-sense intuitions but on the meaning of causal dependence which is required by science itself; for example, water has 0 capacity to begin to exist by itself and depends on the formation of hydrogen at the earlier history of the universe.

It might be objected that the series as a whole has the capacity for causally explaining the beginning of existence of its members because every member would have a cause, and that the beginning of each entity is adequately accounted for by a previously existing entity in an infinite regress.[24]

However, the problem is that every cause of every member is dependent on prior causes and every prior cause is dependent. The objector might reply that this is not a problem because every prior cause has a cause—which is dependent, yes, but it has a cause. In reply, every member having a prior dependent cause is useless for solving the problem that all the members are dependent, just like every wagon having a prior dependent wagon is useless for solving the problem that all the members are dependent. In order for a cause to bring about the next member the cause has to begin to exist first, just as in order for a wagon to pull along the next member the wagon has to begin to move first, but the problem is that in both cases it is dependent on prior member all of which are dependent. Moreover, as noted earlier, every cause of every member and every previously existing entity in such a causal series suffers from the problem of having 0 capacity for explaining the beginning of entities, which is analogous to the Debtors' Scenario, where every previous member suffers from the problem of having 0 capacity for explaining the beginning of money in the chain. There needs to be a source somewhere.

Sceptics might object by claiming that this is not a problem, given that in an infinite regress 'there is always at least one member in existence' to explain why there are subsequent entities which begin to exist. Sceptics might argue that this is the point of disanalogy with the Debtors' Scenario: in an infinite regress of debtors, there is no one who has money, but in an infinite regress of causes, there is always at least one member in existence.

In reply, it is trivially true that, if an infinite regress exists, then there is always one member existing. However, such a series cannot exist because the problem of vicious regress remains: how does any of its member begin to exist in the first place? It is dependent on prior member every one of which is dependent. In other words, the postulation that there is always at least one member in existence relies on the assumption that that existent member(s) begins to exist (if none of the members begin to exist, then it is not the case that there is always at least one member in existence), and I have explained that the problem with this assumption is that none of the members prior to that existent member has the capacity for explaining the beginning of entities. This is the point of analogy to the problem that no one has the capacity for explaining the beginning of money in a series of debtors. In order for there to be money in the series, someone must begin to have money, but no one has the capacity for explaining the beginning of money in the series. Likewise, in a series whereby every member has a beginning of existence, in order for there to be something in existence already, something must begin to exist, but the problem is that no one has the capacity for explaining the beginning of entities in the series.

One might object that the things we observe did not begin to exist; rather, they are merely rearrangements of pre-existent matter-energy. For example, the amino acids existed before they became a part of my body.

In response, the fact that my body involves a rearrangement of pre-existing matter-energy does not deny the fact that various events such as new arrangements of the pre-existent matter-energy did begin to exist (e.g. the event of fertilization has a beginning), and that each event (an event is something with a beginning) depends on prior causes (as shown in Chap. 3). My argument does not require beginning to exist from nothing; it only requires that an infinite regress of dependant events/changes is not the case, and therefore there is a first change brought about by an initially changeless and independently existing First Cause (which can bring about the first change by having libertarian freedom, as explained in Chap. 6).

One might object that according to the static theory of time it is always the case that I began to exist (say) in 1975 and I always existed in 1975.[25] However, even if the tenseless theory of time is true, it is still the case that later events (say my existence in 1975) is dependent on earlier events (say, the existence of my parents prior to 1975, because if they did not exist prior to 1975 I would not have begun to exist in 1975). It should be noted that there are two different senses of 'always exist': (1) existing forever without beginning and therefore doesn't require a cause (see the view of the Oxford School in Chap. 6); (2) having a beginning but a tenseless fact at a particular duration. To appeal to a static theory of time would be to refer to the second sense, but according to the second sense, later events are still dependent on earlier events. Hence, this does not remove the problem with a series of dependent events all of which suffer the same problem of dependence. The solution to this problem requires something that has no beginning and therefore can be the uncaused First Cause.

One might object by suggesting that perhaps matter do not begin to exist and atoms are in continuous motion bringing about water, myself, and so on, in which case 'beginning of existence' is just our way of conceptualizing this motion. According to the conservation of momentum (momentum= mass × velocity), to keep something moving (i.e. changing position), no external cause is needed; thus, one might think that this implies it is false that every change/event has a cause. Carroll writes:

> Aristotle's argument for an unmoved mover rests on his idea that motions require causes. Once we know about conservation of momentum, that idea loses its steam … the new physics of Galileo and his friends implied an entirely new ontology … 'Causes' didn't have the central role that they once did. The universe doesn't need a push; it can just keep going. (Carroll 2016, p. 28)

The conservation of momentum implies that there is no need for external cause; nevertheless, the momentum is itself the cause for the continuing movement. Suppose a moving object M is at location $x_1$ at time $t_1$ and $x_2$ at $t_2$. If M has zero momentum at $x_1$ it would not move to $x_2$, but if M has a certain momentum it would move to $x_2$; thus, the momentum is the cause.[26]

Moreover, the beginning of the state at $t_2$ is dependent on the beginning of the state at $t_1$ having momentum, which is dependent on the beginning of a prior state having momentum. Given the viciousness of an infinite regress of dependent states as explained earlier, there must still be a beginningless and initially changeless First Cause with libertarian freedom. (If there is a *changeless* beginningless entity, that would not require a cause, but that entity would be irrelevant for accounting for the events we observe, unless [as explained in Chap. 6] that entity has libertarian freedom to bring about the first event leading to the series of events we observe; this implies that it would be a First Cause Creator who is *initially* changeless)

Finally, it should be noted that the argument from the viciousness of dependence regress is compatible with time being a continuum of points (which nevertheless has been demonstrated above to be false). Even if there is an infinite number of points between $t_1$ and $t_2$, they sum up together into an interval $t_1$–$t_2$ which is still finite in earlier than extension. Therefore, an entity lasting this interval has a beginning, and hence (according to the causal principle established in Chap. 3) would depend on a cause. As argued above, there cannot be an actual infinite regress of dependent entities with beginnings; therefore, there cannot be a beginningless infinite regress of entities/events each of which has a beginning. Rather, there must be a First Cause which is beginningless and not composed of an infinite series of events/changes, but rather is beginningless and initially eventless/changeless.

In conclusion, the above argument can be summarized as follows:

1. An infinite regress of dependent entities is a vicious regress.
2. An event/change is an entity that has a beginning at the state of having gained or lost a property. (Definition)

3. Whatever begins to exist depends on a cause. (Chap. 3)
4. An event/change is a dependent entity. (From 2 and 3)
5. An infinite regress of events/changes is an infinite regress of dependent entities. (From 4)
6. An infinite regress of events/changes is a vicious regress. (From 1 and 5)

What is required is a First Cause which is beginningless and (initially) eventless/changeless and therefore can exist independently. (It will be shown in Chap. 6 that this beginningless entity cannot be an impersonal universe, since it must have libertarian freedom in order to cause the first event from an initially eventless/changeless state. Thus, the impersonal universe itself would have a beginning, which was freely brought about by a beginningless personal First Cause.)

## 5.6 Can a First Cause Be Avoided by a Causal Loop?

Gott and Li (1998) proposed that a First Cause may be avoided by the suggestion that the temporal series of events at the beginning of the universe is a small time loop, thus allowing it to create itself similar to the way a time traveller could travel to the past and become his/her own mother.

However, such a closed causal loop is contrary to the Generalized Second Law of Thermodynamics (Wall 2013a, 2013b) and faces the following problems.

For a causal loop in dynamic time, the members of a series of events come to be one after another cycles after cycles. Given that an actual infinite regress is not the case (as argued in previous sections), the number of earlier cycles would be finite, and thus there would still be a first cycle with a first event and a First Cause.

A causal loop in static time—in which A requires B to bring about its beginning, B requires C to bring about its beginning, and C requires A to bring about its beginning—is viciously circular. It would be similar to a scenario in which railway wagon A requires wagon B to bring about its motion (i.e. by pulling it), wagon B requires wagon C to bring about its motion, and wagon C requires wagon A to bring about its motion. It is evident that such a viciously circular setup—in which the state of each of the entities in a causal loop is supposed to be dependent on another entity within the loop—would not work. Likewise, in a loop that is supposed to avoid a first cause, the beginning of our universe is required to provide causally necessary conditions for the beginning of existence of other entities within a closed loop, while the beginning of our universe itself requires the existence of these other entities. Such a vicious circular setup would not work as well.[27] It faces a similar problem with that of a vicious dependence regress explained previously; that is, since every member of such a causal series has 0 capacity for explaining why there are entities which begin to exist, all the members would have 0 capacity for explaining why there are entities which begin to exist, because 0 + 0 + 0 … = 0.

Against circles of causes, Oppy (2019a, p. 13) argues: 'It is a fundamental causal principle that, if one thing is a cause of a second thing, and that second thing is a cause of a third thing, then the first thing is a cause of the third thing. However, if there could be a circle of causes … then it could be that there are things that are causes of themselves.' But nothing can be a cause of itself, since causes by definition are causally prior to their effects (ibid.; Oppy's argument depends on the transitivity of causation, which I have defended above in Sect. 5.5).

Another objection has been raised by Rasmussen (2018) with reference to the possibility of a scenario which a static closed loop gives rise to. This scenario involves a time traveller whose older self was able to go back in time to teach his/her younger self how to build a time machine and his/her older self knew how because his/her younger self had been told (Lewis 1976, pp. 148–149) Rasmussen (2018) remarks:

> Consider that there is no knowledge of how to build a time machine in our world. That's because no one figured it out (and we can assume for sake of illustration that it could be figured out). Yet, the same is so in the above scenarios: no one figured out how to build a time machine. Thus, no causally relevant difference explains how such knowledge exists in the loop

and infinite regress scenarios but not ours. (For replies to objections, for example, by David Lewis and others, see Loke 2017a, chapter 4).

## 5.7 Conclusion

While the 'Standard' Big Bang model affirms that matter-energy began to exist at the initial cosmological singularity of the Big Bang, over the years, alternative models of the Big Bang have been developed, and some of them have tried to avoid postulating a first event. However, such models face scientific objections related to the Generalized Second Law of Thermodynamics, acausal fine-tuning, or having an unstable or a meta-stable state with a finite lifetime (Craig and Sinclair 2009, pp. 179–182; Bussey 2013; Wall 2013a, 2013b). Bounce cosmologies which postulate that the universe was born from an entropy-reducing phase in a previous universe and the entropy reverses at the boundary condition (Linford 2020) have been proposed to avoid some of these problems. However, these cosmologies neglected the problem of causal dependence at the interface. While it has been suggested that the universes to either side of the interface might be interpreted as the simultaneous causes of each other (Linford 2020, p. 24), this violates the irreflexivity of causation and amounts to a vicious circularity.

Moreover, there are at least five philosophical arguments against an infinite regress of causes and events which have been proposed in the literature, and any one of these would suffice to establish the conclusion:

1. The argument from the impossibility of concrete actual infinities
2. The argument from the impossibility of traversing an actual infinite
3. The argument from the viciousness of dependence regress
4. The argument from the Grim Reaper paradox
5. The argument from Methuselah's diary paradox

I have defended some of the above-mentioned arguments in this chapter, and demonstrated it is not the case that there is a causal loop which avoids a First Cause. Therefore, there is a first event and a First Cause. In the next two chapters, we shall discover what the First Cause is.

# Notes

1. I thank Oners for suggesting this objection.
2. Linford notes that 'Craig takes the alignment between the direction of time and the entropic arrow to be nomologically necessary (i.e. necessary relative to our world only) because of the second law of thermodynamics. … On the account that Craig endorses, we may have empirically discovered that the passage of time is correlated with entropy increase, but this correlation does not reflect anything deep about the direction of time, and situations can arise in which this correlation becomes broken' (2020, pp. 18–19).
3. It should be noted that the number of events in the future cannot be a potential infinite if the static theory of time is true. See Loke (2017a, chapter 2).
4. Sean Carroll, 'Did the Big Bang Break the Laws of Thermodynamics?' https://www.youtube.com/watch?v=TGs4C60FR68. Accessed 30/3/2020.
5. Against the involvement of God, Hawking claims that, because there was no time before the Big Bang, nothing caused the Big Bang, 'because there was no time for a cause to exist in'. For response, see Chap. 6.
6. For replies to other arguments for an infinite past which have been offered by some scientists and refuted by others, see Loke (2017a, chapter 2).
7. See https://www.youtube.com/watch?v=0X6ism4-KKw; from 11 mins onwards.
8. Concerning the significance of the term 'experience', see the discussion on the static theory of time below.
9. I thank Tong Shih Ping for sketching this proof.
10. I thank Tim Maness for mentioning this point at the 2020 AAR conference.
11. Supposing that he/she is wholly focused on the process of successive addition rather than (say) daydreaming.
12. Concerning the objection that the person would instead experience infinite + finite=infinite, this is based on a misunderstanding of the argument; see the discussion on Morriston (2021) above.
13. I thank Richard Swinburne for raising this question.
14. I thank Anthony66 for suggesting this objection.
15. There are objections to such a claim; see the citation from Craig below.
16. For replies to other points Leon (2011) made, see Loke (2014a).

17. I thank Oliver Davies for raising this point.
18. I thank Don Page for raising this question.
19. This does not imply that the necessity of a particular causally necessary condition. It has been argued that there are counterexamples to transitivity of causation (McDonnell 2018). Those purported counterexamples do not affect the novel argument, because they are disanalogous. For example, there are cases whereby x causes $y_1$ causes z but $y_1$ is not required in the sense that $y_2$ can also cause z. Whereas my argument is saying that z depends on a cause regardless of whether it is $y_1$ or $y_2$. Some other purported counterexamples understood causes as causally sufficient conditions, whereas the concept of causation essential to my argument is causally necessary conditions.
20. I thank Mark Harris for raising this concern.
21. Schaffer did not think that his analogy can be used for causal series, but for grounding of 'source of reality', claiming that 'But a caused entity qua caused entity still has intrinsic reality unto itself. Caused entities do not inherit their reality from their causes. Indeed, a caused entity may also be fundamental, and thus ontologically subsistent in its own right' (ibid.). In reply, my argument here shows that entities in a causal series with beginning of existence are dependent on earlier entities just as debtors are dependent on others to give them money, and thus the analogy works. Trogdon (2018, p. 191) modified Shaffer's argument to argue for inheritance of causal capacity, but his argument does not show (as my argument does) that there is a First Cause with divine properties.
22. Cf. Morganti (2015), who objects to Shaffer by claiming that, rather than entities transferring reality to what they ground, it may be that reality emerges from grounding chains instead. However, money would not emerge when no entity has the capacity for explaining the beginning of existence of money!
23. I thank Brian Wong for raising this objection.
24. I thank Nick Morris for raising this objection.
25. I thank Alex Malpass for suggesting this objection.
26. This is similar to the impetus theory discussed in Feser (2013, pp. 249–250), in which an 'impetus' is imparted to an object by whatever initiated its inertial motion, and which continuously actualizes its potencies with respect to spatial location. Feser notes two problems with the impetus theory. First 'a finite object (such as the baseball of our example) can only have finite qualities. And yet, an impetus, in order to

have local motion ad infinitum as its effect, would at least in that respect be an infinite quality.' In reply, one can think of the effects the impetus cause on future events as a potential infinite, which is finite at any moment. The second problem is 'an impetus would continually be bringing about new effects and thus (as a finite cause) itself be undergoing change; and in that case, we have only pushed the problem back a stage, for we now need to ask what causes these changes in the impetus itself.' The answer is the previous state of the impetus.

27. One might ask whether there could be a causal loop with only one entity causes itself to exist with no beginning point in the loop. In reply, such a loop is incoherent: if there is no beginning point, nothing is actually brought into existence; that is, nothing is caused.

## Bibliography

Aguirre, Anthony. 2007. Eternal Inflation, Past and Future. https://arxiv.org/abs/0712.0571.

Almeida, Michael. 2018. *Cosmological Arguments*. Cambridge: Cambridge University Press.

An, Daniel, Krzysztof Meissner, Pawel Nurowski, and Roger Penrose. 2020. Apparent Evidence for Hawking Points in the CMB Sky. https://arxiv.org/abs/1808.01740.

Barrow, J., and F. Tipler. 1986. *The Anthropic Cosmological Principle*. Oxford: Oxford University Press.

Baum, L., and P. Frampton. 2007. Turnaround in Cyclic Cosmology. *Physical Review Letters* 98: 071301.

Bliss, Ricki. 2013. Viciousness and the Structure of Reality. *Philosophical Studies* 166 (2): 399–418.

Bojowald, M., G. Date, and G. Hossain. 2004. The Bianchi IX Model in Loop Quantum Cosmology. *Classical and Quantum Gravity* 21: 3541.

Bricmont, Jean. 2017. *Making Sense of Quantum Mechanics*. Cham: Springer Nature.

Bussey, Peter. 2013. God as First Cause – A Review of the Kalām Argument. *Science & Christian Belief* 25: 17–35.

Butterfield, Jeremy, and C. Isham. 1999. On the Emergence of Time in Quantum Gravity. In *The Arguments of Time*, ed. J. Butterfield. Oxford: Oxford University Press.

Cameron, Ross. 2008. Turtles All the Way Down: Regress, Priority and Fundamentality. *The Philosophical Quarterly* 58: 1–14.

———. 2018. Infinite Regress Arguments. In *The Stanford Encyclopedia of Philosophy*, ed. Edward Zalta. https://plato.stanford.edu/archives/fall2018/entries/infinite-regress/.

Carroll, Sean. 2016. *The Big Picture*. London: Oneworld.

Chan, Man Ho. 2019. Is the History of Our Universe Finite? *Theology and Science* 17 (2): 248–256.

Copan, Paul, and William Lane Craig, eds. 2017. *The Kalām Cosmological Argument. 2 Vols*. New York: Bloomsbury Academic.

Craig, William Lane. 2000a. *The Tensed Theory of Time: A Critical Examination*. Synthese Library 293. Dordrecht, the Netherlands: Kluwer Academic Publishers.

———. 2000b. *The Tenseless Theory of Time: A Critical Examination*. Synthese Library 294. Dordrecht, the Netherlands: Kluwer Academic Publishers.

———. 2018. The Kalām Cosmological Argument. In *Two Dozen (Or So) Arguments for God*, ed. Jerry Walls and Trent Dougherty. Oxford: Oxford University Press.

———. 2020. Explaining the Applicability of Mathematics. https://www.reasonablefaith.org/writings/question-answer/explaining-the-applicability-of-mathematics.

Craig, William Lane, and Sean Carroll. 2015. *God and Cosmology: The Existence of God in Light of Contemporary Cosmology*. Augsburg: Fortress Publishers.

Craig, William Lane, and James Sinclair. 2009. The Kalām Cosmological Argument. In *The Blackwell Companion to Natural Theology*, ed. William Lane Craig and J.P. Moreland. Chichester: Wiley-Blackwell.

———. 2012. On Non-Singular Space-times and the Beginning of the Universe. In *Scientific Approaches to the Philosophy of Religion*, ed. Yujin Nagasawa. New York: Palgrave Macmillan.

Curiel, Erik. 2019. Singularities and Black Holes. In *The Stanford Encyclopedia of Philosophy*, ed. Edward Zalta. https://plato.stanford.edu/archives/spr2019/entries/spacetime-singularities/.

Daintith, John, ed. 2009. *A Dictionary of Physics*. Oxford: Oxford University Press.

Dretske, Fred. 1965. Counting to Infinity. *Analysis* 25: 99–101.

Ellis, George. 2007. Issues in the Philosophy of Cosmology. In *Philosophy of Physics*, ed. J. Butterfield and J. Earman. Amsterdam: Elsevier.

Ellis, George, U. Kirchner, and W. Stoeger. 2004. Multiverses and Physical Cosmology. *Monthly Notices of the Royal Astronomical Society* 347: 921–936.

Feser, Edward. 2013. *Aristotle on Method and Metaphysics*. London: Palgrave.
Giberson, Karl, and Mariano Artigas. 2007. *Oracles of Science: Celebrity Scientists Versus God and Religion*. New York: Oxford University Press.
Gott, Richard I.I.I., and Li-Xin Li. 1998. Can the Universe Create Itself? *Physical Review D* 58: 023501–023501.
Hagar, Amit. 2014. *Discrete Or Continuous? The Quest for Fundamental Length in Modern Physics*. Cambridge: Cambridge University Press.
Halper, Phillip. 2021. The Kalām Cosmological Argument: Critiquing a Recent Defence. *Think* 20: 153–165.
Hartle, James, and Stephen Hawking. 1983. Wave Function of the Universe. *Physical Review D* 28: 2960–2975.
Hawking, Stephen. 1988. *A Brief History of Time*. London: Bantam.
———. 2018. *Brief Answers to the Big Questions*. New York: Bantam Books.
Huggett, Nick, and Christian Wüthrich. 2013. Emergent Spacetime and Empirical (in)coherence. *Studies in History and Philosophy of Modern Physics* 44: 276–285.
Hume, David. 1779/1993. Dialogues Concerning Natural Religion. In *Dialogues and Natural History of Religion*, ed. J.A.C. Gaskin. Oxford: Oxford University Press.
Jow, Dylan, and Douglas Scott. 2020. Re-evaluating Evidence for Hawking Points in the CMB. *Journal of Cosmology and Astroparticle Physics* 3. https://doi.org/10.1088/1475-7516/2020/03/021.
Koons, Robert. 2014. A New Kalām Argument: Revenge of the Grim Reaper. *Noûs* 48: 256–267.
Leon, Felipe. 2011. Moreland on the Impossibility of Traversing the Infinite: A Critique. *Philo* 14: 32–42.
Lewis, David. 1976. The Paradoxes of Time Travel. *American Philosophical Quarterly* 13: 145–152.
Linde, Andrei. 1994. The Self-Reproducing Inflationary Universe. *Scientific American* 271: 48–55.
Linford, Dan. 2020. The *Kalām* Cosmological Argument Meets the Mentaculus. *The British Journal for the Philosophy of Science*. https://doi.org/10.1093/bjps/axaa005.
Loke, Andrew. 2010. Divine omnipotence and moral perfection. *Religious Studies* 46: 525–538.
———. 2012. Is an Infinite Temporal Regress of Events Possible? *Think* 11: 105–122.
———. 2014a. A Modified Philosophical Argument for a Beginning of the Universe. *Think* 13: 71–83.

———. 2014b. No Heartbreak at Hilbert's Hotel: A Reply to Landon Hedrick. *Religious Studies* 50: 47–50.

———. 2016a. On Finitism and the Beginning of the Universe: A Reply to Stephen Puryear. *Australasian Journal of Philosophy* 94: 591–595.

———. 2016b. On Beginningless Past, Endless Future, God, and Singing Angels: An Assessment of the Morriston-Craig Dialogue. *Neue Zeitschrift für Systematische Theologie und Religionsphilosophie* 58: 57–66.

———. 2016c. On the Infinite God Objection: A Reply to Jacobus Erasmus and Anné Hendrik Verhoef. *Sophia* 55: 263–272.

———. 2017a. *God and Ultimate Origins: A Novel Cosmological Argument. Palgrave Frontiers in Philosophy of Religion Series.* Cham, Switzerland: Springer Nature.

———. 2017b. A Reply to Peter Lyth on Whether an Infinite Temporal Regress of Events Is Possible. *Think* 16: 77–81.

———. 2021a. The Kalām Cosmological Argument and divine omniscience: an evaluation of recent discussions in Sophia. *Sophia* 59: 651–656.

Luna, Laureano, and Jacobus Erasmus. 2020. A Philosophical Argument for the Beginning of Time. *Prolegomena* 19 (2): 161–176.

Malpass, Alex. 2019. *Successive Addition.* https://useofreason.wordpress.com/2019/02/21/successive-addition/. Accessed 14 April 2020.

———. 2021. All the Time in the World. *Mind*, fzaa086. https://doi.org/10.1093/mind/fzaa086.

McDonnell, N. 2018. Transitivity and Proportionality in Causation. *Synthese* 195: 1211–1229.

Moore, A.W. 2001. *The Infinite.* London: Routledge.

Moreland, J.P., and William Lane Craig. 2003. *Philosophical Foundations for a Christian Worldview.* Downers Grove: InterVarsity Press.

Morganti, M. 2015. Dependence, Justification and Explanation: Must Reality Be Well-Founded? *Erkenntnis* 80: 555–572.

Morriston, Wes. 2013. Doubts About the Kalām Argument. In *Debating Christian Theism*, ed. J.P. Moreland, Chad V. Meister, and Khaldoun A. Sweis. Oxford: Oxford University Press.

———. 2021. Infinity, Time, and Successive Addition. *Australasian Journal of Philosophy.* https://doi.org/10.1080/00048402.2020.1865426.

Nomura, Yasunori. 2012. Static Quantum Multiverse. *Physical Review* 86: 083505.

Oppy, Graham. 2006a. *Philosophical Perspectives on Infinity.* Cambridge: Cambridge University Press.

———. 2019a. The Universe Does Not Have a Cause. In *Contemporary Debates in Philosophy of Religion*, ed. Michael Peterson and Raymond Van Arragon. Chichester: Wiley Blackwell.

Penrose, Roger. 2010. *Cycles of Time: An Extraordinary New View of the Universe*. London: Bodley Head.

Pitts, Brian. 2008. Why the Big Bang Singularity Does Not Help the Kalām Cosmological Argument for Theism. *The British Journal for the Philosophy of Science* 59: 675–708.

Poplawski, N.J. 2010. Radial Motion into an Einstein-Rosen Bridge. *Physics Letters B* 687: 110–113.

Pruss, Alexander. 2018. *Infinity, Causation, and Paradox*. Oxford: Oxford University Press.

Rasmussen, Joshua. 2018. Review of God and Ultimate Origins. *European Journal for Philosophy of Religion* 10: 189–194.

Reichenbach, Bruce. 2021. Cosmological Argument. In *The Stanford Encyclopedia of Philosophy*, ed. Edward Zalta. https://plato.stanford.edu/archives/win2016/entries/cosmological-argument/.

Russell, Bertrand. 1969. *History of Western Philosophy*. London: Allen and Unwin.

Schaffer, Jonathan. 2016. Grounding on the Image of Causation. *Philosophical Studies* 173: 49–100.

Sorabji, R. 2006. *Time, Creation and the Continuum*. Ithaca, NY: Cornell University Press.

Steinhardt, P., and N. Turok. 2005. The Cyclic Model Simplified. *New Astronomy Reviews* 49: 43–57. Preprint: http://arxiv.org/abs/astro-ph/0404480.

Susskind, Leonard. 2012. Was There a Beginning?. https://arxiv.org/abs/1204.5385.

Tegmark, Max. 2015. Infinity Is a Beautiful Concept – And It's Ruining Physics. https://www.discovermagazine.com/the-sciences/infinity-is-a-beautiful-concept-and-its-ruining-physics.

Trogdon, Kelly. 2018. Inheritance Arguments for Fundamentality. In *Reality and Its Structure: Essays in Fundamentality*, ed. Ricki Bliss and Graham Priest. Oxford: Oxford University Press.

Van Bendegem, Jean Paul. 2011. The Possibility of Discrete Time. In *The Oxford Handbook of Philosophy of Time*, ed. Craig Callender. Oxford: Oxford University Press.

Veneziano, G., and M. Gasperini. 2003. The Pre Big Bang Scenario in String Cosmology. *Physics Reports* 373: 1–212.

Vilenkin, Alexander. 1982. Creation of Universes from Nothing. *Physics Letters B* 117: 25.

———. 2015. The Beginning of the Universe. *Inference: International Review of Science* 1 (4) https://inference-review.com/article/the-beginning-of-the-universe. Accessed 22 April 2015.

Wall, Aron. 2013a. The Generalized Second Law Implies a Quantum Singularity Theorem. *Classical and Quantum Gravity* 30 (16): 165003. Preprint: arXiv:1010.5513v4 [gr-qc].

———. 2013b. Corrigendum: The Generalized Second Law Implies a Quantum Singularity Theorem. *Classical and Quantum Gravity* 30 (19): 199501.

Wall, Aaron. 2014. Did the Universe Begin? IV: Quantum Eternity Theorem. http://www.wall.org/~aron/blog/did-the-universe-begin-iv-quantum-eternity-theorem/. Accessed 20 January 2017.

Waters, Ben. 2013. Methuselah's Diary and the Finitude of the Past. *Philosophia Christi* 15 (2): 463–469.

Zarepour, Mohammad. 2020. Infinite Magnitudes, Infinite Multitudes, and the Beginning of the Universe. *Australasian Journal of Philosophy*. https://doi.org/10.1080/00048402.2020.1795696.

# 6

# What the First Cause Is

## 6.1 Introduction

In the previous chapter, I have shown that there exists a First Cause of our existence. But what is this First Cause? Is it God, or part of the universe as postulated by Hawking (see below)? A number of formulations of the Cosmological Argument have arrived at the conclusion of an Ultimate Ground (Deng 2019) or a Necessary Being (e.g. Weaver 2016) without showing that the necessary being/ultimate ground is God. Others have claimed that naturalistic accounts of ultimate origin fare at least as well as theistic accounts (Oppy 2009, 2010, 2013a, b), and that

> whatever causal structure for the universe is supposed by the theist can be replicated by the naturalist ... Thus if the free action of God is supposed to be the indeterministic action of a necessary being, the naturalist is free to propose that the universe had an initial state which was itself necessary and indeterministically caused the organized cosmos we experience. (Pearce 2017; following Oppy 2013b)



A. Loke, *The Teleological and Kalam Cosmological Arguments Revisited*, Palgrave Frontiers in Philosophy of Religion, https://doi.org/10.1007/978-3-030-94403-2_6

I shall address these concerns by demonstrating that the First Cause of the universe has the properties of being beginningless, initially changeless (here, 'initial' refers to the first in the series of states [ordered causally], not first the series of changes/events/temporal series), transcendent, immaterial, has libertarian freedom, is enormously powerful and highly intelligent, and therefore worthy of being called the Creator of the Universe. The conclusion follows from premises 6–12 of the KCA-TA and is as follows:

6. Since the First Cause is the first, it is uncaused.

7. Since whatever begins to exist has a cause (Causal Principle), the First Cause is beginningless.

8. Since every change is an event which has a beginning as something/part of a thing gains or loses a property, and since the first change (= first event) does not begin uncaused (given the Causal Principle), the first change (= first event) is caused by a First Cause which is initially changeless. (From 5 and 7)

9. Since the First Cause is initially changeless, it is transcendent and immaterial (i.e. it is distinct from the material universe and is the cause of the universe).

10. In order to cause an event (Big Bang or whatever) from an initial changeless state, the First Cause must have

10.1. the capacity to be the originator of the event in a way that is undetermined by prior event, since the First Cause is the first, and

10.2. the capacity to prevent itself from changing, for otherwise the First Cause would not have been initially changeless and existing beginninglessly without the event/change.

10.1 and 10.2 imply that the First Cause has libertarian freedom.

11. In order to bring about the entire universe, the First Cause is enormously powerful.

12. (+ the Teleological Argument:) In order to bring about a universe with its fine-tuning and order, the First Cause is highly intelligent.

13. A First Cause that is uncaused, beginningless, initially changeless, transcendent, immaterial, has libertarian freedom, and is highly intelligent and enormously powerful is a Creator of the Universe.

14. Therefore, a Creator of the universe exists.

I shall now discuss each of the premises in turn. Among the important contributions of this chapter is a reply to the objections posed by the works of Stephen Hawking (including the objections found in his final book published in 2018), which are of great interest in philosophy of religion debates and science and religion dialogues, as well as the discussion of the relationship between the KCA, the doctrine of *creatio ex nihilo*, and various views and theories of time. The debate between the relational view of time and the substantival view of time continues, just as the debate between the dynamic theory of time and the static theory of time continues. It is beyond the scope of this book to settle the debate. Suffice to note that the KCA-TA defended in this book is compatible with any of these views and theories. I shall explain this point further in what follows, focusing on the relational view and the static theory (because these generate more issues for the KCA which need to be addressed) and commenting on the substantival view and dynamic theory whenever necessary.

## 6.2 The First Cause Is Uncaused, Beginningless, and Initially Changeless

Premise 6 'since the First Cause is the first, it is uncaused' follows by definition of the word 'first' and the word 'cause' and 'uncaused' as defined in Chap. 2 (see further, Chap. 8). Premise 7 follows from the Causal Principle 'whatever begins to exist has a cause' established in Chap. 3.

Concerning premise 8, a 'change' is understood as an event that has a beginning at the state of having gained or having lost a property. Thus, a beginningless change is impossible. Even if events are not discrete, they are still distinct, otherwise they would be changeless. Now it has been shown in Chap. 5 that an infinite regress of changes is not the case—and this is true regardless of what dimensions of time there are. Therefore, there is a first change. Given the Causal Principle and the fact that a change is something that has a beginning as explained in Chap. 5, the first change would have a cause. The cause (X) of the first change (Y) cannot have been caused by another cause (W), for otherwise W causing X would be a change that is prior to Y, in which case Y would not have been the first change. Therefore, given that an infinite regress of changes is not the case, there is a first change which is caused by an uncaused cause, and this uncaused cause would be the first in the series of causes. Now this First Cause cannot be a change prior to the first change (otherwise the first change wouldn't be the first!); thus, the First Cause must be changeless initially. This implies that the First Cause is not an event such as the Big Bang.

Quentin Smith (1996, p. 179) has raised an objection by claiming that only events are causes, and therefore there cannot be a cause for the first event. However, on the one hand, there has been no compelling argument offered to show that causes must be events; one can defend an alternative ontological analysis according to which causality does not have to be a relation between events; rather, the causes can be underlying substances such as agents (Craig 2000c; see the discussion on agent causation in Chap. 3 and below).[1] On the other hand, given the arguments that an infinite regress of events is impossible (Chap. 5) and the argument for the Causal Principle that whatever begins to exist has a cause (see Chap. 3), there must be a first event which is caused by a non-event (e.g. a substance). Given that an event is defined as a change and that the First Cause is initially changeless, the uncaused First Cause is not an event prior to another event. Rather, the First Cause was in an initially changeless state causally prior to bringing about an event, and gained a property (i.e. changed) as it brings about an event; that is, it changed simultaneously with the bringing about of a change. (Thus, there are two events X and Y which happened simultaneously: the change (X) to the First Cause as it brings about Y, and X is concomitant to Y.) This conclusion follows from the previous premises which have shown that the First Cause is beginningless whereas the first event has a beginning, which implies that the beginningless First Cause exists initially without the first event.

It might be objected that, while this view makes sense on a dynamic view of time (given which one can say that the first event 'comes into being'), it seems to be in conflict with the static [B-] theory of time given which the first event does not come into being but exists tenselessly at the first time $t_1$ alongside God (suppose God is the First Cause). If that is the case, how can God be initially changeless? How can God exist without the first event of the universe? (Craig 2000b, p. 221).

There are three possible responses to this question, and I shall explain that all three of them are defensible and any one of them would suffice to answer this question.

Concerning the first possibility, even if we assume the static theory of time, one can say that the timeline of events does not represent all of reality. There is an aspect of reality that is without any event or time, and this can be called the initial state of reality. An aspect of God's being exists in this eventless/changeless and timeless state ('outside of the time line')—that is one way to understand the First Cause being initially changeless and timeless—while another aspect of God's being exists within the timeline and causally interact with the events in the timeline (e.g. it is simultaneous with the first event and brought about the first event). Given the Causal Principle, the beginning of the first entity of the spacetime block would require a cause just like all the beginning of other later entities, and one could say that there is an aspect of God's being that exists at the first moment of time and simultaneously brought about the first entity of a block that exists at that time, even though there would be no earlier time at which that first entity did not exist. On this view, to say that the First Cause existed initially without the first event means that there is an aspect of reality (an aspect of the First Cause) which is outside of the timeline and is beginningless and without change/event. (One might worry that this response contradicts strict Classical Theism, which affirms the doctrine of divine simplicity; I shall explain below that Classical Theism is unwarranted.) We know that distinctions of changes exist in the present portion of reality, and if there is a prior aspect of reality in which such distinctions are absent, that would be changeless state, which I have shown the First Cause is initially in.

Craig objects that the above response concerning the creation of the first event reduces the doctrine of creation to tenseless ontological dependence and thereby emasculates *creatio ex nihilo* (Craig 2000b, p. 221). However, if *creatio ex nihilo* is understood as affirming that the universe has a beginning and does not have a material cause but has an efficient cause (God), then the above response does not contradict this. The difference between the above response and the view that God merely sustains the universe in being is that the latter view is compatible with the universe not having a beginning (i.e. being a spacetime block that is an actual infinite in earlier-than extension), but the above response denies this.

The second possibility is 'to posit a hyper-time in which God brings into being the whole four-dimensional block universe at a moment of hyper-time' (Craig 2000b, p. 221; Craig objects that affirming this view would be extravagant; however, my point here is that this possibility is not excluded by the supposed evidences for B-theory). Such a hyper-time would be an A-theoretical time, which implies that the A-theory is fundamentally correct. While this view is not what B-theorists usually affirm, it is consistent with the evidences (e.g. based on the theory of relativity) which B-theorists cite for their theory, for on this view it remains the case that our spacetime is a four-dimensional block and that all the moments within the block are equally real relative to one another. If valid,[2] the evidences they cite for their theory only imply that an event in the spacetime block does not come into being and go out of being relative to another earlier or later event in the block, but it does not imply that the block itself never comes into being (although this is usually assumed).

The third possibility is to affirm that God initially exists in a form of time prior to the first moment of the block and is thus earlier than it, just like hydrogen and oxygen exist prior to water inside the block and cause it. Just as water is causally dependent on hydrogen and oxygen coming together to form it, the block is dependent on God (the First Cause). This view is proposed by the so-called Oxford School, which includes John Lucas, Richard Swinburne and Alan Padgett. Padgett writes that 'God is in himself temporal in some ways' (1992, p. 126); his view is 'in harmony with the Biblical witness about God and his eternity' (ibid.), which implies being without beginning, and that God is not in any measured time (ibid.) because He is not subjected to the law-like regularities of nature which allow for the periodic processes that underlie isochronic clocks and hence are essential to the measurement of time (p. 127). Applying Padgett's view to the state before the first event of the spacetime block, this would imply that God is in the dimension of time which is not divisible by periodic processes involving events; it is non-metric and unlimited in the earlier-than direction. This unlimited initial state itself exists an earlier-than direction relative to the first event, and thus is prior to the first event in that sense. This view assumes a substantival view of time which affirms that time can exist independently of change. Given this view, the First Cause could have been in an initially changeless state with an actual infinite past extension (i.e. without an 'edge' in the earlier-than direction), causally and temporally antecedent to the first change. In this way, God (the First Cause) could have existed beginninglessly before creation in an undifferentiated, non-metric time and God would not be dependent on such a time because such a time would be a property of God (Padgett 1992). This view does not face the problems with postulating an actual infinite number of earlier durations (see Chap. 5), since the earlier extension of time is undifferentiated. The KCA does not rule out an infinite past if this is understood according to the substantival view of time and that the earlier extension of time is undifferentiated; it merely rules out an infinite regress of causes and changes/events. Given the arguments in the previous chapter there must still be a first event/change even if substantival view of time is correct. Craig and Sinclair (2009, p. 192n.100) note that 'the Kalām argument strictly demonstrates only that metric time had a beginning. Perhaps the cause exists changelessly in an undifferentiated time in which temporal intervals cannot be distinguished.' On this view, the First Cause existed literally and eventlessly before creation, but there was no moment, say, 1 hour or 1 million years before creation (ibid.). Even though, according to the substantival view, in the absence of change, time would still exist as a substance ('the container'), in the absence of change there would be no metric. That is why the Oxford School would say that God exists in unmetricated time prior to His free act of creating the universe (Swinburne 1993, pp. 208–9). With the act of creation 'God freely creates a universe with intrinsic laws of nature that serve as a metric for the physical time of that universe' (Mullins 2015, p. 36).

By contrast, on a relational view of time which defines time as a series of changes/events ordered by 'earlier-than' and 'later-than' relations, an initially changeless First Cause would be initially timeless and hence does not exist 'earlier' than the first event.[3] While the relational view of time is inconsistent with the view of the Oxford School, it is consistent with the Hybrid view according to which God (the First Cause) 'exists timelessly sans creation and temporally at and subsequent to the moment of creation' (Craig and Sinclair

2009, p. 189). The coherence of this Hybrid view has been defended in previous publications by William Lane Craig, and constitutes a major contribution to the discussion on the relationship between God and time. It should be noted that what Craig means is that the First Cause is timeless *without* agent-causing the first event at $t_1$, and temporal *with* agent-causing the first event at $t_1$. There is no contradiction with this view since 'timeless' and 'temporal' have different references on this view (see further, below). According to this view, there is a first moment and a beginning where God's existence in time is concerned. However, this does not imply that 'God's existence has a beginning' simpliciter (contra Leftow 2005, p. 66; 2010, p. 281), because God's existence is not limited to existence in time only; rather, God exists timelessly 'sans creation'. God's existence per se does not have a temporal boundary, since He has a timeless phase which is absent from (say) Oppy's view of the initial state of the universe (see Chap. 3). While Craig has defended this Hybrid view on the assumption of dynamic and relational view of time, I have tried to show that it can also be defended on the assumption of static and substantival view of time as well (see, for example, the discussion on the 'first possibility' above).

One might object by claiming that timeless means existing for zero seconds, which would imply non-existence. This is a misconception, for timeless does not mean existing for zero seconds. A second is a measurement of the temporal dimension; it has a beginning and is defined as a sixtieth of a minute of time, and 'zero seconds' by definition implies being shorter than one second within the measurement of the temporal series. Whereas according to the Hybrid view the First Cause is without beginning and initially changeless and timeless, that is, existing without the temporal dimension initially; therefore, it is not appropriate to use 'zero seconds' to refer to it.

It is also inappropriate to think of this view as involving some 'causal point' prior to the beginning of time,[4] because a point assumes a dimension whereas there was no dimension and hence no point at the initially changeless state which was beginningless and does not require a cause. It is wrong to think that something is changeless if and only if it remains unchanging over an extended time interval. Changeless simply means the absence of change, and since change requires extended time interval, the absence of time interval would also imply the absence of change. Thus, changelessness does not require a time interval; rather, changelessness is also compatible with timelessness.

In his earlier works, Stephen Hawking proposed that the initial state of the universe consisted of a timeless (no boundary) state (Hartle and Hawking 1983; Hawking 1988). This initial state can be understood as a beginningless impersonal First Cause from which all things came, and which avoids the need for a Creator. (It is similar to Craig's hybrid view explained above, except that instead of God it is the universe itself which has a first moment where its existence in time is concerned and yet the universe is beginningless because it does not have a temporal boundary since it has a timeless phase.) Others have offered timeless interpretations of quantum gravity (e.g. Barbour 1999; Deutsch 1997; Anderson 2012) and/or claimed that time and space could have emerged from a timeless and spaceless natural state (e.g. Arkani-Hamed and Trnka 2014; Oriti 2014; Cao et al. 2017; Carroll 2019; Huggett and Wuthrich forthcoming).

However, none of the above can be regarded as established given the lack of a well-established theory of quantum gravity and the problem of underdetermination of scientific theories noted in Chap. 1. Thus, it is not the case that the above scientists have shown that it is possible for the universe to be initially changeless/timeless. On the contrary, Oriti (2014, p. 187) notes 'the ongoing, tentative work of theoretical physicists on models that, most likely, will turn out to be incorrect or only partially understood in the future').

Moreover, most (if not all)[5] of the above concern the measurement of time and not time itself. They do not address the issue of a beginningless and initially changeless/eventless state as it is defined in the context of the Kalām Cosmological Argument (KCA). In other words, they are actually addressing a different problem which does not rebut the conclusion that follows from the premises of the KCA.

For example, regarding the Wheeler–DeWitt equation used by Barbour, Hawking et al., physicist Aron Wall argues that the Wheeler–DeWitt equation does not imply timelessness; rather, it concerns the measurement of time. Wall (2014) writes: 'when we say that the wave-function doesn't change with time, what this really means is that the choice of time coordinate is arbitrary', not that time is an illusion or that it does not exist. "Time' needs to be measured relative to some physical clock. There is no absolute '$t$' coordinate relative to which everything else moves' (ibid.).

Regarding Hawking's use of the so-called imaginary time, while imaginary numbers are used to represent the time coordinate in relativity theory, this does not imply that the mathematical concept has a counterpart in physical reality. As Craig (1990) observes citing Eddington, the use of imaginary numbers for the time coordinate 'can scarcely be regarded as more than an analytical device' (Eddington [1920], p. 48). Imaginary time was merely an illustrative tool which 'certainly do[es] not correspond to any physical reality' (Eddington [1920], p. 181). It has no concrete meaning (similar to an imaginary number such as $\sqrt{-1}$ which has no concrete meaning) and therefore merely used by Hawking as a mathematical trick for avoiding a singularity. As Erasmus (2018, p. 146) explains,

> Wick rotation 'is little more than a convenient mathematical trick' (Isham 1997, p. 399) and imaginary time 'is introduced only for computational convenience' (Vilenkin 2006, p. 182). Consequently, we should not interpret the tunnelling and no-boundary proposals realistically and, thus, the quantum creation hypothesis cannot be a true description of reality.

Concerning using 'imaginary time' to 'change time into (timeless) space', Barrow observes that

> physicists have often carried out this 'change time into space' procedure as a useful trick for doing certain problems in ordinary quantum mechanics, although they did not imagine that time was really like space. At the end of the calculation, they just swop [sic] back into the usual interpretation of there being one dimension of time and three … dimensions of … space. (Barrow 1991, pp. 66–7)

However, in the Hartle–Hawking model, 'Hawking simply declines to re-convert to real numbers. If we do, then the singularity re-appears' (Craig 2000c, p. 228). Since the Hartle–Hawking model does not convert imaginary numbers (which are used instrumentally rather than realistically as explained above) back to real numbers, his model should be understood instrumentally rather than realistically; that is, it does not correspond to any concrete reality.

Indeed, Hawking himself confesses, 'I … am a positivist who believes that physical theories are just mathematical models we construct, and that it is meaningless to ask if they correspond to reality, just whether they predict observations' (Hawking 1997, p. 169). Since the Hartle–Hawking model is intended to be understood in an anti-realistic manner, the model does not intend to describe what reality is or what reality possibly is—and indeed the model cannot do so because, as explained above, imaginary time does not correspond to physical reality. In that case, Hawking's model would not achieve Hawking's intended purpose of justifying the claim that—in reality—'the beginning of the universe … doesn't need to be set in motion by some god' (Hawking and Mlodinow 2010, pp. 134–135). Moreover, it would not rebut the KCA which Craig, myself, and others have presented, because the reasons we have offered in support for the cosmological argument imply that the conclusion of the argument (i.e. there is a Creator of the universe) should be taken in a realist manner.

Additionally, Hawking's proposal ignores the 'zero-point energy' which entails that the initial state is metastable (Gott and Li 1998, p. 38). Craig (2018, p. 401) observes that on Hawking's model, the initial state of the universe 'cannot exist literally timelessly, akin to the way in which philosophers consider abstract objects like numbers to be timeless or theologians take God to be timeless. For this region is in a state of constant flux, which, given the Indiscernibility of Identicals, is sufficient for time.' Boddy et al. (2016) note that vacuum fluctuations are a feature of all quantum systems which ultimately arise as a consequence of the Heisenberg uncertainty principle; even though such fluctuations are not regarded as dynamical because they exist even in 'stationary states'; nevertheless, they give rise to phenomena such as the Lamb shift or Casimir effect. (It should also be noted that the so-called stationary state is called stationary because the probability density does not depend on time; nevertheless, the wavefunction itself is not stationary but continually changes.) What this implies is that there is gaining/losing of properties, which is how change is defined in the context of KCA. Thus, quantum system is not changeless/eventless/timeless as these terms are defined in the context of KCA. To rebut the KCA, the objector has to rebut its premises or its validity; the objector should not dodge the argument by defining the key terms such as 'change/time' and 'changelessness/timelessness' differently. (As explained in Chap. 1, to object to an argument by using an alternative definition would be to miss the point of the argument.)

One might speculate that perhaps there is a spaceless void which has been generating bubbles of universes by quantum fluctuations since eternity. This speculation implies that there has been an actual infinite number of changes (each generation is a change), but I have shown in Chap. 5 that an infinite regress of changes is not possible; thus, this speculation is refuted. (Even if Hawking is able to modify his model such that there is no infinite regress of changes and that there is a first change, that is, a first event, his model would still fail to explain the beginning of the first event because, as explained in Sect. 6.4, the first event must have been brought about by an initially changeless First Cause with libertarian freedom, that is, a Creator.)

In his final book *Brief Answers to the Big Questions*, published posthumously after his death, Hawking tried to explain why he thought that 'the simplest explanation is that there is no God. No one created the universe and no one directs our fate. This leads me to a profound realisation: there is probably no heaven and afterlife either' (Hawking 2018, p. 38).

He made the astonishing claim that 'the laws of nature itself tell us that not only could the universe have popped into existence without any assistance, like a proton, and have required nothing in terms of energy, but also that it is possible that nothing caused the Big Bang. Nothing' (ibid., p. 35).

What made Hawking thought that the laws of nature tell us that it is possible that nothing caused the Big Bang? He began by claiming that 'When the Big Bang produced a massive amount of positive energy, it simultaneously produced the same amount of negative energy. In this way, the positive and the negative add up to zero, always. It's another law of nature' (ibid., p. 32).

However, the claim that the positive and negative energy add up to zero does not imply that the positive and negative energy began to exist uncaused. As noted in Chap. 2, one *can* still ask what made the energy and the laws of nature to be the way they are (indeed, given the Causal Principle defended in Chap. 3, one *should* still ask this question; see below).

Hawking also claimed that at the subatomic level 'conjuring something out of nothing is possible. At least, for a short while. That's because, at this scale, particles such as protons behave according to the laws of nature we call quantum mechanics. And they really can appear at random, stick around for a while and then vanish again, to reappear somewhere else' (Hawking 2018, p. 33).

However, he failed to mention that at the subatomic level quantum particles do not come into existence from absolutely nothing; rather, as noted in Chap. 2, quantum particles are manifestations of pre-existent quantum fields which act according to pre-existent quantum laws.

Hawking seemed to have anticipated the above problems when he asked, 'but of course the critical question is raised again: did God create the quantum laws that allowed the Big Bang to occur?' (Hawking 2018, p. 34). Hawking (2018, p. 37) goes on to say:

> As we travel back in time towards the moment of the Big Bang, the universe gets smaller and smaller and smaller, until it finally comes to a point where the whole universe is a space so small that it is in effect a single infinitesimally small, infinitesimally dense black hole. And just as with modern-day black holes, floating around in space, the laws of nature dictate something quite extraordinary. They tell us that here too time itself must come to a stop.

One might ask where the black hole and the laws of nature came from. Hawking went on to claim that there cannot be a Creator who made these, because

> You can't get to a time before the Big Bang because there was no time before the Big Bang. We have finally found something that doesn't have a cause, because there was no time for a cause to exist in. For me this means

> that there is no possibility of a creator, because there is no time for a creator to have existed in. (Ibid., p. 38)

In response, first, even if there is no physical time before the universe, this does not imply that there is no metaphysical time before the universe in which the Creator could have existed in. As noted earlier, my argument is consistent with a substantival view of time according to which God exists before creation in an undifferentiated, non-metric time, causally and temporally prior to the first event. Concerning the advantage of this view (which is also known as 'relative timelessness'), Craig (2011) notes that it may be helpful for those people who stumble at the idea of God's creating the universe (or the Big Bang) because they assume (unjustifiably in Craig's view) that causes must be prior to their effects in time, and there is no time prior to the Big Bang. Craig replies:

> I'm inclined to say, with most philosophers, I think, that causes need not exist temporally prior to their effects. But for those who are hung up on this difficulty, relative timelessness provides a neat way out: God does exist temporally prior to causing the Big Bang—not in physical time, to be sure, but in His own time, the time in which God Himself endures. (Ibid.)

Second, even if one rejects the above response because one rejects the substantival view of time and embraces a relational view of time instead, there is an alternative response which works on a relational view of time. This alternative response begins by questioning the assumption that underlies Hawking's claim 'there is no time for a creator to have existed in.' One can ask, 'why does God need to exist in time?' Hawking's statement begs the question against a transcendent Timeless Creator who can exist outside of time initially. Earlier on, Hawking (2018, p. 34) attempts to provide a justification for his assumption by claiming that 'our everyday experience makes us think that everything that happens must be caused by something that occurred earlier in time'. However, this claim does not provide adequate justification for the assumption that a cause must be in such a temporal relation with its effect. As Craig argues, the notion that causes always stand in temporal relations with their effects can be treated merely an accidental generalization of our daily experiences, 'akin to Human beings have always lived on the Earth, which was true until 1968. There does not seem to be anything inherently temporal about a causal relationship' (Craig and Sinclair 2009, pp. 188–9). Likewise, Reichenbach (2021) argues that one need not require that causation embody the Humean condition of temporal priority, but may treat causation conditionally or as a relation of production. Hawking (2018, p. 38) also argues: 'time didn't exist before the Big Bang so there is no time for God to make the universe in.' However, God does not need to make the universe in time. Rather, God can be conceived of as being timeless without the universe and in time with the universe, and He brought about the universe together with time (Craig and Sinclair 2009).

Hawking might object that causes only exist within a time context, but there is no time context prior to the Big Bang.

In reply, one should distinguish between the label with the entity labelled. The entity which we call the First Cause is labelled as a 'cause' because it brought about the first event, but this does not mean that the entity cannot have existed in an initially changeless state without bringing about the first event, and entered into time as it brought about the first event. Thus, the Cause of the universe can be initially timeless, and in that initially timeless state it has the capacity (libertarian freedom) to bring about the first event in time (see further, Sect. 6.4.3).

On the one hand, we must be careful not to beg the question against the existence of such an initially timeless Cause, one that is causally but not temporally prior to the universe. On the other hand, a Modus Tollens argument has been offered for the Causal Principle 'whatever begins to exist has a cause' in Chap. 3, and this argument implies that the Causal Principle would hold regardless of whether time exists before the universe began. Given Hawking's claim that there was no time before the Big Bang, this implies that the universe has an (initially) timeless Cause.

Following Morriston (2002b, p. 240), Hawking might object by claiming that his principle 'everything that happens must be caused by something that occurred *earlier* in time' seems to enjoy the same empirical support as the Causal Principle 'everything that begins to exist has a cause', so why does one reject his principle as an accidental generalization while accepting the Causal Principle?

Two points may be said in response. First, Craig explains that 'the univocal concept of 'cause' is the concept of something which brings its effects, and whether this involves causes standing in temporal relations is an incidental question just as whether it involves transformation of already existing materials or creation out of nothing is an incidental question' (Craig and Sinclair 2009, pp. 188–9, 195). Second, Hawking's principle is based on *inductive* generalization of 'our everyday experience' (Hawking 2018, p. 34), and inductive generalizations are susceptible to the fallacy of accidental generalization. Whereas the Modus Tollens argument defended in Chap. 3 is a *deductive* argument and its premises are not based on inductive generalization but are based on conceptual analysis and denying a particular consequent. Hence, it is not susceptible to the fallacy of accidental generalization.

## 6.3 Transcendent and Immaterial

By transcendent I mean 'beyond or above the range of normal or physical human experience' (*Oxford English Dictionary*). By immaterial I mean fundamentally unlike matter-energy as we know it. (One might imagine a First Cause having spatial extension but is initially changeless; however, this would still be different from matter-energy as we know it, which is constantly changing, as explained below.)

Now is has been established previously that the First Cause is initially changeless. Such a First Cause would be beyond the range of normal human experiences of physical reality which is characterized by change, and hence such a First Cause would be transcendent.

Moreover, it would also be distinct from the physical universe which is constantly changing. For according to quantum field theory, the universe is a continuous fluctuating field. Additionally, as noted previously, according to quantum physics, physical entities constantly fluctuate at the quantum level as described by the Heisenberg uncertainty principle (Boddy et al. 2016).

By contrast, the First Cause is not a series of changes/events describable by physical laws; rather, it is initially changeless (and beginningless) and brought about the first event and these physical laws. To insist on calling such a First Cause as a *physical* (or natural) entity would be to use the word '*physical*' to refer to something *very different from what physics* tells us about the physical world, which is inappropriate. Thus, it is more appropriate to call this entity non-physical or immaterial.

Moreover, physical entities do not have 'the capacity to be the originator of an event in a way that is un-determined by prior event, and the capacity to prevent itself from changing' which a First Cause must have, as explained below.

One might object that he/she cannot conceive of a First Cause that is immaterial, spaceless, and timeless, that is, something that has no spatial and no temporal extension, which seems to be non-existence. Three points may be said in response.

First, the lack of extension does not imply non-existence. The key issue is how existence should be understood. While Aristotelian substantivalism invokes the maxim 'to exist is to exist in space and time' (Earman 1995, p. 28), the problem with this view is that space and time are not themselves located in space and time (Moreland and Craig 2003, p. 189). Others may think that to exist is to be physical, but the problem with this view is that disembodied existence is surely conceivable, and it begs the question against an immaterial God (ibid., p. 190). Existence is better defined as 'either the belonging of some property or the being belonged to by a property' (ibid., p. 191). Moreland and Craig (ibid.) explain:

> Things that exist have properties. When something such as Zeus fails to exist, there is no object Zeus that actually has properties. Since unicorns could have existed, this means that the property of being a unicorn could have belonged to something. It would also account for existence itself existing because the belonging-to (exemplification, predication) relation is itself exemplified (a nonfictional, real tiger named Tony and the property of being a tiger both enter into this belonging relation) and the belonging-to relation exemplifies other features (e.g., it has the property of being a relation that belongs to it). (Ibid., p. 191)

Second, one can conceive of immateriality, spaceless, and timeless as the negations of materiality, space, and time. The negation of a meaningful term is meaningful. Materiality is meaningful. Therefore, the negation of materiality is meaningful, and that is what immateriality means. Likewise, space and time are meaningful terms. Therefore, the negation of space and the negation of time are meaningful terms, and those are what spaceless and timeless mean.

One might object that either something is extended or not extended, and if it is not extended it is a point. However, this reasoning neglected the possibility of spacelessness. A point is something in space, whereas spacelessness is not a point in space. Likewise, timelessness is not a point in time. Rather, a timeless and spaceless First Cause would be something that is not in a temporal or spatial dimension and does not have temporal and spatial extension, and it is not non-existence because it has properties, such as the property of causal power which brought about the first event. Having causal power means having the capacity to bring about something; it does not mean/imply/require having spatial or temporal extension.

Third, it has been explained earlier that my argument is consistent with an alternative substantival view of time, according to which the First Cause exists before creation in an initially changeless state in an undifferentiated, non-metric time. According to this view, the First Cause may be conceived of as being temporally (and perhaps also spatially) extended, thus resolving the difficulty. Even though the First Cause may be conceived as being extended in this sense, it remains the case that the First Cause should not be regarded as the universe (understood as the totality of physical reality) or as a part of the universe, for it remains the case that the First Cause is initially changeless whereas physical things are in constant change, as explained above. Moreover, as argued below, in order to bring about the first event from an initially changeless and beginningless state, the First Cause must have libertarian freedom, which is characteristic of a personal Creator rather than an impersonal physical reality behaving in accordance with natural laws.

## 6.4 The First Cause Has Libertarian Freedom

### 6.4.1 How Could the First Cause Bring about the First Event from an Initially Changeless State

It has been explained above that the First Cause was beginningless and initially changeless; that is, it was in a state where it was not gaining or losing any property. One should ask how such a First Cause could bring about the first event from an initially changeless state.

It should be noted that there is a distinction between 'not' and 'cannot'; initially changeless does not mean 'cannot change'; rather, it means 'not-changing initially'. When the First Cause brings about a change, that is, an event, the First Cause itself would undergo a change, that is, a change from 'existing without the event' to 'existing with the event'. But how could that happen?

Could the First Cause be initially changeless due to necessity and initiated the first change out of necessity, that is, in a deterministic, fixed, law-like way? If that were the case, the necessity that initiated the first change would have to overcome the necessity that imposed the initial changelessness, and if it can do so, it would have done so necessarily and the First Cause would not have been initially changeless and the first change would have been coexistent with the First Cause. But this cannot be the case because, as shown in previous sections of this chapter, the First Cause is initially changeless and the first change has a beginning whereas the First Cause does not; thus, they cannot be coexistent. Thus, it cannot be the case that the First Cause was initially changeless due to necessity and initiated the first change out of necessity.

Could the First Cause be a quantum system which initiated the first change contingently? This would be similar to Bohr's interpretation of quantum fluctuation according to which, although the quantum field is a necessary condition, the fluctuation of the field happened indeterministically. Oppy (2009, Footnote 8) claims that there is no relevant difference between appealing to indeterminism in physical systems and non-deterministic agent causation in this case. As an example, consider the following scenario:

> Suppose the laws of nature are such that a 'primeval atom' with no internal structure might decay, generating a Big Bang and the universe as we know it. Before it did decay nothing happened. We may suppose that the laws of nature can be formulated to describe this primeval atom as having existed for an infinite time with an unchanging infinitesimal probability of decay per second.[6]

(In this scenario, infinite time without anything happening should be understood in accordance with a substantival view of time which postulates that time can exist without change.)

In reply, on the one hand, while many types of events have been claimed to be subjected to quantum indeterminacy (e.g. radioactive decay), it is not true that quantum physics has proven that an event can begin to exist indeterministically and contingently, given the viability of deterministic interpretations of quantum mechanics such as Bohm's pilot-wave theory (see Chap. 2) and the possibility of other deterministic theories. On the other hand, a quantum system is constantly changing as explained previously, whereas the First Cause is initially changeless, as argued previously. Therefore, a quantum system cannot be the First Cause. (To elaborate, I shall explain below that the First Cause must not only have the capacity for initiating the first event, but also the capacity for preventing itself from changing. In the case of a quantum system, there is no such preventive condition; that is why fluctuations are constantly happening and therefore a quantum system cannot be in a state which is beginningless and initially changeless.)

Concerning the scenario mentioned above, the postulation 'with an unchanging infinitesimal probability of decay per second' is incoherent, since if (according to the scenario) 'nothing happened' in that state before decay, then there would be no measure of time and hence no 'second'. Additionally, while it has been argued that there are instances of time-delayed causation which indicate that not all instances of causation are simultaneous (Grünbaum 1994), in no instance is there a delay of infinite time as the scenario postulates.

On the contrary, Aguirre and Kehayias note (2013): 'It is very difficult to devise a system—especially a quantum one—that does nothing "forever", then evolves. A truly stationary or periodic quantum state, which would last forever, would never evolve, whereas one with any instability will not endure for an indefinite time.' (Even though Aguirre and Kehayias are arguing against a particular model, namely, the Emergent Universe Scenario, the point they are making is generalizable to those models that postulate something doing nothing 'forever'. Halper [2021, p. 160] notes that Aguirre has argued elsewhere that the universe may be eternal into the past [Aguirre 2007], but in that model, namely, the Eternal Inflation Model, it is not the case that something 'does nothing forever'. Rather, that Eternal Inflation Model affirms a beginningless and continuous changing scenario, that is, an inflation that continues forever globally. This implies an actual infinite regress of changes and is refuted by the arguments in Chap. 5.)

Concerning those models in which something 'does nothing forever', Chan (2019, p. 251) explains the problem is that

> In a stable state, the 'decay life time' would be infinite. Without any external causes, this state would exist forever. However, in an unstable state, the initial state would change to other state in a finite time and the 'decay life time' is finite ... If the initial state of our universe is a stable state, no Big Bang would occur because this state would exist forever without the Big Bang. Since we have the Big Bang based on observations, our initial state must not be a stable state. If the initial state is an unstable state, Big Bang would occur but the time for this initial state must be finite. This implies that a beginning must exist in the initial state because of its finite life time.

One might ask, if time is composed of chronons with a smallest duration (say) of Planck time dimension—an extended simple—would it be the case that particles are changeless within that dimension, and if that is the case would that not imply that they are changeless and then changed with the next duration? This case however is disanalogous to the First Cause because the chronon has a beginning and these particles within the chronon are caused to change with the next duration by prior events or things and thus have prior causes, whereas the First Cause is beginningless and has no prior cause. Likewise, quantum states transition through a zero point is from an event to another event, it is not the same as initiating the first event from a beginningless and initially changeless (i.e. eventless) state.

The objector might ask whether, even though the First Cause is not a quantum system, could there be some other form of impersonal entity (one might call it an 'Initial Natural Thing', Oppy 2019b, p. 229) which exists necessarily, beginninglessly, and initial changelessly and initially timelessly as the First Cause, from which the first event indeterministically arose. In this case, the first event could have begun to exist without sufficient condition, that is, have a probabilistic cause (Rasmussen and Leon 2018, p. 64; Pearce 2017; following Oppy 2013b).[7] On this view, the first event is explained by the initial state of the impersonal entity which exists necessarily and which follows probabilistic natural laws.

In reply, first, I have argued previously that an initially changeless First Cause would be immaterial and thus not describable by science, whereas the 'Initial Natural Thing' is supposed to be natural and thus describable by science. There is no scientific basis for such a natural thing. It is science fiction.[8]

More seriously, the objector's postulation is still plagued by the problem similar to what physicists Aguirre, Kehayias and Chan noted above. Even though their point concerns infinite earlier durations rather than timelessness, nevertheless both infinite earlier durations and timelessness share a point of commonality, namely, both are beginningless, that is, not having any limit in the earlier than direction. As I shall explain further below, that is what is relevant for my argument, given which their point is relevant for illustrating a problem which I shall go on and develop into an argument below. My argument is not dependent on the current state of cosmology but is based on the analysis of the necessary conditions for an event to begin from a beginningless and initially changeless First Cause.

To elaborate, the beginning of the first event would imply a change to the First Cause, as it brings about and exists in a new (causal) relation with the first event. If the First Cause is an impersonal entity and the first event arose indeterministically from it (or if the First Cause is a system of tension of opposites)[9], that is, if there is a (non-epistemic) probability (between 0 and 1) of the first event arising from the initial state of such an entity, this would imply that the initial state of such an entity is unstable. That is, it has a disposition for changing with the beginning of the first event. An impersonal entity would not be able to 'hold back' its disposition; this implies that it would exist in the initial state for only a finite duration, rather than beginninglessly and (initially) changelessly/timelessly. This is a problem because the premises of the KCA have shown that the First Cause exists beginninglessly and (initially) changelessly/timelessly. Thus, the impersonal unstable entity cannot in fact be the First Cause, contrary to supposition.

One might ask, 'suppose the indeterministic first cause has a 50% probability of bringing about the first event. Would it not be the case that in half of the possible worlds, the first cause would never change and exists in a timeless state?'

In reply, we know that the First Cause *did* change in the actual world to bring about the first event of our world. The point about impersonal first cause being timeless in possible worlds is irrelevant because the KCA only needs to prove that there is a personal First Cause in the actual world by ruling out the First Cause being impersonal in the actual world. Moreover, in order for such an impersonal first cause to remain unchanging beginninglessly in some possible worlds, it would have to be unlimitedly stable, which means it would not have been able to change and bring about the first event in the actual world. Thus, such an impersonal first cause cannot be the cause of first effect in the actual world; this implies that it cannot in fact be the First Cause, contrary to supposition.

It has also be explained previously that it is not the case that the First Cause initiated the first change out of necessity, that is, in a deterministic, fixed, law-like way with a probability of 1, for in that case the First Cause would not have been initially changeless and the first change would have been coexistent with the First Cause. On the other hand, if the First Cause exist beginninglessly and changelessly and is impersonal, then it would be unlimitedly stable. There would be no likelihood/propensity/tendency/disposition for change. In other words, the probability of the first event would be 0, which means it would not have happened. (I have argued previously that quantum systems are not changeless; my point here is that, *even if* a quantum system is initially changeless, it would not change because in that case it would be eternally stable. I have also argued in Chap. 3 that it is not the case that the first event began uncaused.)

The only solution to the above problem is a beginningless First Cause with libertarian freedom, that is, a personal agent with control over its action and hence having freedom to change from a beginningless and initially changeless state. The objector might ask, 'what is the probability of such a First Cause bringing about the first event?' In reply, an agent free choice is evidently different from (say) coin throwing where there is some definite objective probability of landing heads; it is open to proponents of agent causation to deny that agent acts have objective probabilities (Buchak 2013). It has been argued that control act is not a chance event (Lowe 2008, p. 195). While others have argued that a finite, conditioned agents such as mere humans are often affected by volitional tendencies and preferences such that their free action is characterized by objective probabilities (O'Connor 2016), there is no good reason to think that the First Cause of the universe would be subjected to such limitations and conditioning. On the contrary, the foregoing discussion indicates that there is a First Cause with such absolute control that the first event is not a probabilistic or deterministic event. In other words, the First Cause is a personal agent with the power to control itself by having the following two capacities:

1. The capacity to initiate the first change/event, for the first change cannot be caused by another entity since the First Cause is the First.
2. The capacity to prevent itself from changing initially and hence maintain its stability in the initially changeless state, that is, the capacity to prevent the capacity to initiate the first change from initiating it initially, for otherwise the First Cause would not have been initially changeless and existing beginninglessly without the first change. (The capacity to control itself and prevent itself from changing differentiates indeterministic libertarian freedom from indeterministic quantum system [suppose for the sake of the argument that quantum physics is truly indeterministic; I offered an argument against this in Sect. 3.3]; the latter lack this capacity and hence is constantly changing, although it is still indeterministic in the sense that the results can be different even though the prior condition is the same.)

As I shall explain further below, having the above two capacities implies that the First Cause has libertarian freedom, and hence is a personal agent. The causation of the first event is therefore due to freedom rather than the result of deterministic causation describable by an impersonal law of nature. Moreover, it has been noted that the indeterministic theories of freedom which have been offered fall into three main groups: non-causal theories, event-causal theories, and agent causal theories (Clarke and Capes 2013). Now the cause of the first event cannot be a prior event (since the first event is the first), and thus event-causal theories of libertarian freedom are not relevant here. Non-causal theories are also not relevant, given the causal principle defended in this book. The only relevant theory of libertarian freedom is agent causal theory which affirms that the agent is the ultimate source of the free event. Moreland (2017, p. 302) notes that 'advocates of libertarian agency employ a form of personal explanation that stands in contrast to a covering law model'.

Oppy (2009) objects that agent causation is controversial and 'it is not a secure foundation upon which to rest a convincing argument for the existence of God'.

In reply, I do not posit agent causation as a foundation to rest my argument, nor did I build in the concept of libertarian freedom into the indeterminancy of the First Cause. Rather, the conclusion of Libertarian freedom is deduced based on the kind of indeterminancy that is required to bring about the first change from an initially changeless beginningless state, and the conclusion of agent causation is arrived at deductively from the preceding premises of my argument on which the argument rests. In other words, the conclusion that the First Cause has libertarian freedom is not assumed but deduced from the premises of the KCA; that is, the KCA shows that the First Cause exists beginninglessly and the first effect (first event) has a beginning, and in order for this to be the case the First Cause must have libertarian freedom, as I have explained previously.

In response to the objection that the notion of a Divine agent cause of the initial singularity is obscure, Moreland (2017, pp. 306–307) notes:

> We understand exercises of power primarily from introspective awareness of our own libertarian acts, and we use the concept of action so derived to offer third-person explanations of the behaviour of other human persons.

> There is nothing obscure about such explanations for the effects produced by other finite persons … In fact, naturalists like John Searle, John Bishop and Thomas Nagel all admit that our basic concept of action itself is a libertarian one.

The possession of libertarian freedom implies that the First Cause is not an impersonal entity such as an initial singularity (contra Oppy 2019a, p. 22). Rather, the First Cause is a Creator God.

### 6.4.2 Should We Call It Libertarian Freedom?

It might be objected that one should not call the two capacities mentioned above libertarian freedom, because libertarian freedom is associated with a mind with the capacity for reasoning and decision making, but it has not yet been shown that the First Cause has other properties of a mind with the capacity for reasoning and decision making; in particular, it has not yet been shown that the First Cause brought about the first event purposefully rather than accidentally. The two capacities could be something else (call it Blark power) not involving agency or decision making.[10]

To address this objection, one can argue that, to demonstrate that something *x* has property *y*, one only needs to demonstrate that *x* has the properties sufficient for *y*, one does not need to demonstrate that *x* has the properties associated with *y*. For example, SETI (Search for Extra-Terrestrial Intelligence) researchers can reasonably conclude that Extra-Terrestrial Intelligence exists if they pick up a certain signal under certain circumstances. Even if our understanding of the intelligence that is capable of producing that signal is associated with *human* intelligence, the association with humans is not essential to the definition of intelligence.

Likewise, it can be argued that having the above mentioned two capacities explained above are sufficient for having libertarian freedom. First, it should be noted that the First Cause of the universe was not caused to bring about the first event by some prior causes nor prevented from doing so by outside forces; thus, it is truly free in this sense. Second, the only notion of freedom which has those two capacities is libertarian freedom.

According to the other notion of freedom, that is, compatibilism, the events are determined by prior events and there is nothing with (1) the capacity to initiate change, and also (2) the capacity to prevent itself from changing. Thus, compatibilist freedom is not what the First Cause has; only libertarian freedom follows from the deduction of the KCA. Third, as explained in Sect. 6.4.1, the only notion of causation we have which is consistent with a first cause having the two capacities I mentioned is agent causation having libertarian freedom; thus, it is not ad hoc to call the First Cause an Agent. On the other hand, we have no prior notion of what 'Blark' means; thus, it is ad hoc to use it.

It might be objected that simply having a notion of libertarian freedom in no way establishes its reality, just as having a notion of unicorns does not establish its reality, and that it is still being disputed whether human beings has libertarian freedom (which I am ascribing to the First Cause).[11] However, this objection is based on fallacious reasoning. No one has yet demonstrated that a unicorn exists, but that does not imply that no one can discover that a unicorn exists in the future. If one day someone discovered a white horse-like beast with a single large, pointed, spiralling horn projecting from its forehead, we would say a unicorn has been discovered. One does not have to demonstrate that unicorns exist first before they discover a real-world entity with the characteristics of unicorns. Rather, having a pre-existing notion/concept of unicorn would be sufficient. Likewise, to ascribe Libertarian freedom and agent causation to a First Cause with the relevant characteristics, having a pre-existing notion/concept of Libertarian freedom and agent causation would be sufficient. An instance of *x* does not need to exist first in order for us to discover an instance of *x*; rather, having a pre-existing notion of *x* would be sufficient. I have already shown using the KCA-TA that the First Cause has the characteristics which fit our pre-existing notion of libertarian freedom, thereby demonstrating that libertarian freedom does exist (in the First Cause, at least).

It might be objected that agent causation is not a feature of physics. In reply, that is because physics does not offer a complete explanation of all reality; indeed, physics itself requires deductive and inductive reasoning the justification of which is philosophical. On the one hand, there is no proof that physics offers a complete explanation of all reality. On the other hand, I have offered a proof that the First Cause brought about the first event via libertarian freedom which is characteristic of agent causation. My argument is coherent and consistent with everything currently known in science. The objection that my argument is not consistent with science is based on the assumption that science offer a complete explanation of everything at all moments, which is a philosophical assumption, not a scientific one (i.e. it is the assumption of scientism, not science), and it is a fallacious philosophical assumption as explained in Chap. 1.

It might be objected it is obscure how libertarian freedom works. Nevertheless, as explained in Chap. 1, the conclusion of a sound argument (i.e. a deductively valid argument with true premises) must be true, regardless of whether we know of other details like further explanations concerning how it works, and I have already explained why my argument for a First Cause with libertarian freedom is sound. It should be noted that physics itself admits some of the lawful relationships among physical entities are brute facts having no further explanations' (Koons and Bealer 2010, p. xviii). Indeed, the impossibility of infinite regress of explanations implies that on any worldview there would be brute facts, and I have explained only a First Cause with libertarian freedom can be the brute fact to terminate the causal regress.

In summary, given the three points mentioned above, it is justified to conclude that the First Cause is an agent having libertarian freedom in virtue of having those two capacities, and it is not necessary to demonstrate that the First Cause has other properties of a mind with the capacity for reasoning and decision making. Nevertheless, I shall provide evidences for the latter as well by arguing that the First Cause brought about the first event purposefully rather than accidentally in Chap. 7. This will be accomplished by completing my defence of the Teleological Argument and combining it with the KCA to demonstrate that the First Cause is an intelligent designer of the universe.

### 6.4.3 Is the First Event Random?

A libertarian free act does not entail that the act is un-determined and random. While such a free act is not determined by prior events (and thus is indeterministic in this sense; see below), it is nevertheless determined by a personal agent who is the cause of the action, and the agent freely willed the action rather than randomly, and the agent can will in accordance with reason. On the other hand, calling the first event random does not explain how the first event could have begun from an initially changeless first cause; as explained previously, only libertarian freedom can explain this.

To elaborate, libertarian free acts are indeterministic but not uncaused. As Randolph Clarke and Justin Capes explain, on agent-causal theories, a free act (or some event internal to such an act) must be caused by the agent; and it must not be the case that either what the agent causes or the agent's causing that event is causally determined by prior events. Thus, an agent is in a strict and literal sense an originator of the free act. This combination of *indeterminism and cause* (origination) is thought to capture best the idea that, when we act freely, a plurality of alternatives is open to us and *we determine*, ourselves, which of these we pursue. In response to the objection that the explanatory role of reasons seems to be excluded, Clarke and Capes (2013) suggest an account in which a free action is caused by the agent *and* non-deterministically caused by agent's recognizing certain reasons for which she acts. Acting for a reason does not mean that the person has a reason which determined her choice for a reason (contra Levy and McKenna 2009, p. 121). Rather, as Lowe (2008, pp. 181–190) explains, acting for a reason means that the reason for which the agent acted is simply the reason which the agent *chose* to act upon. Being 'responsive' to a reason for acting in this manner is not being *determined* to act in a certain way by that reason. Thus, indeterminism and causality can both be affirmed, and it is not a random act given that reason is involved.

One might object that, if every beginning has a cause, then the beginning of the event which is 'an agent's causing an event' has a cause (Rowe 2003, p. 73), which appears to generate an infinite regress of causes. Craig replies that 'Partisans of agent causation typically say that the agent's causing some effect is not an event requiring a cause, either because it is not itself an event, but just a way of describing an agent's causing an event, or if it is an event, then it is not further caused' (Craig and Sinclair 2009, p. 194n. 101, citing O'Connor 2000, Chap. 3). Libertarian freedom does not posit an infinite regress or random creation without a cause; rather, the agent is the First Cause of the free act (no regress) and he/she acts for a reason (not random).

It might be objected that there cannot be deliberation in timelessness and hence the decision would be random. In reply, the conclusion does not follow, because the decision can still be made for a reason. For the First Cause could be an omniscient Mind who is aware of all propositions in an initially timeless changeless state and therefore does not need time to think about those reasons. The word 'thought' essentially refers to something X in the mind, and X can be ideas that one is aware of. Moreover, there is no contradiction in saying that something M has a changeless (i.e. timeless) awareness of ideas (i.e. thought) and their logical relations. Therefore, it is not true to say that time must exist first in order that the First Cause can have a thought.

One might object that, if God (suppose God is the First Cause) has reasons for creation (e.g. bless creatures), then the decision to create is made as a result of those reasons, and the decision would be determined by those reasons and hence is not free but occurred by necessity.

The answer is that those reasons can be understood as a necessary condition but not a sufficient condition for the decision, which can therefore still be caused and free. The intention can be one which is freely chosen. Thus, suppose (for example) God—because of His perfect goodness and love—freely created a universe with humans who have significantly morally valuable freedom for His loving purpose of wanting to bless these creatures with the knowledge of Himself who is the Good.[12] In this case, having reasons to bless creatures does not imply that He has to bless creatures, neither does it imply that God could not have refrained from creating initially. The reasons for creation are not coercive, there might also be reasons for not creating, there may well be goods (related to creating and not-creating) which are incommensurable, and even among equal value options, there may be variation (Pruss 2016). Hence, God did not create out of necessity. According to the Christian tradition, God by definition (ex hypothesi) is a free agent who is perfect and therefore has no need; a perfect agent would not experience any insufficiency and hence would have no need to express Himself in creative acts or self-glorification. Rather, He created out of perfectly free love for creatures, and in this way manifested His perfection, that is, His glory. The creation (which has beginning) by a First Cause (which has no beginning) is therefore an evidence of His perfection.

One might ask whether those reasons would be the First Cause(s) given that those reasons are the necessary conditions of the decision. In reply, against the idea that an action done on the basis of a reason is caused by that reason, Pruss (2018, pp. 184–185) argues:

> We can understand a reason as a mental content or a thinkable favoring an action. A reason is thus something abstract. But in addition to the mental contents or thinkables, there are the token thinkings that realize these contents. It is not the reasons considered as abstract thinkables that are causes of an agent's actions. Rather, it is the token thinkings that realize these thinkables that are the causes of an agent's actions.

Pruss goes on to say that, while there are infinitely many reasons on the basis of which God created as He did, this does not imply that there are infinitely many concrete token thinkings in the mind of God given that 'multiple thinkables can be realized in a single act of thinking … when one believes the moon is round and gray, one thereby also believes that it is round and that it is gray. Likewise, multiple reasons can be realized in a single act of thinking' (ibid.). The reasons are abstract, they do not begin to arise in the Mind of God but are being aware of by the Mind in the initially changeless and beginningless state. 'For a reason' is the aim of the choice. The thinking of these reasons is a necessary condition but not a sufficient condition for a rational free choice; thus, by itself it does not determine the choice. Rather, the thinking of these reasons (the final cause) and free will of the Agent (God) (the efficient cause) are what brought about the first event, and 'the exercise of God's free will' is merely descriptive of this. God had thoughts in the sense of being aware of them in the initially changeless state, but was not choosing to bring about the first event initially. When He freely chose to bring about the first event, time began.

### 6.4.4 Libertarian Freedom and Time

As noted earlier, being initially changeless does not mean it is not able to change, just as someone not carrying out an action initially does not mean he/she is unable to act. Here, 'initial' refers to the first in the series of states (ordered causally), not first the series of changes/events/temporal series. With regard to the Hybrid view, there is no contradiction in saying that a First Cause is initially timeless, and then entered into time when it acted. 'Enter' is a temporal concept, as the First Cause brought about the first change (=first event) time also began to exist, and the First Cause entered into time as it brought about the first event (temporal causation). Therefore, in this case there is an initially atemporal cause with temporal causation. On this view the First Cause does not come before all else in time. Rather, the first state of the First Cause existed without time as explained above. The First Cause can freely move out of the timeless state and bring about time. On this view, the temporal event of the universe beginning is not caused by the First Cause in its timeless phase, rather, the First Cause is in time as it causes that event. Libertarian freedom is not a temporal concept; it is a capacity. While 'change' necessarily involves time, the 'ability to change' does not. A timeless agent with libertarian freedom may be without change initially, but having the capacity to bring about the first event, and when he does so, change and time would begin. The First Cause changes and enters into time with the exercise of the freedom to create the universe. Now Mullins (2020, p. 226) has raised the following objection to Craig's hybrid view:

> A change is things' being one way at a particular moment, and then being different at the next moment. If time exists if and only if change exists, then it would seem that time cannot exist without there being a series of moments. This has a counterintuitive entailment—there is no time at the first moment because there is no change at the first moment.

In reply, as noted in previous chapters, a change essentially involves a thing or part of a thing gaining or losing one or more properties; it does not have to involve 'being one way at a particular moment, and then being different at the next moment', rather, it can involve 'being one way at a particular state without the dimension of time, and being different at another state with the dimension of time'. Thus, the First Cause can be in an initially changeless state in which it was not gaining or losing property, and as it brings about the universe, there is a gain of a new property and hence a change together with the first moment and time.

The claim that 'to be able to change, one must exist within a time matrix' is inaccurate, for a timeless agent with libertarian freedom may be initially without change but having the capacity to start changing, and when he does so time would begin. To ask 'how long was this cause changeless for? a millisecond? five minutes? etc.' would be to ask a meaningless question given the relational view of time, according to which in the absence of change there is no time, whereas 'millisecond, five minutes, etc.' involve a measurement of time. To claim that 'time would still pass' is to assume change, since 'pass' is a change. Hence, time would not pass if nothing else exist except something that is initially changeless.

Changeless means absence of change; there is nothing in the notion of this absence itself that requires an extent (temporal or otherwise); the notion of 'no extent and no change (i.e. no gaining or losing of properties at the initial state)' is perfectly coherent. The problem is that many people are too used to thinking in temporal terms and subconsciously asking 'changeless for how long', which of course begs the question against the timeless-sans-creation view by presupposing temporal extent ('for how long').

As an analogy for the Hybrid view, one may think of a situation (call this Situation X) in which nothing else (e.g. no clock, no time dimension) exists except a motionless person who exists (initially) changelessly without beginning: he has the ability to move, but as long as he does not actually move there is no change and no time. (On this view, it is false to say that the person is motionless at $t = 0$, for $t = 0$ implies a time dimension, but on this view there is no time dimension in that motionless state.) When he moves and performs an act, that itself is a change, that is, a temporal causation, and that is what bringing about temporality means. When the person causes the effect he would no longer be motionless. Thus, it is not the case that the man is both moving and motionless simultaneously. There is therefore no contradiction.

Hence, it is wrong for Wielenberg (2020, p. 3) to state, 'But Craig also says that the first cause must be timeless; otherwise, how could it have the power to create time itself?' Actually, what Craig means is that the First Cause is timeless without 'agent-causing of B at t1', and temporal with 'agent-causing of B at t1'. There is no contradiction.

Concerning Craig's illustration of 'a man sitting changelessly from eternity … could freely will to stand up', Wielenberg (2020, p. 3) writes: 'But now suppose that (i) the man causes the effect of standing up while he is sitting.'

Craig can reply that when the man causes the effect he would no longer be sitting. Thus, it is not the case that the man is both seated and fully upright simultaneously. There is therefore no impossibility. (Wielenberg may be presupposing a beginning point. If so, see Chap. 5, where I discuss Craig and Sinclair's [2012, p. 100] rejection of the idea that having a beginning requires having a beginning *point* because it lands one in the ancient Greek paradoxes of motion.)

Wielenberg (2020, p. 3) writes: 'Similarly, on Craig's view, the temporal event of the universe beginning is caused by God in His timeless phase.' But this is mistaken. On Craig's view, the temporal event of the universe beginning is *not* caused by God in His timeless phase; rather, God is in time as He causes that event.

Contrary to Wielenberg (2020, p. 3), this view does not imply 'the causal inertness of God in His timeless phase', for in that timeless phase God possesses the causal power to bring about the first event which He refrained from exercising in that timeless phase, and which was exercised at the first duration of time. Having that power (which He refrained from exercising) in that timeless phase distinguishes God from (say) abstract objects which are timeless but have no such power—that is why we say that abstract objects are causally inert. Neither is it accurate to characterize this view as saying that 'a temporal being caused the universe' (ibid.) simpliciter. Rather, according to this view the universe is created by a God who is timeless without creation and temporal with creation

Wielenberg also claims that the view that God caused the beginning of time has the problem of implying that God's exercise of causal power (GA) is a temporal event 'causally prior to the beginning of time, which is impossible, since it would make the existence of time a prerequisite for an event that is causally prior to the beginning of time and hence would require time to be causally prior to itself' (2020, pp. 4–5).

In reply, instead of saying that 'God's exercise of causal power (GA) is an event that caused the beginning of time', one can say 'God's exercise of causal power' (GA) is just a way of describing an agent (God) causing an event/change (the beginning of the universe in this case), and the beginning of time (the first moment) is concomitant to the event ('the beginning of universe'). Hence, God's act of creation does not depend on the pre-existence of time (a moment at which He creates); rather, the existence of time is dependent on God's act of creation.

Wielenberg (ibid., p. 7) claims that the intrinsic change of God entailed by GA implies that GA is an event which is both caused and uncaused. However, this is a non-sequitur. As noted earlier, rather than saying that GA is an event, GA can just be a way of describing God causing an event, and this can entail an 'intrinsic change' as follows: The first state of not-causing exists without change initially, and thus is timeless (on a relational view of time). As God causes the first change, this entails the second state of causing which is concomitant to God causing the first change, and the difference between the first and second state is an 'intrinsic change' which is not-uncaused but is concomitant to (and simultaneous with) God causing the first change. There is no uncaused event in the above scenario.

Thus, one can coherently affirm:

1. God is initially timeless.
2. God's exercise of causal powers brings about the initial state of the universe.
3. As God exercises His causal power, time begins.
4. The universe is caused by God in His temporal state.

It might be asked, 'since there is a succession of distinct states (initial changelessness followed by change), would it be coherent to state that God's timeless state does not temporally precede the existence of the universe? How are we to make sense of the notion of the succession of states not being a temporal sequence?'[13]

In reply, the First Cause being changeless-sans-first-event and changes with the first event does not imply a temporal succession of two states, because according to the Hybrid view the initial changeless state is not a state in time but timeless, that is, without a temporal dimension which only exists with the first event. Hence, this is not a case of succession of two temporal states. One can make sense of the notion of the succession of states not being a temporal sequence by thinking of time as involving a dimension and/or change, and according to the Hybrid view in the original state there is neither. In this way the First Cause can be causally prior but not temporal prior to the first event.

One might ask, 'if there is no time separating the timeless First Cause and the first event, then the two must coexist. In that case, how can it be that the First Cause is timeless sans (without) the first event?'

In response, on the Hybrid view, the difference in properties between timelessness (which is beginningless) and time (which has a beginning) implies that the timeless First Cause and the first event do not coexist, and that the First Cause can be timeless without the first event. God was (1) initially changeless without creation—there was no event in that state and no universe as well; (2) God changed with the bringing about of the beginning of the universe, in which state the universe existed alongside God. There is distinction with a difference between (1) and (2), and it shows that it is not the case that the universe and God coexisted.

One might object that for $x$ to change is for $x$ to have property $p$ at $t_m$ that $x$ does not have at $t_n$, and therefore it is impossible that timeless entities change. However, proponents of the Hybrid view can argue that 'for $x$ to change is for $x$ to have property $p$ at $t_m$ that $x$ does not have at $t_n$, *or for x to have p in timelessness that x does not have at t*', and thus there is no incoherence there.

It might be objected that, while it is easy to conceive of how the First Cause can have libertarian freedom on a dynamic theory of time, it is difficult to conceive this on a static theory of time. In reply, Craig argues that static time is compatible with human libertarian freedom; if that is so, it would be compatible with the Divine First Cause having libertarian freedom as well. Craig (2015) explains that

> the B-theory does not imply that events which lie in our future are causally determined with respect to antecedent event. Indeed, some such event could be wholly undetermined by antecedent causes. On any standard

> definition of libertarian freedom, therefore, such an event could be a genuinely free choice.

He also argues that, 'on a B-theory of time, although we cannot change the future, we can act in such a way that if we were to act in that way, the future would be different' (ibid.).

Likewise, one can argue that, on a B-theory of time, God can refuse to act in such a way that, if He were to refuse to act in that way, the universe would not have existed at $t_1$.

One might object: if God is initially changeless, then His willing of the universe must be without beginning, in which case the universe should also be without beginning, but this contradicts the KCA which argues that the universe has a beginning (Morriston 2000).

Citing J.P. Moreland, Craig replies that it is insufficient for P to have merely the intention and power to bring about R; rather, there must also be a basic action on the part of P, a free undertaking which took place simultaneously with the first effect in time. Craig concludes the failing of Morriston's objection is that in speaking of God's willing that the universe exists, he does not differentiate between God's timeless intention to create a temporal world and God's undertaking to create a temporal world. Once we make the distinction, we see that creation ex nihilo is not an instance of state–state causation (Craig 2002).

In short, one should note the distinction between God's *intending* to create a universe and His *undertaking* it, that is, His bringing about that intention. Given this distinction, Craig argues that it's possible for God to eternally intend to bring about the universe, and then to freely and spontaneously undertaking to create it a finite amount of time ago. Thus, the universe was freely brought about by the Divine Agent who has libertarian freedom, and the 'undertaking' is for the purpose of accomplishing something; therefore, it is not random.

One might object that the distinction between God's intending to create a universe and His undertaking it does not exist if there are no actual distinct differences intrinsic to God in the initially changeless state.[14] This objection confuses between conceptual distinction and distinction-in-the-concrete. Conceptually, there is a distinction between God's intending to create a universe and His undertaking it—these two mean different things. However, concretely, such a distinction does not exist in the being of God in His initially changeless state. Nevertheless, His capacity for the exercise of libertarian freedom existed in that state, and when He exercised this capacity He undertook the creation of the universe and the distinction began to exist concretely.

Leon (in Rasmussen and Leon 2018, p. 63) objects that the distinction between deciding and undertaking that decision arises 'in three main types of case: When you do not yet know what you will decide to do; when a decision the time for carrying out your decision has not yet arrived; and when you have weakness of will that (at least temporarily) prevents you from carrying out your decision', but none of these conditions applies to an omniscient, timeless, omnipotent, and morally perfect God. Likewise, Morriston argues that 'An omnipotent being cannot suffer from weakness of will. An omniscient being cannot change its mind. And a timeless being cannot meaningfully be said to "delay" undertaking to carry out its intentions. So it is very hard indeed to see how God's eternal will to create can fail to be sufficient for His undertaking to do so' (Morriston 2002a, p. 107).

However, the three main types of cases that Leon explained can be regarded as accidental to humans but not essential to the distinction between deciding and undertaking that decision. An omniscient, timeless, omnipotent, and morally perfect God could have known what He would decide to do with the beginning of time, while also willed to initially refrain from undertaking creation because it is consistent with divine perfection to be initially changeless (the conclusion of initial changelessness follows from the premises of the Kalām as argued previously). In that case there is no weakness of will that prevents God from carrying out His decision to create; rather, God not carrying out His decision is due to His will to initially refrain. There is no changing of mind, since the divine mind has always planned to initially refrain and to create at the first moment. This refraining is not a delay (since delay involves time but there is no time in the state of refraining) but rather an exercising of the capacity to prevent itself from changing initially, as argued previously.

It might be objected that a decision seems to be an action that only makes sense in time.[15]

In reply, the term 'decide' refers to the end result of consideration of reasons. While humans require time to consider and make up their mind because human mental capacity is limited, a superior Being who does not suffer from this limitation would not have this requirement. There is no contradiction in saying that a superior Being has an initially changeless (i.e. initially timeless) awareness of ideas and reasons (i.e. thought), their logical relations, and the resulting freely made decision. Therefore, it is not true to say that time must exist first in order that the Superior Being can have a thought. What is essential to the consideration of ideas and a decision is logical sequence, not temporal sequence, which, although necessary for humans, is unnecessary for a Superior Being who can be aware of all logical sequences timelessly. God decides in an initially timeless state (He does not require time to make up His mind) and acts with the beginning of time; therefore, it is not the case that God decides and acts at the same time. These two are not 'at the same time' because God decides in an initially timeless state, not at the same time with the action.

## 6.4.5 Contradiction with Classical Theism

It might be objected that the conclusion that the Divine First Cause changed with the exercise of libertarian freedom is inconsistent with the doctrines of a strong notion of divine immutability, essential divine timelessness, and divine simplicity, which has been held by many Christian theologians (e.g. Augustine, Anselm, and Aquinas) and is known as Classical Theism.

Classical Theism however has been rejected by many theologians today for being contrary to the Scripture and philosophically untenable (Mullins 2015). I have argued in Loke (2014, 2018) that there is insufficient philosophical, theological, or Scriptural justification for a strong notion of divine immutability, essential divine timelessness, and divine simplicity; that these views are not required for Perfect Being Theism; and that these views face difficulties concerning the doctrine of the Incarnation, which is of central importance for understanding the divine (and human) in Christian Theology. Additionally, these views face difficulties concerning the doctrine of creation. For in order for a universe with beginning to be caused by a God without beginning, this would require God to refrain from using His active powers in the beginningless-state-sans-the-universe, and use His powers at creation. This implies that God is not Pure Act, since Pure Act entails intrinsic essential changeless-ness, whereas refrain-use implies that the First Cause is not essentially changeless but only initially changeless as explained in previous sections. By contrast, a beginningless (first) cause which (as Aquinas claimed) is Pure Act would (contrary to Aquinas) bring about a beginningless universe. However, this consequence is contrary to orthodox Christian doctrine that the universe has a beginning, and this consequence is also contradicted by the conclusion established previously that the First Cause brings about a first event which has a beginning whereas the First Cause has no beginning. Moreover, since God is the Creator and since creation involves God using His powers, which He does not use in the initially changeless state, this would imply that God's internal properties do change with creation.

Through his argument from motion Aristotle had concluded that for any motion to occur there must be some unmoved mover, that is, God, who, being fully actual, cannot change because He has no potentiality not already fully realized (Aristotle, *Metaphysics* 12.5–9). Thomists have similarly argued that changes involve the actualization of potentials (Feser 2017, p. 26) and are explained by a hierarchical causal series (the cup of coffee is held up by the desk, which is help up by the floor, which is held up by the foundations, which is held up by the earth…) with a first member 'without any potential for existence requiring actualisation. This is pure actuality … uncaused cause, Aristotle's "Unmoved Mover" … unactualized actualizer' (ibid., p. 27). Concerning Aquinas' Fifth Way, it has also been argued that, since its purpose is to explain the teleological potential that is present in all things, the explanation cannot have such a potential in itself, but must be Pure Act (Newton 2014, p. 576).

In reply, while one can agree that changes involve the actualization of potentials and that there is a First Cause of a hierarchical causal series, this does not imply that such a First Cause must have no potentiality not already fully realized (by 'potential' I am referring to 'active power in the state of not-being-used', i.e., the state of refraining from using active power, see above). Rather, a beginningless First Cause having libertarian freedom to freely actualize its own potential (e.g., to use its active power to create the universe, and hence is the first cause of this actualization and explains why that potential is actualized rather than not-actualized) would terminate the hierarchical causal series just as well. On this view, the First Cause was initially changeless sans the first event, but it has the un-actualized potential to bring about the first event, and as it actualizes its potential to bring about the first event, it also actualizes its potential to sustain the things that are brought about by the first event. (This does not mean that the First Cause actualized all its potentials at the first event; rather, it is possible that the First Cause could have other potentials, such as [as Christian theologians affirm] the potential for Incarnation [see Loke 2014], which was actualized at a later time. The possibility of such a view implies that it is not necessary the case that the First Cause must be a Pure Act.) While the Thomist assumes that something cannot actualize its own potential, a libertarian agent who is a beginningless First Cause can do that, and this does not involve something bringing about its own existence since the First Cause is beginningless and eternally existed. On this view the change in God's properties was brought about by God Himself as He brought about the first event and continues to freely choose to sustain the world in existence. God can change His initially state of changelessness given that that state is not essential to divine nature (I argue for this in Loke 2014, chapter 5) and given that God has libertarian freedom. Affirming that God can change properties that are non-essential to divine nature does not imply that God can change properties essential to divine nature (such as properties of being uncaused, beginningless, omnipotent, etc.).

## 6.5 The First Cause Has Tremendous Power

The enormity of the power of the First Cause is indicated by the enormity of the effect down the causal chain, namely, the entire universe. Scientists have discovered that the sun which illuminates our earth is merely one of the over 200 billion of stars in our galaxy. Even if we could travel at the speed of light—about 300,000 kilometres per second—it would take about 100,000 years to travel from one end of the galaxy to the other. More astounding still is the fact that our galaxy is merely one of the over 100 billion galaxies in existence, many of which have hundreds of millions of stars. And this is merely the currently observable universe; the actual universe is much larger than this. The universe is truly awesome, and as shown in previous chapters, these billions of stars and galaxies ultimately came from a First Cause who is the Creator of the universe.

Against the First Cause having enormous power, one might object that one cannot make such deduction of the degree of power from the effect. For example, while we can infer from the effects that the atomic bomb causing the destruction of Hiroshima in 1945 was tremendously powerful, it might be argued that the bomb was the end result of a process leading from less powerful entities, such as the tiny elements of uranium and the little 'bullet-like' mechanism shot in to the uranium to start the reaction.[16] Sceptics might object that the bullet, which has little energy, is analogous to the First Cause.

Further reflection reveals that the above example is disanalogous to the First Cause of the universe in the following way. The tremendous power of the atomic bomb is due to the mass of the uranium, which contains a lot of energy given the conversion of mass into energy in accordance to $E = mc^2$. While the bullet brought about the conversion of the mass of the uranium into energy, the existence of the mass-energy of the uranium was not brought about by the bullet. Whereas the existence of the entire universe with its tremendous amount of mass-energy was ultimately brought about by the First Cause, which therefore has tremendous power to bring about all these.

As noted in Chap. 2, some cosmologists have proposed the Zero Energy Universe theory according to which the net energy of the universe is zero. One might think that, if that is true, then there is no reason to think that the First Cause would be required to possess tremendous power to bring about zero energy.

However, this is a misconception. I have explained in Chap. 2 that, even if Zero Energy Universe theory is true, one still has to ask what made the energy and the laws of nature to be the way they are. The First Cause must still be enormously powerful in order to be able to make the tremendous amount of positive and negative energy to be the way they are—out of zero energy! While humans with limited powers require pre-existing matter-energy to work from in order to create (say) an atomic bomb and the feeble bullet trigger requires pre-existent uranium to start the nuclear reaction, the First Cause does not require pre-existing matter-energy in order to bring about a series of events that resulted in the billions of stars and galaxies as well as the negative energy of gravity and the amazing laws of nature. This is an indication that the power of the First Cause far surpasses ours; indeed, it far surpasses anything else we know.

## 6.6 Conclusion

I have defended premises 6–11 of KCA-TA and show that the First Cause is not a series of changes (= events) describable by physical laws; rather, it is initially changeless (premise 8) and brought about the first event with the physical laws. It is also distinct from the physical universe which is constantly changing according to quantum field theory, and which does not have 'the capacity to be the originator of an event in a way that is undetermined by prior event, and the capacity to prevent itself from changing', which a First Cause must have (premise 10). I have explained and defended the claim that these two capacities imply that the First Cause has libertarian freedom.[17] Thus, the First Cause cannot be part of the physical universe as postulated by Hawking's no-boundary proposal, which as explained above is unproven and scientifically flawed. Rather, as shown by premises 6–11 of KCA-TA, the First Cause is uncaused, beginningless, initially changeless, has libertarian freedom, and is enormously powerful, that is, a transcendent immaterial Creator of the Universe. With regard to the relationship between the First Cause and time, I have shown that both the Hybrid view and the view of the Oxford School are defensible; any one of them would be sufficient for the conclusion of this book. I have also shown that the conclusion of KCA, as well as the doctrine of *creatio ex nihilo*, is consistent with the relational view of time and the substantival view of time, and it is also consistent with the dynamic theory of time and the static theory of time. Thus, for the purposes of this book it is not necessary to settle the debates between these views and theories. I personally think that there are other philosophical reasons for thinking that the static theory of time is false, but the point here is that, regardless of which of these view or theory is true, there must still be a First Cause which is uncaused, beginningless, initially changeless, has libertarian freedom, and is enormously powerful. The conclusion that the First Cause is a Creator who brought about the first event purposefully rather than accidentally can be further strengthened by considering the evidences of fine-tuning and order of the universe, which have been explained in Chap. 4. I shall complete my demonstration that the First Cause is a Designer in the next chapter.

## Notes

1. See also Alfred Freddoso's comparison of Suarez's analysis of causation with contemporary theories in Freddoso's *Introduction to Suarez* (2002).
2. A-theorists such as Craig deny that these purported evidences support the B-theory; see Craig (2000a, 2000b).
3. Some philosophers have argued that time could continue to exist even if all events were to cease (Shoemaker 1969). I think this argument can be rebutted, but rebutting it will take us too far afield; in any case, Craig and Sinclair (2009, p. 192) notes that the arguments of Shoemaker 'are inapplicable in the case at hand, where we are envisioning, not the cessation of events, but the utter absence of any events whatsoever'.
4. As Oppy did during the debate: https://www.youtube.com/watch?v=a8NrTv-Durc&t=129s.
5. Except perhaps Oriti (2014), who proposes geometrogenesis, that is, the coming of spacetime into being with the physical condensation of the 'spacetime atoms'. As noted above, Oriti acknowledges that this view is not well-established.
6. Koons (2014, pp. 261–262), attributing it to an anonymous referee. Koons replies by pressing a dilemma: 'either there is an intrinsic metric to the pure passage of time, or not. If there is, then the infinite past is actually divided into an infinite number of periods, contrary to the conclusion of the Reaper paradox. If there is no intrinsic measure of time, then the imagined scenario is impossible, since it supposes an extended period during which absolutely nothing happens' (ibid.). However, the second horn of the dilemma presupposes a relational view of time. The opponent could deny this and hold to a substantival view of time, according to which there can be time without change or process.

7. While Oppy has called it the Initial Singularity, he writes that ""Initial Singularity" is just a convenient label for whatever it is that exists in the initial state of natural reality. It would work equally well to use, instead, the label "Initial Natural Thing"' (Oppy 2019b, p. 229).
8. I thank William Lane Craig for helpful input here.
9. I thank Andres M for suggesting this.
10. I thank Vaal for raising this objection.
11. I thank Vaal for raising this objection.
12. I discuss God's reasons for creation in *Evil, Sin and Christian Theism* (Loke 2022).
13. I thank John Pascal for raising this question.
14. I thank Mediator media for raising this objection.
15. I thank Louigi Verona for raising this objection.
16. I thank Vaal for raising this objection.
17. This conclusion also provides a response to Kant's First Antinomy; for details and replies to other objections concerning the properties of the First Cause, see Craig (1979); Loke (2017a, chapter 6).

## Bibliography

Aguirre, Anthony. 2007. Eternal Inflation, Past and Future. https://arxiv.org/abs/0712.0571.

Aguirre, Anthony, and John Kehayias 2013. Quantum Instability of the Emergent Universe. Cornell University. https://arxiv.org/abs/1306.3232.

Anderson, E. 2012. Problem of Time in Quantum Gravity. *Annalen der Physik* 524 (12): 757–786.

Arkani-Hamed, Nima, and Jaroslav Trnka. 2014. The Amplituhedron. *Journal of High Energy Physics* 2014: 30. https://doi.org/10.1007/JHEP10(2014)030.

Barbour, J. 1999. *The End of Time*. Oxford: Oxford University Press.

Barrow, J. 1991. *Theories of Everything: The Quest for Ultimate Explanation*. Oxford: Oxford University Press.

Boddy, K.K., S.M. Carroll, and J. Pollack. 2016. De Sitter Space Without Dynamical Quantum Fluctuations. *Found Phys* 46: 702–735.

Buchak, L. 2013. Free Acts and Chance. *Philosophical Quarterly* 63: 20–28.

Cao, ChunJun, Sean M. Carroll, and Spyridon Michalakis. 2017. Space from Hilbert Space: Recovering Geometry from Bulk Entanglement. *Physical Review D* 95: 2.

Carroll, Sean. 2019. *Something Deeply Hidden: Quantum Worlds and the Emergence of Spacetime*. New York: Dutton.

Chan, Man Ho. 2019. Is the History of Our Universe Finite? *Theology and Science* 17 (2): 248–256.

Clarke, Randolph, and Justin Capes. 2013. Incompatibilist (Nondeterministic) Theories of Free Will. *The Stanford Encyclopedia of Philosophy*, ed. Edward Zalta. http://plato.stanford.edu/archives/spr2013/entries/incompatibilism-theories/. Accessed 20 January 2014.

Craig, William Lane. 1979. Kant's First Antinomy and the Beginning of the Universe. *Zeitschrift für Philosophische Forschung* 33: 553–567.

———. 1990. What Place, Then, for a Creator?: Hawking on God and Creation. *British Journal for the Philosophy of Science* 41: 480–486.

———. 2000a. *The Tensed Theory of Time: A Critical Examination*. Synthese Library 293. Dordrecht, the Netherlands: Kluwer Academic Publishers.

———. 2000b. *The Tenseless Theory of Time: A Critical Examination*. Synthese Library 294. Dordrecht, the Netherlands: Kluwer Academic Publishers.

———. 2000c. Naturalism and Cosmology. In *Naturalism: A Critical Analysis*, ed. William Lane Craig and J.P. Moreland. London: Routledge.

———. 2002. Must the Beginning of the Universe Have a Personal Cause? *Faith and Philosophy* 19: 94–105. http://www.reasonablefaith.org/must-the-beginning-of-the-universe-have-a-personal-cause-a-rejoinder#ixzz2YAV5QeN5. Accessed 13 January 2010.

———. 2011. #232 The Metric of Time. https://www.reasonablefaith.org/question-answer/P220/the-metric-of-time.

———. 2015. Q&A #450: Does the B-theory of Time Exclude Human Freedom? http://www.reasonablefaith.org/does-the-B-theory-of-time-exclude-human-freedom#ixzz4O4einczH. Accessed 15 October 2016.

———. 2018. The Kalām Cosmological Argument. In *Two Dozen (Or So) Arguments for God*, ed. Jerry Walls and Trent Dougherty. Oxford: Oxford University Press.

Craig, William Lane, and James Sinclair. 2009. The Kalām Cosmological Argument. In *The Blackwell Companion to Natural Theology*, ed. William Lane Craig and J.P. Moreland. Chichester: Wiley-Blackwell.

———. 2012. On Non-Singular Space-times and the Beginning of the Universe. In *Scientific Approaches to the Philosophy of Religion*, ed. Yujin Nagasawa. New York: Palgrave Macmillan.

Deng, Duen-Min. 2019. A New Cosmological Argument from Grounding. *Analysis*. https://doi.org/10.1093/analys/anz071.

Deutsch, David. 1997. *The Fabric of Reality: The Science of Parallel Universes and Its Implications*. London: Penguin.

Earman, John. 1995. *Bangs, Crunches, Whimpers, and Shrieks: Singularities and Acausalities in Relativistic Spacetimes*. New York: Oxford University Press.

Eddington, A. 1920. *Space, Time and Gravitation*. Reprint edition: Cambridge Science Classics. Cambridge University Press, 1987.

Erasmus, Jacobus. 2018. *The Kalām Cosmological Argument: A reassessment*. Cham: Springer Nature.

Feser, Edward. 2017. *Five Proofs of the Existence of God*. San Francisco: Ignatius Press.

Gott, Richard I.I.I., and Li-Xin Li. 1998. Can the Universe Create Itself? *Physical Review D* 58: 023501–023501.

Grünbaum, Adolf. 1994. Some Comments on William Craig's 'Creation and Big Bang Cosmology'. *Philosophia Naturalis* 31: 225–236.

Halper, Phillip. 2021. The Kalām Cosmological Argument: Critiquing a Recent Defence. *Think* 20: 153–165.

Hartle, James, and Stephen Hawking. 1983. Wave Function of the Universe. *Physical Review D* 28: 2960–2975.

Hawking, Stephen. 1988. *A Brief History of Time*. London: Bantam.

———. 1997. The Objections of an Unashamed Positivist. In *The Large, the Small, and the Human Mind*, ed. R. Penrose and M. Longair. New York: Cambridge University Press.

———. 2018. *Brief Answers to the Big Questions*. New York: Bantam Books.

Hawking, Stephen, and Leonard Mlodinow. 2010. *The Grand Design*. New York: Bantam Books.

Huggett, Nick, and Christian Wuthrich. forthcoming. *Out of Nowhere: The Emergence of Spacetime in Quantum Theories of Gravity*. Oxford: Oxford University Press.

Isham, C. J. 1997. Creation of the universe as a quantum process. In *Physics, philosophy, and theology: A common quest for understanding* (pp. 375–408) (3rd ed.), ed. R. J. Russell, W. R. Stoeger, & G. V. Coyne. Vatican City State: Vatican Observatory.

Koons, Robert. 2014. A New Kalām Argument: Revenge of the Grim Reaper. *Noûs* 48: 256–267.

Koons, Robert and George Bealer, eds. 2010. *The Waning of Materialism*. Oxford: Oxford University Press.

Leftow, Brian. 2005. Eternity and Immutability'. In *The Blackwell Guide to the Philosophy of Religion*, ed. W. Mann. Malden, Blackwell.

———. Eternity. In *A Companion to Philosophy of Religion*, ed. C. Taliaferro, P. Draper, and P. Quinn, 2nd ed. Chichester: Wiley.

Levy, Neil, and Michael McKenna. 2009. Recent Work on Free Will and Moral Responsibility. *Philosophy Compass* 4: 96–133.

Loke, Andrew. 2014. *A Kryptic Model of the Incarnation*. London: Routledge.

———. 2017a. *God and Ultimate Origins: A Novel Cosmological Argument. Palgrave Frontiers in Philosophy of Religion Series*. Cham, Switzerland: Springer Nature.

———. 2018. Review of Timothy Pawl's in Defense of Conciliar Christology: A Philosophical Essay. *Faith and Philosophy* 34: 114–119.

———. 2022. *Evil, Suffering and Christian Theism*. London: Routledge.

Lowe, E.J. 2008. *Personal Agency: The Metaphysics of Mind and Action*. Oxford: Oxford University Press.

Moreland, J.P. 2017. Libertarian Agency and the Craig/Grünbaum Debate About Theistic Explanation of the Initial Singularity. In *The Kalām Cosmological Argument*, ed. Paul Copan and William Lane Craig. New York: Bloomsbury Academic.

Moreland, J.P., and William Lane Craig. 2003. *Philosophical Foundations for a Christian Worldview*. Downers Grove: InterVarsity Press.

Morriston, Wes. 2000. Must the Beginning of the Universe Have a Personal Cause? *Faith and Philosophy* 17: 149–169.

———. 2002a. Craig on the Actual Infinite. *Religious Studies* 38: 147–166.

———. 2002b. Causes and Beginnings in the *Kalām* Argument: Reply to Craig. *Faith and Philosophy* 19: 233–244.

Mullins, R.T. 2015. *The End of the Timeless God*. Oxford: Oxford University Press.

———. 2020. The Divine Timemaker. *Philosophia Christi* 22: 207–233.

Newton, William. 2014. A Case of Mistaken Identity: Aquinas's Fifth Way and Arguments of Intelligent Design. *New Blackfriars* 95: 569–578.

O'Connor, Timothy. 2000. *Persons and Causes: The Metaphysics of Free Will*. Oxford: Oxford University Press.

———. 2016. Probability and Freedom. *Res Philosophica* 93: 289–293.

Oppy, Graham. 2009. Cosmological Arguments. *Nous* 43: 31–48.

———. 2010. Uncaused Beginnings. *Faith and Philosophy* 27: 61–71.

———. 2013a. *The Best Argument Against God*. London: Palgrave Macmillan.

———. 2013b. Ultimate Naturalistic Causal Explanations. In *The Puzzle of Existence*, ed. Tyron Goldschmidt. New York: Routledge.

———. 2019a. The Universe Does Not Have a Cause. In *Contemporary Debates in Philosophy of Religion*, ed. Michael Peterson and Raymond Van Arragon. Chichester: Wiley Blackwell.

———. 2019b. Review of P. Copan and Craig, W. The Kalām Cosmological Argument Volume Two: Scientific Evidence for the Beginning of the Universe. *European Journal for Philosophy of Religion* 11: 225–229.

Oriti, Daniele. 2014. Disappearance and Emergence of Space and Time in Quantum Gravity. *Studies in the History and Philosophy of Modern Physics* 46: 186–199.

Padgett, Alan. 1992. *God, Eternity, and the Nature of Time*. New York: St. Martin's.

Pearce, Kenneth. 2017. Foundational Grounding and the Argument from Contingency. In *Oxford Studies in Philosophy of Religion*, vol. 8. Oxford: Oxford University Press.

Pruss, Alexander. 2016. Divine Creative Freedom. *Oxford Studies in the Philosophy of Religion* 7: 213–238.

———. 2018. *Infinity, Causation, and Paradox*. Oxford: Oxford University Press.

Rasmussen, Joshua, and Felipe Leon. 2018. *Is God the Best Explanation of Things: A Dialogue*. Cham: Springer Nature.

Reichenbach, Bruce. 2021. Cosmological Argument. In *The Stanford Encyclopedia of Philosophy*, ed. Edward Zalta. https://plato.stanford.edu/archives/win2016/entries/cosmological-argument/.

Rowe, William. 2003. Reflections on the Craig-Flew Debate. In *Does God Exist? The Craig-Flew Debate*, ed. Stan Wallace. Aldershot: Ashgate.

Shoemaker, S. 1969. Time Without Change. *Journal of Philosophy* 66: 363–381.

Smith, Quentin. 1996. Causation and the Logical Impossibility of a Divine Cause. *Philosophical Topics* 24: 169–191.

Suarez, Francisco. 2002. *On Creation, Conservation, and Concurrence*, Trans. Alfred Freddoso. South Bend.: St. Augustine's Press.

Swinburne, Richard. 1993. God and Time. In *Reasoned Faith*, ed. Eleonore Stump. Ithaca: Cornell University Press.

Vilenkin, Alexander. 2006. *Many Worlds in One*. New York: Hill and Wang.

Wall, Aaron. 2014. Did the Universe Begin? IV: Quantum Eternity Theorem. http://www.wall.org/~aron/blog/did-the-universe-begin-iv-quantum-eternity-theorem/. Accessed 20 January 2017.

Weaver, C. 2016. Yet Another New Cosmological Argument. *International Journal for Philosophy of Religion* 80 (1): 11–31.

Wielenberg, Erik. 2020. Craig's Contradictory Kalām: Trouble at the Moment of Creation. *TheoLogica*. https://doi.org/10.14428/thl.v4i3.55133.

# 7

# Ultimate Design

## 7.1 Introduction

In Chap. 4, I have presented the evidences of fine-tuning and order of the universe, demonstrated that the following are the only possible categories of hypotheses concerning 'fine-tuning and order': (i) Chance, (ii) Regularity, (iii) Combinations of Regularity and Chance, (iv) Uncaused, and (v) Design, and ruled out (i), (ii), and (iii). In this chapter, I shall rule out (iv) Uncaused, defend the conclusion of Design against scientific, philosophical, and theological objections, and demonstrate the superiority of the design inference used in this book compared with alternative approaches.

## 7.2 Against the 'Uncaused' Hypothesis

It has been suggested that 'fundamental laws might be brute facts, meaning that they have no explanation at all' (Sober 2019, p. 37). The Hartle–Hawking model which has been discussed in Chap. 6 is an example of a cosmological model in which the laws of nature (in this case, the laws of



A. Loke, *The Teleological and Kalam Cosmological Arguments Revisited*, Palgrave Frontiers in Philosophy of Religion, https://doi.org/10.1007/978-3-030-94403-2_7

quantum gravity) exist as an uncaused brute fact in an initially timeless and beginningless state. Concerning fine-tuning, Oppy (2013) suggests the possibility that the properties of the naturalistic initial state of the initial singularity lay in certain appropriately narrow ranges, which guaranteed that it is metaphysically necessary that subsequent natural causal reality would be life-permitting.[1] To explain why the physical entities are sustained in an orderly manner, one might appeal to a non-causal explanation of some sort in terms of deeper metaphysical principles. For example, Lange (2009) tries to explain why the laws are true by appealing to the (purported) fact that no matter how things had started out, the laws would still have been true, and then explaining why that counterfactual is true, by saying that no matter how things had started out, that counterfactual itself would still have been true, and so on *ad infinitum* (he calls this 'the lawmaker's regress') (p. 146). If it is a law that *p*, then various subjunctive facts explain why *p* is the case, and for each of these subjunctive facts, various further subjunctive facts explain why it is the case, and so forth. All of those subjunctive facts help to make it a law that *p* (p. 149). Each of the subjunctive facts that helps to constitute a law's necessity is itself metaphysically necessary, its necessity constituted by other subjunctive facts that help to constitute the law's necessity (p. 155). Others have suggested what makes laws metaphysically necessary are essential properties of the natural kinds (Ellis 2001) or dispositional properties (Mumford 2004).

There are at least two problems with such views.

First, all such models in which the laws of nature are brute fact cannot work because, as explained in Chap. 6 while discussing the Hartle–Hawking model, an infinite regress of events is not possible and in order for the first event to begin, it must be caused by an initially changeless First Cause with libertarian freedom. In other words, the first event must have been brought about freely, and not in a law-like way which guaranteed that it is metaphysically necessary that subsequent natural causal reality would be life-permitting (cf. Oppy 2013). Thus, the First Cause which caused the first event cannot be part of the physical universe which is constantly changing and does not have libertarian freedom, and therefore it cannot be a naturalistic initial state postulated by Oppy (2013). Rather, as shown by premises 6–11 of the KCA-TA, the First Cause is uncaused, beginningless, initially changeless, has libertarian freedom, and is enormously powerful, that is, a transcendent immaterial Creator of the Universe. This implies that the physical universe cannot be the uncaused First Cause; rather, it has a first event, which implies it has a beginning and therefore (according to Causal Principle) has a cause, and hence its properties of being fine-tuned and highly ordered would also have a cause. We need to ask why, after the First cause brought about first event (regardless of whether this is the first event *of our universe* or the first event of something else), it eventually resulted in a fine-tuned and highly ordered universe.

Second, Frederick notes that, while the sceptic might claim that the laws of nature are metaphysically necessary, this does not answer the question of how it could be necessary that unthinking, mindless things always accord with natural laws (Frederick 2013, pp. 272–273). While Lange, Ellis, Mumford et al. attempt to provide a *non-causal* explanation of what makes the laws metaphysically necessary, this does not answer the question of how it could be necessary that unthinking, mindless laws always accord with such an explanation, in such a way that the order within the universe can be described by sophisticated mathematical equations which indicate a high degree of ordering. In other words, their explanation does not answer how it could be necessary that the subjunctive facts (Lange) or the physical entities have stable essences (Ellis) or dispositions (Mumford) that persist throughout time which enable them to behave in ways describable by such mathematical laws. Likewise, saying that it is just the nature of physical entities to behave in such an orderly manner does not answer how the nature of unthinking, mindless things could be such that they (almost) always accord with natural laws describable by mathematical equations (e.g. Schrodinger equation, Dirac equation, etc.), such as the highly intricate order of quantum mechanics which scientists observe from moment to moment.

Leslie (1989) asks us to consider a hypothetical scenario in which 'particles regularly formed long chains which spelled out 'GOD CREATED THE UNIVERSE', this then being shown to result inevitably from basic physics' (p. 109). It would be unconvincing to object that this is not evidence of design by claiming that the laws of nature are metaphysically necessary and are brute facts or that this is the only universe that we have observed. The point of this hypothetical scenario is that it is likewise unconvincing to object that our fine-tuned and highly ordered universe is not evidence of design by claiming that the laws of nature are metaphysically necessary and are brute facts or that this is the only universe that we have observed.

Some physicists seem to have thought of the laws of physics as the uncaused cause of the universe.[2] The problem with this view is that, as explained in Chap. 2, a law of physics is not a concrete thing but merely a description of behaviour of concrete physical things, and descriptions by themselves do not make things happen one way or another. Therefore, the laws of physics cannot be the uncaused cause of the universe. Something else is needed; that is, a concrete First Cause is required to make the universe in accordance with the descriptions of the laws of physics, and to be able to do that the First Cause would have to be intelligent as well (like an architect making a house in accordance with the description in the blueprint).

Leslie (1989) however has attempted to offer an alternative explanation for the lawfulness of the universe by saying that it is a prerequisite for having a good universe, and that there is a teleological explanatory principle that favours goodness, a view which he traced back to Plato in which reality is structured after the Form of the Good on which all existent things owe their being. He claims that the abstract ethical requirement that the good exist has 'creative power' partially to determine (or simply constrain) which possible world exists. Leslie (2016, p. 51) states that "The Good is "what gives existence to things"". In answer to the question whether the abstract ethical requirement would be too purely abstract to act creatively, Leslie (1989, p. 169) writes:

> Well, if by 'being purely abstract' you just mean 'having no practical power' then you entirely beg the question against Neoplatonism when you classify ethical requirements as always 'purely abstract'. Surely requirements for the existence of things are not at all clearly realities of the wrong sort for bringing things into existence. (The abstract truth that two and two make four, or the fact that quadratic equations cannot ride horses, would in contrast be realities quite wrong for this task.)

In other words, what Leslie meant by abstract ethical requirement is not what modern philosophers mean by abstract when they refer to (say) 2 + 2 = 4, that is, things with no causal power (see Chap. 3). Rather, what Leslie meant by abstract ethical requirement is something that have causal powers (indeed, he uses the term 'creative power'). Thus, what he meant by 'abstract' is really what modern philosophers would call 'concrete' causes. (Rosen 2020 notes that 'Plato's Forms were supposed to be causes *par excellence*, whereas abstract objects are generally supposed to be causally inert in every sense.') Hence, what Leslie calls an 'ethical requirement' is actually what modern philosophers would call a concrete necessary existing First Cause that has creative power to bring about universes. However, as explained above, such a First Cause would have to have libertarian freedom and intelligence in order to bring about the first event resulting in a fine-tuned and ordered universe. Hence, such a First Cause would be a Creator God.

Against calling this ethical requirement 'God', Leslie (2016) argues that the ethical requirement is that which accounts for why a world-creating deity exists (p. 54). While noting the strong tradition that God's existence is necessary because God is eternal, Leslie (1989, p. 168) objects that 'the eternal may not be necessary at all; it is logically possible that a thing should simply happen to exist eternally.' Leslie's argument is similar to the Leibnizian and Thomist Cosmological Arguments which claim that, even if the universe is eternal in the sense of having no beginning, it is not necessarily existent[3] and would still require a Necessary Being or a sustaining First Cause to explain its existence. Proponents of these arguments would claim that being beginningless is a necessary condition but not a sufficient condition for necessary existence.

The Leibnizian and Thomist Cosmological Arguments are controversial and it is beyond the scope of this book to settle the controversies here. The following points would suffice to address the issues that are relevant here.

First, it has been explained in Chap. 6 that the Thomist idea of Pure Act is not justified, for one can hold to the alternative view of a First Cause having libertarian freedom to freely actualize its own potential and this would terminate the hierarchical causal series. Second, the postulation of a First Cause that is both beginningless and not being sustained in existence would terminate both the temporal causal series and a hierarchical causal series. Given a First Cause (call this God) that is both beginningless and not being sustained in existence, it is not it is logically possible that God 'happens to exist' given that 'happens' (occurs, comes into being)[4] involves a beginning whereas God (the First Cause) is beginningless. Existing without a beginning and not being sustained in existence implies that God was not brought about; that is, He is uncaused. This is not a case of making its own quality justify its own existence. In fact, it would be fallacious to think of something beginningless as being dependent on its own property of beginningless for its existence, since beginningless is merely a description of the way it has existed (see Chap. 3). The question 'what makes the First Cause beginningless?' is illogical, since being beginningless and unsustained implies that it is uncaused and that nothing makes it this way. Likewise, even if (as Leibniz argues) the First Cause has other properties in addition to beginninglessness which explains why it exists necessarily, it would be fallacious to think of the First Cause as being dependent on that property, since that property would merely be a description of the way it has existed. In any case it should be noted that, as demonstrated above, what Leslie calls an 'ethical requirement' with creative power is really 'a Creator God' rather than 'a property of God which explains why God exists necessarily'. However, if by ethical requirement Leslie intends to refer to what modern philosophers would call an abstract object, then as noted in Chap. 3, abstract objects merely describe relations or possibilities, or are merely exemplifiable by things; they do not make things happen and have no creative power to bring about the first event. Thus, in any case the conclusion that a Creator God exists cannot be avoided.

## 7.3 In defence of Design

As shown by the logically exhaustive list in Chap. 4, the only remaining category of hypotheses is (v) Design. In what follows, I shall reply to various objections against the likelihood of Design.

Philosopher Willem Drees claims that introducing a god as an explanatory notion only shifts the locus of the question: Why would such a god exist (Drees 1996, pp. 267–269)? Likewise, Dawkins has asked the infamous question, namely, if the laws of nature are designed by a God, then who designed this God (Dawkins 2006, p. 188)? Dawkins thinks that consideration of this question renders the existence of God unlikely. He writes:

> The whole argument turns on the familiar question 'Who made God?', which most thinking people discover for themselves. A designer God cannot be used to explain organized complexity because any God capable of designing anything would have to be complex enough to demand the same kind of explanation in his own right. God presents an infinite regress from which he cannot help us to escape. This argument … demonstrates that God, though not technically disprovable, is very very improbable indeed. (Dawkins 2006, p. 109)

In reply, the assumption that complexity by itself requires a designer ('any God capable of designing anything would have to be complex enough to demand the same kind of explanation in his own right') is false. The reason is as follows: 'Design' is a causal notion; '*x* is designed' means that *x* is *caused* to be what it is in accordance with the purposes of a designer. Now it is important to note that the often-held assumption that 'everything has a cause' is false.[5] What the Modus Tollens argument for the Causal Principle in Chap. 3 has shown is that everything *that begins to exist* has a cause. However, if something is without beginning and is not being sustained in existence, then it was not brought about by a cause; it didn't come from nothing nor from anything (since 'brought about' either implies a beginning of existence or being sustained in existence). Such a thing is uncaused, which implies nothing designed it. As explained in previous chapters, the KCA demonstrates that an infinite causal regress is impossible and that there is a beginningless and uncaused Divine First Cause of the universe. The word 'God' is used to refer to the First Cause, which (as explained in Chaps. 3 and 6) is beginningless, initially changeless,[6] and exists uncaused and necessarily and hence undesigned and not fine-tuned, regardless of whether God is complex or simple. Whereas (as explained in Chaps. 3 and 6) physical entities have beginnings and they change continually; therefore, they exist contingently and require an explanation for why they behave in an ordered way.

To elaborate, note that what Dawkins means by organized complexity is something that is composed of a variety of parts arranged in a highly specific manner (Dawkins 1986, Chap. 1). The word 'arranged' implies a beginning to the formation of the arrangement of the parts. It is evident that our physical universe is composed of parts that can be separated from one another, and that these parts can be arranged (e.g. separate pieces of wood can be arranged to form a table). However, a First Cause (God) which is beginningless and initially changeless is not formed by the arrangement of parts, since arrangement implies a beginning and a change whereas the First Cause is beginningless and initially changeless. Therefore, even if Dawkins argues that the ideas in God's mind are parts of God's mind and that God is complex in this manner, it would still remain the case that God does not need a designer because His complexity is of a different sort. That is, as Glass (2012, p. 50) observes, God's mind is not composed of a variety of parts that are arranged together to form the mind of God. This view does not require the notion of divine simplicity (the view that God has no part whatsoever); I have argued in Loke (2014, 2018) that there is insufficient philosophical, theological, or Scriptural justification for this notion. The word 'part' can simply mean that which in some way falls short of being the whole of that entity; this does not imply that the parts are caused or that the parts had been put together to make up the whole. Neither does it imply that the parts are independent and separable. I have argued elsewhere that God's mind can be conceived of as an undivided intuition (Loke, forthcoming). Postulating that the being of God has parts does not violate divine aseity, because one can deny that there is a dependence of the whole on the parts, since the parts and the whole in this case are uncaused and the parts are not prior to the whole.

On the other hand, physical entities have beginnings and they change continually; therefore, they exist contingently and require an explanation for why they behave in an ordered way. Consider the hypothetical scenario by Leslie (1989) noted earlier: a scenario in which 'particles regularly formed long chains which spelled out "GOD CREATED THE

UNIVERSE," this then being shown to result inevitably from basic physics' (p. 109). One might ask whether a designer would not be required if the basic physical laws in this case are beginningless, unsustained and metaphysically necessary brute facts. In reply, as explained in Chap. 2, a law of nature is not a concrete thing but merely a description of events. However, each formation of the long chain as well as each event which ground such a formation has a beginning, and thus (on the basis of the Causal Principle) has a cause. Therefore, these events are not necessary but contingent; that is, they are dependent on their causes, such that later events would have been different if earlier events are different. Moreover, as explained in Chap. 5, a series of events cannot be infinite in the earlier-than direction; therefore, it cannot be beginningless. An atheist might suggest that perhaps the series of events of our universe originated from a physical entity (say) an initial singularity which has no parts and is initially changeless, rather than a Creator. However, as explained in Chap. 6, in order for an initially changeless entity to bring about the first event, it must have libertarian freedom. Furthermore, for it to bring about a series of events that result in a high degree of specified complexity such as the mathematically describable order and fine-tuning, it would require intelligence because other alternative explanations would not work as argued in earlier chapters. Therefore, the initial entity has to be a Creator and Designer.

Oppy has also objected that, since (according to proponents of the KCA) God (the First Cause) could have freely chosen to make a physical world in which it was not the case that highly ordered mathematical theories apply, the existence of a physical world in which such theories apply is a brute contingency on this theistic view just as it is on a particular naturalistic view. Thus, this theistic view does not provide a superior explanation than naturalism for our highly ordered universe, for 'when we get to free choice, and you think, "Why this rather than that?", there's no explanation now to be given of why you ended up with one rather than the other' (Oppy, in Craig 2020). Craig replies that 'On theism, the applicability of mathematics to the physical world is a contingency, but it is not a brute contingency (a "happy coincidence"). It has an explanation in the free decision of a transcendent, personal Designer' (ibid.). Oppy would object that 'Why God freely chose X (our highly ordered universe) rather than not-X' is still a brute contingency.

In reply, first, while on the theistic view it is a brute contingency why God freely chose X, it is not uncaused, because the choosing of X is caused by God (see further, below); thus, it does not violate the Causal Principle established in Chaps. 2 and 3. Whereas to postulate our highly ordered universe began uncaused (which Oppy has suggested, see Chap. 3) would violate the Causal Principle.

Second, while it is a brute contingency why God freely chose X (our highly ordered universe) rather than not-X, it is nevertheless chosen for a reason (e.g. to manifest His wisdom) and involves design by a highly intelligent designer who has the capacity to bring about and thus explain the existence of our highly ordered universe. Whereas to postulate our highly ordered universe began uncaused does not explain why our universe is highly ordered, since 'began uncaused' imply the denial of any such capacity. In other words, on the Design hypothesis, the high degree of ordering of our universe can be accounted for given a highly intelligent Creator who has the capacity to bring about such a high degree of ordering, even if the reason for creation is not a sufficient condition and even if 'why create rather than not create' is not fully accounted for by the reason but is an act of free choice and brute contingency. Whereas on Oppy's hypothesis discussed in Chap. 3, there is no capacity for explaining the high degree of ordering of our universe. (As an analogy, SETI [Search for Extra-Terrestrial Intelligence] researchers can reasonably conclude that Extra-Terrestrial Intelligence exists if they pick up a certain signal under certain circumstances, given their knowledge that an Intelligence with the relevant capacity is required to produce the signal. This conclusion should be accepted even if we do not know why the ETI choose to produce [rather than not-produce] the signal [for all we know, this may be a brute contingency due to the libertarian free choice of the ETI], and even if we do not yet have independent evidence for ETI producing the signal.)

In response to the above two objections, Oppy might defend the alternative possibility that our highly ordered universe did not begin uncaused but instead arose indeterministically from a metaphysically necessary, impersonal, and highly ordered initial state of reality in accordance with necessary natural laws which are indeterministic.

In reply, first, I have argued above in Sect. 6.4.1 that the problem with this postulation is that the metaphysically necessary initial state of reality is initially changeless, immaterial, and has libertarian freedom, and therefore it is not impersonal.

Second, the hypothesis of an uncaused and un-designed Mind as the First Cause does not face the problem which besets the hypothesis of an uncaused and un-designed universe. That is, the former hypothesis can satisfactorily explain how mindless physical entities can *consistently* behave in an orderly manner which can be described by mathematical equations, and how it can consistently manifest a uniformity and rationality which human rationality can discern and systematize. Whereas the latter hypothesis, being mindless, cannot explain these satisfactorily, as argued previously. As Rasmussen and Leon (2018, pp. 104–105) elaborate using the notion of intentionality:

> Fundamental reality has intentional powers, which themselves do not depend upon fine-tuned material conditions. Intentional powers allow the foundation to aim for interesting ends, such as an evolution leading to a complex creature who can make a princess drawing. With intentional powers at the foundation, we have a mechanism to explain why the world unfolds toward something beyond merely dots of dust. This mechanism provides a probability pump, which renders organized complexity far more probable/expected. Of course, a mind that itself depends on material complexity would only relocate the problem; its existence would then be just as surprising (i.e., improbable) as the material complexity we are seeking to explain. For this reason, a foundational mind would, by hypothesis, be a mind that exists prior to material complexity. The foundational mind does not depend on organized complexity; rather, it provides the ultimate explanation of all organized forms.

Moreover, to postulate our universe arose indeterministically from a metaphysically necessary, impersonal, and highly ordered initial state of reality does not explain why our universe is fine-tuned. As argued in Chap. 4, the fundamental principles or laws of nature do not uniquely determine a fine-tuned universe (and avoid the Boltzmann Brain problem, etc.), and 'physics is blind to what life needs. And yet, here we are' (Lewis and Barnes 2016, p. 181).

Against the existence of an Immaterial Mind, it might be objected that our experiences of human minds indicate that they do not exist apart from the body. This claim has been challenged by other scholars using various arguments for substance dualism, including the evidences of near-death experiences (Loose et al. 2018). In any case, the claim is based on a limited sampling of human minds on earth; it does not show that an immaterial mind cannot exist anywhere else in the universe or apart from the universe. The association of physical brains with minds can be regarded as an accidental property akin to human beings have always lived on the Earth, which was true until 1968. Out-of-body experiences are intelligible notions, even if one does not believe them. Likewise, a timeless immaterial mind is an intelligible notion and not self-contradictory, and indeed most philosophers throughout history have no problem conceiving it, and this is the reason for thinking that mind-physical dependence is 'accidental' to the notion of a mind.

One might object that, given that the minds which we know of (e.g. human minds) are in time, the view that there can be an initially timeless Divine Mind is special pleading. In reply, special pleading is an informal fallacy wherein one cites something as an exception without justifying the special exception. Saying that God's thoughts are initially timeless is not a case of special pleading because there are at least two justifications for it, namely:

1. The premises of the KCA-TA, from which it follows that an initially timeless Creator and Designer exists.
2. God's thoughts can be fundamentally similar to ours in the sense that they involve intentionality, awareness of logical relations, and so on—there is no need to be in time in order to possess these properties. Likewise, having intelligence means having knowledge, understanding, foresight, wisdom, purpose, and intention; it does not mean/imply/require having spatial or temporal extension. One can think of a Mind having an initially timeless awareness of truths (including truths about highly ordered structures) and which has the capacity to bring about something in accordance with these truths.

Against the conclusion of a Designer, Hume has objected that the 'design' seems to be less than perfect; for example, if the purpose of creating the universe was to allow for life, this universe shows examples of 'imperfect design', such as the presence of natural evil such as tsunamis, hurricanes, and so on that destroy life (Hume 1779/1993, pp. 68–69, 71, 113).

Nevertheless, this objection does not show that the existence of a Designer is unlikely—at least, not on my argument-by-exclusion formulation of the design argument.[7] To illustrate, if one were to discover in the midst of a jungle a factory which has the capacity for making motorcars, one would reasonably conclude that it was designed even if some of the equipment in the factory were faulty and even if all the cars would be destroyed eventually (e.g. due to corrosion of its parts). The reason is because it is unreasonable to think that the components of the factory were fundamentally brought together and assembled by Chance, Regularity, or Combinations of Regularity and Chance, or that the factory began to exist Uncaused, and as explained previously the only remaining explanation is Design. The fact that some of the equipment in the factory were faulty or that the parts are corruptible does not refute this conclusion and could be due to various other reasons. One might think that it is due to an imperfect designer, but it could also be due to another person who came and disrupted the factory after it was built, or it may be due to a perfect designer who allows for these imperfections for his other purposes which we are presently unaware of. Thus, on the one hand, the conclusion that the designer is unlikely or the designer is imperfect does not follow from the presence of imperfections in the factory, because there are alternative explanations which need to be considered and ruled out (and they have not been ruled out). On the other hand, the conclusion that a Designer exists follows from the existence of the factory given that we have ruled out the alternative explanations to Design.

Likewise, even though there are imperfections within the universe, it remains the case that the evidence of fine-tuning and the laws of the universe which are describable by sophisticated mathematical equations indicate the existence of a Designer. The reason is because, as explained in previous sections of this book, it is unreasonable to think that the fine-tuning and order were brought about by Chance, Regularity, or Combinations of Regularity and Chance, or that they are Uncaused, and the only remaining explanation is Design. The fact that there are imperfections within the universe does not refute this conclusion and could be due to various other reasons. One might think that it is due to an imperfect designer, but it could also be due to another person (e.g. fallen angelic beings) who disrupted the designer's creation (Peckham 2018), or it may be due to a perfect designer who allows for these imperfections for his other good purposes which we are presently unaware of.

Goff (2019, p. 107) notes that theists can try to come up with explanations for why God would allow suffering, but he objects that this can seem like special pleading or ad hoc alterations. However, this objection would not work if the explanations given are justified on the basis of reasons/evidences and/or follow from the postulation of theism itself. For example, Goff (ibid.) complains that the observation that life had come about through the gruesome process of natural selection falsified theism. However, he fails to note that it has been argued that choosing to care for the weak, lonely, and vulnerable is a harder thing for humans to do in a Darwinian world, and this makes moral behaviour such as freely choosing to care for those in need to be of great value, and hence God who cares about such moral value chose to create a Darwinian world in which moral behaviour that are of such great value can exist (Peels 2018). Moreover, the wonders of nature, which include the incomprehensible degree of fine-tuning and the 'very advanced mathematics' involved in constructing the universe (Dirac 1963) explained in Chap. 4, indicate that the Designer's wisdom far exceeds ours. Given the evidence for the existence of such a God, 'we should not expect to grasp more than a small fraction of either the goods which lead God to act as God acts (including divine acts of allowing evil) or the constraints that make such divine allowings needful' (Perrine and Wykstra 2017, p. 86). Therefore, even though we may not know the reason why God allows certain instances of suffering, that does not mean there is no good reason which is known to God.

Sober (2019, pp. 51, 67) objects that the last response to the problem of evil, namely, that it is very hard for human beings to understand what God's goals are, would likewise undercut the Teleological Argument, for how then could we know that God would want to create a world that could support intelligent life?

In response, we can know that there exists a Designer who wanted to create a world that could support intelligent life by ruling out all the possible alternative explanations for such a world (viz. (i) Chance, (ii) Regularity, (iii) Combination of Regularity and Chance, and (iv) Uncaused), and this has been accomplished in the earlier parts of this book, but we have not ruled out all the possible good purposes which this Designer (whom we call 'God') might have for allowing imperfections. The Teleological Argument for the existence of God does not require direct access to the purposes which the mind of the Designer (if such a Designer exists) would have—indeed, we have no such access[8]—but only direct access to the world of the phenomena by which we can discover the phenomena of fine-tuning and order and infer that there is a Designer by ruling out the alternative explanations (see further, Sect. 7.6). Whereas the argument from evil against the existence of God requires the proponent of the argument to rule out the purposes which the mind of the Designer (if such a Designer exists) might have in order to rule out the possibility that there might be good purposes for why the Designer (if He exists) might allow suffering, but given the failure to do so, the argument fails. (The literature on the problem of evil is huge and it is beyond the scope of this book to discuss this issue further;[9] for examples of other responses, see Loke 2022a and the sources cited.)

Thus, on the one hand, the conclusion that the designer is unlikely or the designer is imperfect does not follow from the presence of imperfections in the universe, because there are alternative explanations which need to be considered and ruled out (and they have not been ruled out). On the other hand, the conclusion that a Designer exists follows from the existence of fine-tuning and order of the universe, given that we have ruled out the alternative explanations to Design in previous sections of this book. It might be objected that we have not established that the Designer is indeed perfect or morally good, but a proponent of the Teleological Argument can reply that it is not the purpose of the argument to do so (see further, Chap. 8).

It should also be noted that the Teleological Argument does not require the premise that there is order everywhere. For example, think again of the factory in a middle of a messy jungle. Even though there is disorder everywhere around the factory, the presence of the factory would still require an explanation—how did the parts of the factory (e.g. the parts of the assembly line which install the engine, install the hood, install the wheels, etc.) come together to form the factory? Likewise, even though there are apparent chaos and disorder in various parts of the universe, the presence of fine-tuning and mathematically describable order of the universe would still require an explanation, and I have argued that the best explanation is design.

Against God being life-loving, it has been asked why is there so little of life in the universe? 'Why didn't God choose laws that permit life to exist across a much wider range of possible values for their constants?' (Sober 2019, pp. 66–67), such that there is life (say) in Venus, Mars, and so on? Why are most regions in the universe hostile to life?

In reply, on the one hand, we do not know how many living things God actually created to conclude that there is only a little of it, given the possibility that there could be many life forms in faraway regions of the universe or in other spiritual dimensions (in which angelic beings, for example, may dwell). On the other hand, it can be argued that God in His foresight created laws such that there is no evidence of life in Venus, Mars, and so on and then 'suspended these probabilities' by creating life on earth so as to show that He cares for the earth and the living things on it. In any case, even if there are no life anywhere else in the universe, the fact remains that, if the universe is not fine-tuned, there would not be life anywhere in the universe, including planet earth, and it has been argued previously that the best explanation for this is design.

A theological objection to fine-tuning has been raised by Halvorson (2018), who argues that, if God could be expected to create a nice universe, then God could also be expected to set favourable chances for a nice universe, which He did not; therefore, the fine-tuning argument defeats itself. In support of his main claim he writes:

> Consider a sinister game of reverse Russian roulette: your captor hands you a revolver with five chambers filled, and one empty. Now suppose that you pull the trigger, and you hear 'click' … you've survived. What should you conclude? Should you conclude that your captor rigged the game so that

> you wouldn't die? But then why would your captor begin the game by filling five of the six chambers? Why not fill only one … or, even better, don't fill any at all? … In application to the FTA, the analogy is as follows: God created laws such that almost all physically possible universes are lifeless. And yet, the fine-tuning advocate wants us to believe that God designed this 'game' so that we would win. Wouldn't this be a strange way for a deity to operate? Why would God make things hard for himself? (p. 126)

Halvorson's objection raises interesting theological issues concerning fine-tuning. Is a universe fine-tuned for life also fine-tuned for death?

Nevertheless, there are at least two problems with Halvorson's objection.

First, Halvorson's reverse Russian roulette analogy is not quite appropriate; according to his analogy, one out of six of the chambers of the revolver was not filled, and 1/6 (= 0.166 …) is a non-negligible probability. Thus, even though you survived, you might still wonder whether your captor rigged the game because he wanted you to live, or did you survived by chance. Against the former hypothesis, you might ask why would your captor begin the game by filling the rest of the chambers in the first place, and you might conclude that perhaps he just wanted to play the cruel game with you for the fun of it without rigging the game (since there is a non-negligible probability that you survived by chance). However, it has been argued in Chap. 4 and this chapter that all the alternative hypotheses to Design are extremely unlikely and thus have negligible probabilities (far lower than 0.001; in cryptography, negligible probability is typically assigned a value of less than $1/2^{128}$; as noted in Chap. 4, the fine-tuning of entropy alone has been argued to be lesser than that). It would be analogous to the resolver having zillions of chambers, all of which except one was filled. In that case, the fact that you survived would leave you in no doubt that your captor had rigged the game so that you would live, regardless of why your captor began the game by filling the rest of the chambers in the first place.

Second, Halvorson's objection has similarities to the Deist Voltaire's (1764/1901, p. 273) objection against miracles when he claims that ascribing miracles to God would indicate a lack of forethought:

> It is impossible a being infinitely wise can have made laws to violate them … if He saw that some imperfections would arise from the nature of matter, He provided for that in the beginning; and, accordingly, He will never change anything in it.

However, McGrew (2013) notes that Paley (1794/1859, p. 12) and others have replied that God in His foresight would have wanted to set up a universe with regularities that no mere human could abrogate and then suspended them so as to authenticate a revelation.

Likewise, with regard to Halvorson's objection, it can be argued that God in His foresight determined the laws of nature such that almost all physically possible universes are lifeless, and 'suspended these probabilities' by creating a universe that is fine-tuned so as to authenticate a revelation, namely, His General Revelation though His creation (for a theological defence of General Revelation and Natural Theology, see Sudduth 2009; Loke 2019). In other words, God wants life to be naturally unlikely so that we would recognize His hand in designing a life-permitting universe. Moreover, if God had created the natural laws such that life is naturally likely, we might take it for granted that we are alive, whereas the fact that it is naturally unlikely and yet we are alive would make many people feel grateful to be alive. It is widely recognized that gratitude is a virtue and thus it is plausible that God would want to foster it. Concerning Halvorson's question 'Why would God make things hard for Himself?', as a professing Christian, Halvorson should have known that, according to Christian theology, God is willing to make things hard for Himself in order to accomplish His loving purposes for humankind, even to the extent of enduring the suffering of the crucifixion for our sake in order to redeem us.[10]

I shall now show how the logically exhaustive list of hypothesis devised in Chap. 4 is useful for answering a number of objections against the inference of the Teleological Argument.

## 7.4 Reply to Hume's Classic Objections

Many have thought that the Teleological Argument has already been dealt a death blow by philosopher David Hume. Historian of science Jonathan Topham writes, 'It has sometimes seemed inexplicable to historians that natural theology, and particularly the argument from design, continued to be so prevalent in the anglophone world in the wake of Hume's assault' (Topham 2010, p. 66).

Topham summarizes Hume's assault as follows:

1. The central analogy between natural phenomena and human artefacts could not be used convincingly to infer the God of Christianity.
2. The universe was so unlike human productions that the analogy between the two was extremely tenuous.
3. Such analogies were based on so limited a knowledge of the universe; perhaps at other times and in other places nature was even less like a machine.
4. This was the only universe of which anyone had experience, invalidating it as the basis of an inductive inference.
5. No one had had direct experience of the creation of a universe.
6. Even if one allowed that the universe was the product of an intelligent designer, that would only lead to an infinite regress, since the designer's intelligence would require explanation.
7. Since the cause inferred must be proportionate to the effects, such a designer could not be the infinite being of Christian theology.
8. One could not be sure whether there was one designer or many, or,
9. indeed, given the imperfections in nature, whether the designer(s) was incompetent or malevolent.
10. There were other analogies that might be considered to be at least as satisfactory as that between the universe and a machine, such as that between the universe and a living organism. In this case, one might argue that, since all animals were actuated by a soul, God must be the soul of the world; or one might argue that, like a plant, the world had grown from a seed.

11. The appearances of design in nature might reasonably be accounted for as the fortuitous consequences of a chaotic system of matter in motion (ibid.).

While the 11 objections represent an extensive critique of the Teleological Argument, they are far from fatal. With respect to objections 1, 2, and 10, it has been shown in this book that the conclusion of the Teleological Argument does not have to be based on analogy, but can be shown to be based on argument by exclusion. With respect to objections 3, 4, 5, and 11, it has been shown that, even though our knowledge of the universe is limited, that this was the only universe of which anyone had experience, and that no one had had direct experience of the creation of a universe; nevertheless, given the Causal Principle (see Chap. 3) and the conclusion that our universe has a beginning (see Chaps. 5 and 6), the order and fine-tuning that we observe would still require a causal explanation and the best explanation is still design (the conclusion is arrived at by exclusion and not inductively). With respect to objection 6, an infinite regress has already been refuted in Chap. 5, and as explained above in response to Dawkins, a beginningless, uncaused, and intelligent First Cause would not require an explanation. With respect to objection 9, I have argued above that the 'imperfections' do not imply the unlikelihood of Design. With respect to objections 1, 7, and 8, the Teleological Argument is never intended to be a sufficient argument for the Trinity Monotheistic God of Christianity but part of a cumulative case which includes (for example) the historical argument for the claims and resurrection of Jesus (Craig and Moreland ed. 2009; Loke 2017, 2020).

## 7.5 Addressing an Objection to Argument by Exclusion

A sceptic might object that, even if each of the alternatives to Design is improbable, their disjunction is not improbable. For example, consider the outcome of rolling a fair die. Even if the probability of each of the alternatives to 6 (i.e. 1, 2, 3, 4, 5) is rather low (i.e. 1/6), their disjunction is not improbable (i.e. 5/6).

In reply, first, the die example assumes that each outcome has a non-negligible probability: 1/6 (= 0.166 …) is non-negligible and we often do see the outcome of (say) 3 happening naturally. However, it has been argued in previous chapters that each of the alternatives to Design is not the case or extremely unlikely and thus has zero or negligible probability (far lower than 0.001; in cryptography, negligible probability is typically assigned a value of less than $1/2^{128}$).

Second, in the case of rolling a fair die, it can be shown that the outcome of 6 has equal probability to each of the non-6 alternative outcomes. However, it has been explained earlier that, while it has been shown that each of the alternatives to Design has negligible probability, it has not been shown that Design has equally negligible probability.

The die example however is analogous to the case for Design in this sense: the probability of each of the logical alternatives must add up to 1 (1/6 + 1/6 + 1/6 + 1/6 + 1/6 + 1/6 = 1). Likewise, the epistemic probability of each of the five possible categories of explanations— namely: (i) Chance, (ii) Regularity, (iii) Combination of Regularity and Chance, (iv) Uncaused, and (v) Design—must add up to 1. Since each of the four alternatives to Design has negligible probability and that the probability of the disjunction of four negligible probabilities is negligible, it can be concluded that our universe is designed (i.e. the probability of Design has negligible difference from the probability of 1). (It might be objected that one could also reverse the direction of the argument from exclusion, so that [according to the critic] the improbability of design as the explanation should lead us to think that there is a higher probability of non-design explanations than we had previously estimated.[11] However, I have already argued previously that, on the one hand, there is no good reason to think that Design is improbable; on the other hand, the improbability of non-design explanations is well-established.) Even if one disagrees with my assessment that each of the naturalistic alternative hypotheses has 'negligible probability', one can still say that each of these naturalistic alternatives has been shown to be very improbable. For example, even if one assigns to each of the four naturalistic alternatives a probability of 1 in 1000 (which is very generous given the arguments in previous chapters), that still leaves Design with a high epistemic probability of 99.6%. This should warrant assent from a reasonable person.

## 7.6 Response to Difficulties Concerning Determining the Prior Probability that God Design the Universe

My argument from exclusion avoids a difficulty often mentioned against other approaches to inferring design, namely, the difficulty of assigning *prior* probability for Design. Proponents of fine-tuning argument have argued that, while the fine-tuning is improbable under atheism, it is not improbable under theism: '[Since] God is an all good being, and it is good for intelligent, conscious beings to exist, it is not surprising or improbable that God would create a world that could support intelligent life' (Collins 1999, p. 54). It is good for embodied, intelligent, conscious beings to exist because 'intelligent conscious beings can actualize noble values in the world, such as moral values, aesthetic values, and epistemic values' and they can be aware of God and 'can communicate and establish a deep relation of love with God, if God exists' (Chan and Chan 2020, pp. 6–8 citing Swinburne). Halvorson (2018, p. 129) notes that a defender of the Fine-Tuning argument would argue that, while a life-permitting universe is improbable conditional on God's non-intervention, it is probable conditional on God overriding the probabilistic laws of physics, but he objects that 'not many of us—even the theists among us—have a prior probability for the claim that God will intervene in a certain situation'. Sober claims that the likelihoodist formulation of the design argument is the best formulation,[12] but it is beset by the problem of assigning prior probability for Design given the difficulties of knowing the putative designer's goals (pp. 29, 44–45, 62). Moreover, our ground rules of inferring intelligent design are based on our empirical knowledge of *human* intelligence, which may not carry over to hypotheses involving non-human intelligent designers (Sober 2003, p. 38; see also Manson 2020, who argues that God's mind is so different from ours that we cannot judge what God would likely do; thus, the probability that there is a life-permitting universe if God exists should be regarded as *inscrutable*). Likewise, Grünbaum complains that we have no independent evidential access to God's choices and motives. He argues that this is unlike the case of ordinary action-explanations, for example, an unreasonable reprimand of an academic colleague by the department chairman, where we have access to independent evidence as to the content of the agent's motives other than the action taken by the agent. He thinks that, absent such evidence, we should reject the proffered action-explanation as viciously circular (Grünbaum 2000, section 3). The problem of assigning prior probability for Design is further accentuated by the presence of imperfect design (see Sect. 7.5), which atheists argue are evidences against the goodness of the Creator assumed by Collins et al. While theists can reply that this objection fails using the approach of sceptical theism, atheists might reply that the sceptical theism approach highlight the difficulties of knowing the putative designer's goals mentioned by Sober.

Now Barnes (2019; citing Hawthorne and Isaacs 2017, 2018) has used what he calls the *Awesome Theistic Argument* test (ATA) to argue that the kind of inscrutable probability objection raised by Manson to the Fine-Tuning Argument (FTA) fails, as follows:

> Manson contends that the fine-tuning sceptic can limit the extent of their judgement of inscrutability, so that while being unconvinced by the FTA, they could agree that "there would be evidence of God's existence if, for example, the stars miraculously rearranged themselves to spell out the Nicene Creed" (2018: 5). And yet a starry Nicene sceptic could block this argument by claiming that the probability of "We believe in one God, the Father Almighty, Maker of all things visible and invisible …" appearing in the night sky if God exists is inscrutable. This, if anything, is more plausible than declaring that the probability of a life-permitting universe on theism is inscrutable, and yet the conclusion is absurd. If the starry Nicene sceptic would be irrational to block that argument by appealing to inscrutability, then the fine-tuning sceptic must also be irrational.

My argument-by-exclusion formulation of the design argument complements the ATA by showing *why* the conclusion is absurd (see the analogy of 'discovering a factory in the jungle' mentioned in Sect. 7.3), while also avoiding the above objections which beset the likelihoodist formulation of the design argument.

To elaborate, given that the list of categories of hypotheses mentioned previously (viz. (i) Chance, (ii) Regularity, (iii) Combination of Regularity and Chance (iv) Uncaused, and (v) Design) is logically exhaustive, and given that the laws of logic are necessarily true for all entities human or non-human (see Chap. 1), we can argue for the Design hypothesis by exclusion and without vicious circularity and without violating any ground rules. This can be done by arguing that, while the alternatives to design are unlikely, the Design hypothesis is not. Given that all the alternatives to design fail (as has been shown previously), it can be argued using a Modus Tollens argument:

1. If there is no intelligent designer of the universe, the universe would not be fine-tuned and highly ordered given the failure of all the alternative hypotheses (viz. Chance, Regularity, Combinations of Chance and Regularity, Uncaused).
2. The universe is fine-tuned and highly ordered.
3. Therefore, there is an intelligent designer of the universe.

Therefore, we should accept the conclusion of design regardless of whether we have access to independent evidence concerning the content of the agent's motives. Swinburne points out that we can often have strong evidence for a hypothesis that a particular person committed the crime, without having the slightest idea of his reasons for bringing it about in the exact way that he did (Swinburne 2005, p. 924). Likewise, we can have strong evidence for a hypothesis that an event—for example, a magician pulling out a rabbit from the hat—happens as a result of design without knowing how the designer (e.g. the magician) pulls it off. Thus, objections based on our ignorance of the motives or mechanisms of the process of divine creation (e.g. 'we really do not know how God "pulls it off"') fail to rebut the conclusion that the laws of nature are designed. In other words, 'we often are able to tell that an intelligent designer made an object even though we have no idea what that putative designer's goals were' (Sober 2019, pp. 44–45).

Against this, Sober objects by claiming that this inference is an inductive sampling reasoning which

> focus exclusively on the causes we have actually observed; it ignores causes that may have operated before human beings existed, or that may have operated far away in space, or that may have occurred too slowly for human beings to notice. The inductive sampling version of the design argument is biased against theories that postulate unobservable processes. (ibid., p. 29)

In reply, my argument is not based on inductive sampling but based on deduction using a logically exhaustive list of hypotheses which covers all possible hypotheses, regardless of whether they involved entities that exist long ago or far away or processes that are too slow or unobservable, and the conclusion of Design is arrived at by exclusion of the alternative hypotheses based on their essential characteristics. Hence, my argument is not susceptible to Sober's objection.

In summary, my argument by exclusion—based on the logically exhaustive list of hypotheses formulated in Chap. 4—can lead to the conclusion that the universe is designed without having to first assign a prior probability for Design, thus avoiding the objections by critics on this point entirely. In this aspect, my formulation of the design argument is better than the likelihoodist formulation as well as other formulations which are beset by those objections.

Concerning the prior probability of naturalism versus prior probability of theism, I have argued above that, on the one hand, there is no good reason to think that prior probability of theism is low and the prior probability of naturalism is high. While many atheists would subjectively push up the prior probability of naturalism (due to simplicity), the criterion of simplicity is only valid if all else is equal. The theist can use the KCA to argue that all else is not equal and that theism has a higher prior than atheism. Moreover and in any case, the observation by Lewis and Barnes (noted in Chap. 4) that our conclusions would not depend much on the prior probability of the theory if our data is very good implies that the final probability of the constants being 'fine-tuned' by the 'Chance hypothesis' would be very low, and I have argued in Chaps. 4, 5, and 6 that this problem cannot be avoided by all the other alternative hypotheses to Design (Regularity, Combination of Regularity and Chance, and Uncaused). Therefore, we can conclude that the final probability of the Design hypothesis is high.

## 7.7 Reply to Objections Concerning the Range of Explanatory Latitude

Against Swinburne's defence of the Teleological Argument from the order of the universe, Grünbaum (2004, p. 605) objects that, whatever the laws of nature turn out to be, the theist would explain these as *brought about by God; hence, the range of the explanatory latitude of the theistic volitional explanation is too permissive and the supposed evidences (i.e. the laws of nature) provide no check on the validity of the explanatory premises.* Grünbaum complains that the proposed theistic explanation fails to transform scientific brute facts into specifically explained regularities, for contrary to Swinburne's contention, the divine volitional explanation provides no epistemically viable account of why the physical energy conservation law holds, let alone of why the magnitude of the total energy is what it is (ibid., p. 562).

In reply, Grünbaum's objection would only work against Swinburne's version of the argument, which claims that 'The very same criteria which scientists use to reach their own theories lead us to move beyond those theories to a creator God who sustains everything in existence' (Swinburne 1996, p. 2). This claim makes Swinburne vulnerable to the objection that his theistic hypothesis does not make predictions in the same way as scientific theories, and that it does not transform scientific brute facts into specifically explained regularities the way Grünbaum demanded. Likewise, an important reason why several authors have objected to Dembski's eliminative approach (see Chap. 4) by emphasizing the necessity of providing some positive argument in favour of design is because Dembski claims that his theory of Intelligent Design is scientific, and according to these authors' definition, a scientific theory would be expected to make a range of testable predictions (Dawes 2007, pp. 71, 79; Fitelson et al. 1999, p. 487).

However, Grünbaum's objection would not work against the argument from the mathematically describable order of the universe presented in this book. For the argument defended here does not follow Swinburne in claiming to use the very same criteria which scientists use to reach their own theories. Contrary to Dembski, my book does not claim to defend Design as a scientific theory. Instead, I argue in Chap. 1 that science is not the only way to knowledge (contra scientism), that science itself requires the laws of logic, that the laws of logic imply that the conclusion of a deductively valid argument with true premises must be true (regardless of whether it makes testable predictions), and I have explained in the rest of the book why my argument is deductively valid and why the premises are true. Therefore, the conclusion of Design is true.

Additionally, the argument defended in this book is not based on the premise that the laws of nature should be able to be described by one mathematical form rather than the other, but that they should be able to be described by any highly ordered mathematical form at all. It is true that a range of possible laws of nature describable by a range of possible mathematical equations is possible. Nevertheless, given that a particle, for example, could have moved in billions of alternative directions other than consistently in the direction described by any form of mathematical equation (see Chap. 4), the explanatory latitude of the Design hypothesis is still vastly more restricted than the hypothesis that there is no external creative cause. Thus, the observations concerning whether particles do move in the manner describable by mathematical equations would still serve as a check with regard to the evidences for the Design hypothesis, and these observations constantly confirm the evidences for the Design hypothesis. It is true (as Grünbaum argues) that the hypothesis that God exists entails nothing about the numerical value of the energy of the universe being of a certain value E (Grünbaum 2005, p. 935), given that God could have assigned other values (Swinburne 2005, pp. 923–924). Nevertheless, as explained in Chap. 4 and this chapter, the evidence that particles do behave in the manner describable by mathematical equations is still evidence for the conclusion that there is a Designer who, for whatever reason, causes them to behave in this manner, resulting in the numerical value of the energy of the universe being of a certain value E. Swinburne explains it thus,

> But of course the probability that he would choose a particular disjunct is low; and I am not appealing to there being just the amount of energy there is (rather than some other slightly different amount) as confirmatory evidence of the existence of God. But the evidence which I am adducing as evidence of the existence of God confirms the claim that he brought about just the amount there is … Analogously, footprints of a kind that the suspect would have made if he had been at the scene of the crime confirm the hypothesis that he was at the scene of the crime and so put his feet in the exact position when the footprints were found, without it being the case that the prints being at that exact position rather two millimetres to the west has any confirming effect on the hypothesis. (ibid.)

One might complain that, just as the existence of God does not entail that the numerical value of energy in the universe should be E rather than other value, likewise, the existence of God does not entail that the universe exhibits very sophisticated mathematical order, given that God could have chosen to create a universe without this order. Why then should we think that the existence of sophisticated mathematical order is evidence for God?

In reply, the reason why E is not evidence for God is not merely because the existence God does not entail E, but also because there are alternative plausible explanations for E that does not involve a designer. Whereas, in the case of the existence of sophisticated mathematical order, we have already ruled out the plausibility of alternative explanations, and therefore this should be regarded as an evidence for a Designer (God). To elaborate, given the vast number of possible alternative disordered schemes and given that the alternative categories of hypotheses in the logically exhaustive list are unlikely (see Chap. 4), the probability that, without an external intelligent cause, we should observe the ordered scheme which we do observe is extremely low, and this is evidence against the null hypothesis that no external intelligent cause is required (Cf. Grünbaum 2004, p. 599).

Therefore, even though the existence of God does not entail the existence of a highly mathematically ordered physical reality, nevertheless the existence of a highly mathematically ordered physical reality is evidence for God because all the alternative explanations have been excluded.

Against appealing to God as the Creator of natural laws, Mumford (2004, pp. 147–148) complains that how God's decrees come to be manifest in nature remains unexplained. He writes: 'they are essentially supernatural, so how do laws have effects in nature? This is not a compelling model of how laws govern. This relation between laws and the world is a paradigmatic *deus ex machina*.' The latter is illustrated by the classic cartoon of the scientist writing the elaborate theorem on the chalkboard with '*then a miracle occurs*' in step two to fill in for what he could not work out. Others might object that accepting God as a conclusion opens the floodgates to virtually any competing explanation where one can just posit 'the ability to do X' to solve the problem, such as posit the intelligence and power to create a universe to a Magic Beaver.[13]

In reply, my argument does not postulate a Designer as a *deus ex machina*, nor is the conclusion of my argument based on ignorance (my argument is not 'because we don't know how to explain the laws of nature, therefore God'). Rather, the conclusion is based on the analysis of the necessary conditions (e.g. what is required for an initially changeless First Cause to bring about the first event) and follows from deduction and exclusion (we know by deduction that there are only a few possibilities and all the rest have been excluded, therefore God). My argument does not posit a First Cause having libertarian freedom merely as a possible solution among many alternative solutions. Rather, I have explained that a First Cause having libertarian freedom and intelligence follows deductively from the premises I presented. Thus, there is no other possibilities and no floodgates opened to a Magic Beaver for which we have no independent reason or evidence to think is the case. The classic cartoon case is disanalogous because the 'miracle' does not follow from the previous steps of the theorem and is based on ignorance of what should follow from those steps, and this ignorance is open to being filled by all kinds of alternative explanations such as a Magic Beaver to be posited to solve the problem. Whereas my conclusion follows deductively from my premises and is not based on ignorance but on reasons and analysis of the necessary conditions (e.g. what is required for an initially changeless First Cause to bring about the first event). Thus, it is not open to being filled by other explanations because there isn't any other viable logical alternative and there is only one viable conclusion which follows deductively from the premises, namely, the conclusion that the Creator and Designer of the universe exists.

Not knowing how the supernatural affect the natural is not a compelling objection, because our lack of understanding of a relation is not a good reason to reject the existence of the relation. As Koons and Bealer point out, physics itself admits lawful relationships among physical entities that are extraordinarily diverse in nature and, in turn, admits relations of causal influence and law-grounded explanation among these entities. Physics allows, moreover, that some of these lawful relationships are brute facts having no further explanations (Koons and Bealer 2010, p. xviii). Likewise, the relationship between mind and body (and between 'supernatural' and 'natural') could well be a brute fact having no further explanation. Kojonen (2021, p. 64) notes that

> the problem of not being able to provide further details about the mechanism is not necessarily unique to theism: as Dawes (2009, pp. 51–53) notes, in all explanations there comes a point where we reach the level of basic causal powers, and are unable to specify further intermediate mechanisms. To insist on an explanation at such a truly basic level would just lead to an infinite causal regress.

Moreover, SETI (Search for Extra-Terrestrial Intelligence) researchers can reasonably conclude that Extra-Terrestrial Intelligence exists if they pick up a certain signal under certain circumstances, even if they do not yet know the actual mechanism by which the Extra-Terrestrial Intelligence created the signal.

## 7.8 Conclusion

In this chapter, I complete my refutation of the alternative hypotheses to design by offering two considerations against the hypothesis that the fine-tuning and order of the physical universe is Uncaused (the other alternatives have already been refuted in Chap. 4). First, all such models cannot work because, as explained in previous chapters, the physical universe cannot be the uncaused First Cause; rather, it is constantly changing and has a first event, which implies it has a beginning and therefore (according to Causal Principle) has a cause; hence, its properties of being fine-tuned and highly ordered would have a cause.

Second, the 'Uncaused' hypothesis does not explain how it could be the case that unthinking, mindless things consistently accord with natural laws.

I have defended the hypothesis that the best explanation for why unthinking mindless physical entities consistently have such an orderly behaviour is that there is a Mind who determined that they should be like this, by replying to various arguments against the likelihood of Design. For example, in answer to the infamous question 'Who designed God?' (Dawkins 2006, p. 188), I have explained that 'God' refers to the First Cause which is beginningless, initially changeless, uncaused and necessarily existent and hence is un-designed. In reply to the objection from 'imperfections' such as the presence of natural evil (Hume 1779/1993), this objection, even if successful, does not imply that a designer is unlikely, only that the designer is imperfect (moreover, as noted above, various plausible theodicies concerning why a perfect Designer might allow evil have already been offered by scholars; see Loke 2022a). Against the theological objection that, if the universe is fine-tuned, it should not be the case that almost all physically possible universes are lifeless (Halvorson 2018) or that most regions in our universe are hostile to life (Sober 2019, pp. 66–67), it can be argued that God wants life to be naturally unlikely so that we would recognize His hand in designing life.

In conclusion, while the alternatives to design are unlikely, the Design hypothesis is not. Since the list of hypothesis is logically exhaustive as shown in Chap. 4, one can argue for the Design hypothesis by exclusion without having to first assign a prior probability for Design, thus avoiding the objections by critics on this point entirely.

Moreover, my argument does not postulate a Designer as a *deus ex machina*, nor is the conclusion of my argument based on ignorance. Rather, the conclusion is based on the analysis of the necessary conditions (e.g. what is required for an initially changeless First Cause to bring about the first event) and follows from deduction and exclusion. My argument does not posit a First Cause having libertarian freedom merely as a possible solution among many alternative solutions. Rather, I have explained that a First Cause having libertarian freedom and intelligence follows deductively from the premises I presented and that there is no other viable possibility. Hence, there are no floodgates opened to be filled by other explanations because there isn't any other viable logical alternative and there is only one viable conclusion which follows from the premises, namely, the conclusion that the Designer of the universe exists.

## Notes

1. Oppy (2013) also considered the alternative possibility that there are at least some aspects of fine-tuning of natural causal reality that arise contingently at non-initial stages of natural causal reality as the results of the outplaying of objective chance. However, this possibility has been refuted in Chap. 4 when considering the Chance hypothesis.
2. Cosmologist Don Page wrote to me about this in personal correspondence, attributing it to Stephen Hawking.
3. Lewis and Barnes (2016, p. 328) argues that 'The Universe is not a necessary being because "there is nothing necessary about how it is, or that it is, or how it behaves", unlike (say) a triangle which necessarily has 3 vertices. This is why science needs observations; we can't figure out the Universe from our armchairs. We need to go outside and look.'
4. *Merriam-Webster Dictionary.*
5. Concerning the historical circumstances that led to the lamentable prevalence of this false assumption among atheist philosophers (Bertrand Russell et al.), see Clarke (1970).
6. With regard to the concerns raised by the Thomistic Cosmological Argument, the initial changelessness of the First Cause implies that the First Cause does not require a sustaining cause; the subsequent changes can be understood as being initiated and sustained by the libertarian freedom of the First Cause (see Chap. 6).
7. This objection may affect other formulations, such as the likelihood formulation. See Sect. 7.6.
8. Unless the Designer chooses to grant us such an access in some ways.
9. I discuss this issue in greater detail in Loke (2022).
10. Concerning the debate about divine impassibility and a defence of the view that the Second Person of the Trinity suffered in his human nature, see Loke (2014, chapter 4).

11. I thank an anonymous reviewer for suggesting this objection.
12. He constructs a likelihoodist formulation of the design argument as follows:

    Pr (the value of physical constant *x* is in *W* | God set the value of *x* & *W* is narrow) >
    Pr (the value of physical constant *x* is in *W* | a mindless chance process set the value of *x* & *W* is narrow) (p. 62).

13. I thank Vaal for raising this objection.

## Bibliography

Barnes, Luke. 2019. A Reasonable Little Question: A Formulation of the Fine-Tuning Argument. *Ergo* 6 (42).

Chan, K., and M. Chan. 2020. A Discussion of Klass Landsman's Criticisms of the Fine-Tuning Argument. *Theology and Science*.https://doi.org/10.1080/14746700.2020.1755544.

Clarke, Norris. 1970. A Curious Blindspot in the Anglo-American Tradition of Antitheistic Argument. *The Monist* 54: 181–200.

Collins, Robin. 1999. A Scientific Argument for the Existence of God: The Fine-Tuning Design Argument. In *Reason for the Hope Within*, ed. M. Murray. Eerdmans: Grand Rapids.

Craig, William Lane. 2020. Explaining the Applicability of Mathematics. https://www.reasonablefaith.org/writings/question-answer/explaining-the-applicability-of-mathematics.

Craig, William Lane, and J.P. Moreland, eds. 2009. *The Blackwell Companion to Natural Theology*. Chichester: Wiley-Blackwell.

Dawes, Gregory. 2007. What Is Wrong with Intelligent Design. *International Journal for Philosophy of Religion* 61: 69–81.

Dawkins, Richard. 1986. *The Blind Watchmaker*. London: Penguin Books.

———. 2006. *The God Delusion*. London: Bantam Press.

Dirac, Paul. 1963. The Evolution of the Physicist's Picture of Nature. *Scientific American* 208: 45–53.

Drees, Willem. 1996. *Religion, Science, and Naturalism*. Cambridge: Cambridge University Press.

Ellis, Brian. 2001. *Scientific Essentialism*. Cambridge: Cambridge University Press.

Fitelson, Brandon, Christopher Stephens, and Elliot Sober. 1999. How Not to Detect Design. *Philosophy of Science* 66: 472–488.

Frederick, Danny. 2013. A Puzzle About Natural Laws and the Existence of God. *International Journal for the Philosophy of Religion* 73: 269–283.

Glass, David. 2012. Darwin, Design and Dawkins' Dilemma. *SOPHIA* 51: 31–57.

Goff, Philip. 2019. Did The Universe Design Itself? *International Journal for Philosophy of Religion* 85: 99–122.

Grünbaum, Adolf. 2000. A New Critique of Theological Interpretations of Physical Cosmology. *The British Journal for the Philosophy of Science* 51: 1–43.

———. 2004. The Poverty of Theistic Cosmology. *The British Journal for the Philosophy of Science* 55: 561–614.

———. 2005. Rejoinder to Richard Swinburne's 'Second Reply to Grünbaum. *The British Journal for the Philosophy of Science* 56: 927–938.

Halvorson, Hans. 2018. A Theological Critique of the Fine-Tuning Argument. In *Knowledge, Belief, and God*, ed. Matthew Benton, John Hawthorne, and Dani Rabinowitz. Oxford: Oxford University Press.

Hawthorne, John, and Yoaav Isaacs. 2017. Misapprehensions About the Fine-Tuning Argument. *Royal Institute of Philosophy Supplement 81*: 133–155.

Hawthorne, J., and Y. Isaacs. 2018. Fine-tuning Fine-Tuning. In *Knowledge, Belief, and God*, ed. M. Benton, J. Hawthorne, and D. Rabinowitz. Oxford: Oxford University Press.

Hume, David. 1779/1993. Dialogues Concerning Natural Religion. In *Dialogues and Natural History of Religion*, ed. J.A.C. Gaskin. Oxford: Oxford University Press.

Kojonen, Rope. 2021. *The Compatibility of Evolution and Design*. Palgrave Frontiers in Philosophy of Religion Series. Cham: Springer Nature.

Koons, Robert and George Bealer, eds. 2010. *The Waning of Materialism*. Oxford: Oxford University Press.

Lange, Marc. 2009. *Laws and Lawmakers. Science, Metaphysics, and the Laws of Nature*. Oxford: Oxford University Press.

Leslie, John. 1989. *Universes*. London: Routledge.

———. 2016. A Way of Picturing God. In *Alternative Concepts of God*, ed. Andrei Buckareff and Yujin Nagasawa. Oxford: Oxford University Press.

Lewis, Geraint, and Luke Barnes. 2016. *A Fortunate Universe: Life in a Finely Tuned Cosmos*. Cambridge: Cambridge University Press.

Loke, Andrew. 2014. *A Kryptic Model of the Incarnation*. London: Routledge.

———. 2017. *God and Ultimate Origins: A Novel Cosmological Argument. Palgrave Frontiers in Philosophy of Religion Series*. Cham, Switzerland: Springer Nature.

———. 2018. Review of Timothy Pawl's in Defense of Conciliar Christology: A Philosophical Essay. *Faith and Philosophy* 34: 114–119.

———. 2019. Theological Critiques of Natural Theology: A Reply to Andrew Moore. *Neue Zeitschrift für Systematische Theologie und Religionsphilosophie* 61: 207–222.

———. 2020. *Investigating the Resurrection of Jesus Christ: A New Transdisciplinary Approach*. London: Routledge.

———. 2022a. *Evil, Suffering and Christian Theism*. London: Routledge.

Loose, Jonathan, Angus Menuge, and J.P. Moreland. 2018. *The Blackwell Companion to Substance Dualism*. Oxford: Wiley Blackwell.

Manson, N. 2020. How Not to Be Generous to Fine-Tuning Sceptics. *Religious Studies* 56 (3): 303–317.

McGrew, Timothy. 2013. Miracles. In *The Stanford Encyclopedia of Philosophy*, Edward N. Zalta (ed.). Available at http://plato.stanford.edu/archives/spr2013/ entries/miracles/.

Mumford, Stephen. 2004. *Laws in Nature*. London: Routledge.

Oppy, Graham. 2013. Ultimate Naturalistic Causal Explanations. In *The Puzzle of Existence*, ed. Tyron Goldschmidt. New York: Routledge.

Paley, William. 1794/1859. *A View of the Evidences of Christianity*. London: John W. Parker and Son.

Peckham, John. 2018. *Theodicy of Love: Cosmic Conflict and the Problem of Evil*. Grand Rapids: Baker.

Peels, Rick. 2018. Does Evolution Conflict with God's Character? *Modern Theology* 34 (4) https://doi.org/10.1111/moth.12435.

Perrine, Timothy, and Stephen Wykstra. 2017. Skeptical Theism. In *The Cambridge Companion to the Problem of Evil*, ed. Chad Meister and Paul Moser. Cambridge: Cambridge University Press.

Rasmussen, Joshua, and Felipe Leon. 2018. *Is God the Best Explanation of Things: A Dialogue*. Cham: Springer Nature.

Rosen, Gideon. 2020. Abstract Objects. In *The Stanford Encyclopedia of Philosophy*, ed. Edward Zalta. https://plato.stanford.edu/archives/spr2020/ entries/abstract-objects/.

Sober, Elliot. 2003. The Argument from Design. In *God and Design*, ed. N. Manson (pp. 25–53). Routledge. Reprinted in *The Blackwell Guide to Philosophy of Religion*, 2004, ed. W. Mann (pp. 117–147).

———. 2019. *The Design Argument*. Cambridge: Cambridge University Press.

Sudduth, Michael. 2009. *The Reformed Objection to Natural Theology*. London: Routledge.

Swinburne, Richard. 1996. *Is There a God?* New York: Oxford University Press.
———. 2005. Second Reply to Grünbaum. *The British Journal for the Philosophy of Science* 56: 923–924.
Topham, J. 2010. Natural Theology and the Sciences. In *The Cambridge Companion to Science and Religion*, ed. Peter Harrison. Cambridge: Cambridge University Press.
Voltaire. 1764/1901. Philosophical Dictionary. In *The Works of Voltaire*, Vol. 11. New York: E. R. DuMont.

# 8

# Ultimate Designer

## 8.1 Summary of Important Conclusions from Previous Chapters

We have come to the end of our quest concerning the ultimate design of our universe, a quest that has brought together the disciplines of philosophy, science, and religion. The justification for using such a transdisciplinary approach for gaining a fuller understanding of reality has been given in Chap. 1, where it has been shown that scientism, verificationism, and empiricism are untenable. In particular, it has been demonstrated that philosophical considerations are important for scientific theories, and that what is mathematically possible is not concretely possible if it violates certain metaphysical considerations (while what is mathematically impossible is concretely impossible as well). Hence, even if something is mathematically and/or logically possible but metaphysically impossible, it would still be actually impossible. This is significant because it implies that appealing to cosmological models which postulate an infinite regress of events (e.g. Eternal Inflation Model) or logically consistent closed causal loop are incapable of defeating the metaphysical arguments



A. Loke, *The Teleological and Kalam Cosmological Arguments Revisited*, Palgrave Frontiers in Philosophy of Religion, https://doi.org/10.1007/978-3-030-94403-2_8

against the possibilities of infinite regress of events and closed causal loops presented in this book. Instead, those metaphysical arguments stand as a defeater for these cosmological models.[1] In view of the importance of philosophical considerations explained in Chap. 1, cosmologists should not merely construct models of the universe without considering the philosophical arguments against certain models. (The neglect of this point has contributed to the lack of consensus in cosmology; it has been said that one can come up with any theorem to prove that the universe has a beginning or no beginning depending on the assumptions[2]; this book has shown which assumptions should be rejected.)

This book has contributed to the discussion by developing these philosophical arguments in engagement with modern science, and demonstrating that whatever begins to exist has a cause (against recent objections in Linford 2020 and others) and it is not the case that there is an actual infinite regress of events or a causal loop which avoids a First Cause. This implies that there is a first event and a First Cause.

It has often been objected that we cannot observationally confirmed this First Cause caused the universe and that we cannot know the answer concerning ultimate reality because that is far beyond our ordinary experiences. I have replied to these objections by explaining in Chap. 1 that the laws of logic are necessarily true, and that they would hold even at levels of reality far beyond our ordinary experiences, such as at the beginning of time or at the level of timelessness (there cannot be shapeless squares at such levels too). Following the laws of logic, the conclusion of a deductively valid argument from true premises must be true (regardless of whatever realm of reality), and I have explained in previous chapters that the Kalām Cosmological Argument is deductively valid and the premises are true, and therefore the conclusion that there is a First Cause with libertarian freedom is true.

Concerning the Teleological Argument, while many scientists and philosophers have argued that there are evidences of design in fundamental physics, others have objected that there could be alternative hypotheses which have yet to be considered (Ratzsch and Koperski 2019). I have addressed this problem by developing an original deductive argument which demonstrates that the following are the only possible categories of hypotheses: (i) Chance, (ii) Regularity, (iii) Combination of Regularity and Chance, (iv) Uncaused, and (v) Design. I have demonstrated that there are essential features of each category such that, while the alternatives to design are unlikely, the Design hypothesis is not. Since the list is logically exhaustive, the epistemic probabilities of the five categories must add up to 1. Even if one assigns to each of the four alternatives a probability of 1 in 1000 (which is very generous given the arguments presented in this book), that still leaves Design with a high epistemic probability of 99.6%. One can thus argue for Design by exclusion without having to first assign a prior probability for Design (thus avoiding the objection in Sober 2019), and I have shown that my argument avoids the problems that beset alternative forms of design inference.

An objector might complain that the conclusion of the Teleological Argument still falls short of 100 % epistemic certainty. Then again, there isn't 100 % epistemic certainty for most things in life either. For example, there isn't 100 % epistemic certainty that what you are reading now is authored by a human being rather than a randomly typing monkey. The latter hypothesis is logically possible, yet unlikely. The more probable answer ought to be accepted as the true answer; that is, what you are reading now ought to be regarded as the product of an intelligent author, though certainly not as intelligent as the Creator who created such elegant equations of physics and such astonishing degree of fine-tuning! The fine-tuning and order of the universe is therefore a strong evidence for a highly intelligent Creator, and given an epistemic probability of at least 99.6% this ought to be accepted as the true answer.

Thus, the Teleological Argument by itself—in particular, the undeniable evidence of the mathematically describable order of the universe by itself (see for example Steiner's point in Sect. 4.4.1)—is already sufficient for concluding that the universe has a Creator. I have also explained previously that the KCA by itself is sufficient for concluding that the universe has a Creator. Therefore, even if one of these two arguments is refuted, the conclusion that the universe has a Creator can still stand.

I have argued that both arguments are in fact defensible, and that the Cosmological Argument can be used to strengthen the Teleological Argument even further, by answering the question 'Who designed the Designer?' through demonstrating that there is a beginningless and undesigned First Cause with libertarian freedom. On the other hand, the

Teleological Argument can be used to strengthen the Cosmological Argument by providing considerations for thinking that a First Cause brought about the first event intelligently rather than accidentally or naturally. In particular, it has been shown in Chaps. 4 and 7 that it is unlikely that unintelligent cause(s) can bring about a universe in which mindless unthinking physical entities constantly behave in ways that can be predicted by mathematical equations, which can be treated by physicists as expressing a kind of software of the universe (Heller 2013, p. 594). It has also been shown that it is unlikely that unintelligent cause(s) can generate an 'explosion' such as the Big Bang that would bring about the creation of ordered systems (solar system, quantum system, etc.) rather than disorder and debris. Thus, the best explanation is that the present universe is the result of an intelligent Designer who programmed the 'software' and engineered the 'explosion'.

## 8.2 Concerning the God-of-the-Gaps Objection

It might be asked whether the arguments defended in this book are God-of-the-gaps arguments and whether the conclusion that the Creator exists is based on ignorance. When ancient people did not understand certain natural phenomena (such as thunder), they thought that these are caused by the gods (e.g. Thor). As scientific understanding progresses, such religious explanations are replaced by scientific ones. Haught (2004, p. 238) notes that the problem with 'God of the gaps' explanations is that they appeal to God 'at a point in inquiry when there is still plenty of room for further scientific elucidation'. Therefore, it might be objected that the fine-tuning argument is based on current science which might be explained away eventually by a naturalistic answer with the progress of science (Loeb 2014). Carrier claims that 'scientists have consistently found physical explanations for every phenomenon they have been able to thoroughly examine .... There is not a single instance on record of any fact that has been thoroughly examined by scientists that turned out to have no identifiable physical origin' (Carrier 2003).

In reply, the conclusion of the KCA is not based on ignorance. Rather, it is based on reasons. The argument is not 'because we still do not know how to explain the origin of the universe, therefore there is a Creator'. Rather, the argument is, because there are reasons (discussed in previous chapters) for thinking that an actual infinite causal regress and a closed causal loop is not the case, therefore there is a First Cause. It is because there are reasons for thinking that whatever begins to exist has a cause, therefore this First Cause is beginningless. The rest of the properties of this First Cause are likewise derived on the basis of reasons rather than ignorance, as shown above. Moreover, as explained previously, each step of the argument is strictly deductive in nature, for which no alternative explanation is possible, whereas 'a 'god of the gaps' explanation is one on which it is at least possible in principle that some nondivine explanation might be correct' (Feser 2017, p. 271).

Contrary to Carrier, scientists have not found physical explanations for the ultimate origin of our universe. While the progress of science would generate newer understandings of the laws of nature as explanations for the phenomena we observe, as shown by the KCA, the progress of science would not replace a First Cause (Creator) as an explanation for the existence of all things, including the laws of nature themselves which must have come from this First Cause.

Concerning the Teleological Argument, one might attempt to explain away design (using science or otherwise) by appealing to alternative explanations. However, where the mathematical order and fine-tuning is concerned, it has been shown in previous chapters that all the possible alternative explanations (chance, regularity, combinations of chance and regularity, uncaused) would fail as ultimate explanations for these phenomena. Ratzsch and Koperski (2019) note that 'evidence of design in nature does not automatically imply gaps. Design built or "front-loaded" into nature from the very beginning would require no further interventions within the historical flow of nature and therefore no gaps.' Hume et al. have claimed that an infinite regress of causes/events is possible given which there is no beginning for design to be front-loaded into. However, it has been shown in this book that the KCA can be used to strengthen the Teleological Argument by demonstrating that an infinite regress of causes/events is not the case and thus there is a first event, and that this first event cannot have been brought about by a regular/natural/mechanistic/scientific process but by an act of libertarian freedom of the First Cause. (See also the response by Frederick in Sect. 7.1 concerning the mathematically describable order of the universe. As noted earlier, the undeniable evidence of the mathematically describable order of the universe by itself is already sufficient for the Teleological Argument; the discovery of the evidences for fine-tuning only makes the argument stronger.)

Therefore, while the science concerning fine-tuning will be updated in the future, with regard to whatever scientists discover (e.g. a new law of nature), it can still be asked where did that come from (e.g. where did that law of nature come from). The basic logical form of my argument would still remain, and no matter what scientists discover in the future, there must still be a First Cause for that discovery. Even if scientists discover one day that our universe is a digital simulation (Bostrom 2003) or it was created by intelligent being(s) living in another universe (Harrison 1995), we could still ask where did that digital simulation/intelligent being(s) come from; that is, what caused it? If one claims that the intelligent being or the cause of the digital simulation is uncaused, that would imply that the intelligent being/cause of the simulation is the First Cause of our universe and it would also have other properties which the First Cause must have as deduced by the KCA-TA, namely, is beginningless, initially changeless, has libertarian freedom, intelligence, and is enormously powerful. In other words, the conclusion of the KCA-TA that a Creator of the universe exists would still follow.

Even if we live in an illusory world (e.g. in a matrix), the conclusion would still follow. In such a world, it remains the case that the existence of changes and beginning of changes cannot be denied. As Craig (Copan and Craig 2017 vol. 1, p. 67) notes, on the thesis of the mind-dependence of becoming, there is at least the appearance of temporal becoming of the physical world. An illusion or appearance of becoming involves becoming, so that becoming cannot be mere illusion or appearance. Thus, even the radical sceptic who doubts all of his/her perceptions of the world external to his/her mind must still grant the existence of changes in his or her own subjective mental states. While we observe changes through a filter of perception, we cannot deny that changes exist. Given the impossibility of an actual infinite regress of changes, as well as the truth of the

Causal Principle that whatever begins to exist has a cause (the violation of which would entail that his/her subjective experiences would be very different from what they are), the conclusion that an initially changeless First Cause with libertarian freedom and intelligence (i.e. Creator) exists would still follow.

Contrary to Carrier (2003), who claims that all physicists would find a non-naturalistic conclusion to be quite absurd, many of the greatest physicists throughout history (e.g. Newton, Einstein [see Chap. 4, footnote 5) have recognized God as the ultimate explanation for the existence and order of the universe. They do not regard this conclusion as anti-scientific because they do not hold to the fallacious ideas of scientism (see Chap. 1). They recognize that philosophy examines primary causes while science examines secondary causes. Cosmologist William Stoeger offers an account of how science, philosophy, and theology can complement one other concerning ultimate origins:

> Physics and cosmology as sciences are incapable of exploring or directly accounting for the ultimate source of existence and order which philosophy and theology, properly understood, provide. By the same token, philosophy and theology are not equipped to investigate and describe the processes and relationships which contributed to the expansion, cooling and subsequent structuring of the universe on macroscopic and on microscopic scales. Thus, philosophy and theology seek to provide an understanding of the origin and evolution of the universe which is complementary to that which physics and cosmology contribute. (Stoeger 2010, p. 174)

Thus, philosophy can let us know about Divine First Cause while leaving scientists (e.g. cosmologists) to find out the secondary causes concerning mechanisms and to work on understanding the details of the process. The fact that the latter is still unknown and there is no consensus among cosmologists at this time does not contradict the conclusion that the former can be known using philosophical arguments such as the KCA-TA. With regard to the objection that we should always try to find a scientific explanation, the KCA-TA demonstrates that the ultimate explanation cannot be a scientific one, because the first event is brought about by a First Cause with libertarian freedom (premise 8) and not by a mechanism describable by a law of nature. The possession of libertarian freedom by the First Cause implies agent causation and a personal explanation. Moreover, being initially changeless, the First Cause is not a physical entity (such as the universe or multiverse) which is characterized by constant changes. While the progress of science would generate new theories to explain various aspects of the physical world, it would not replace the First Cause (Creator) as the ultimate explanation for why the physical world exists in the first place, as demonstrated by the KCA-TA. Thus, the conclusion of the KCA-TA cannot in principle be overturned by future scientific discoveries. Rather, future discoveries would only enhance our understanding of the wisdom of the Creator through understanding the laws which He had created.

## 8.3 Limitations of the KCA-TA and responses

One might object that that KCA does not rule out other timeless concrete entities existing alongside God, and neither does it prove that there is only one First Cause.

In reply, one can speculate about other entities which may or may not exist, but what needs explanation is the series of changes which we observe within our universe, and I have already explained why an infinite regress of changes is impossible and why this implies that there is an initially changeless First Cause with libertarian freedom. The conclusion that there is a single First Cause is more reasonable than multiple first causes given the widely accepted scientific principle (Ockham's razor) that causes should not be multiplied beyond necessity (Craig and Sinclair 2009, p. 192). This principle is widely used by atheists (e.g. Oppy 2013a), who think that, since thunder (for example) can be sufficiently explained by natural laws, there is no need to postulate a thunder god to explain it; thus, the existence of a thunder god should be rejected. Likewise, theists can argue that, since a single Creator is sufficient to explain the origin of the universe, there is no need to postulate additional creators or other timeless concrete entities.

The conclusion of a single Creator is further strengthened by Sudduth's (2009, p. 210) observation:

> The unity of order throughout the cosmos is evidence for a single cause of this order. If we postulate a single designer, then we would expect to find the same fundamental physical laws governing the behavior of objects over vast distances of space and time in the cosmos. We would also expect to find different particular physical laws explicable in terms of these fundamental physical laws.

Concerning premise 6 'since the First Cause is the first, it is uncaused', it has been noted in Chap. 2 that I am referring to the First Cause of change and that this First Cause is not something that is brought into existence. One might object that such a First Cause might nevertheless be something that is sustained in existence, and thus is caused in the sense of having a sustaining cause. If that is the case, then given the impossibility of infinite regress of sustaining causes or a closed loop, the First Sustaining Cause would be the true First Cause (here, the word 'cause' is used in a different sense, not as a cause of change, but as something that sustains another thing changelessly). Such a sustaining First Cause might not be the entity which brought about the first event (cf. premise 10 of KCA-TA), and it might be impersonal.

Two points may be said in response.

First, while Aquinas had famously argued for a First Sustaining Cause and he was not a proponent of the KCA and did not think that a First Cause of time can be demonstrated, he nevertheless affirmed that there is such a First Cause of time on the basis of Christian tradition and that the First Cause of time is also identical with the First Sustaining Cause. Now there are disputes concerning whether the Thomistic Cosmological Argument is sound, and I have argued in Sect. 6.4 that, if there is a First Sustaining Cause, there is no good reason to think that it is a Pure Act which is distinct from the First Cause of time. On the contrary, I have argued that the First Cause of time can also be the One who sustains all else in existence. Therefore, it would be simpler (following Ockham's razor) to regard the First Sustaining Cause to be identical with the First Cause demonstrated by the Kalām.

Second, atheists who affirm a naturalistic First Cause (e.g. Oppy and Hawking) typically assume that this First Cause is not being sustained in existence by (say) a Thomistic First Cause. For the sake of parity, the theistic proponent of the KCA may assume the same, given that (as explained under the first point above) there is no good reason to think that there is a First Sustaining Cause which is distinct from the First Cause of time.

Another objector to the KCA might suggest the hypothesis that there are two beginningless beings—God and primordial matter—and that God (the First Cause with libertarian freedom) caused the primordial matter to change, hence bringing about the first event of physical reality. In this case, the primordial matter would be the material First Cause without libertarian freedom but it might be enormously powerful (like a powerful bomb waiting to be triggered), while the efficient First Cause has libertarian freedom but may have little power (the trigger of a bomb may have little power in itself). In this way, the conclusion that there is one First Cause with both libertarian freedom and enormous power may be avoided.

Three points may be said in response.

First, the above scenario which is intended to avoid the conclusion of this book faces the problem that the efficient First Cause with libertarian freedom would still need to have enormous power and intelligence in order to form a highly ordered and fine-tuned universe from the material according to his intelligent plan.

Second, the so-called primordial matter would be initially changeless and hence (as argued in Chap. 6) immaterial, given which it is problematic to call it matter.

Third, there is no good reason to think that there are two beginningless beings rather than one. Therefore, it would be simpler (following Ockham's razor) and—in light of points 1 and 2—less problematic to think that there is one beginningless First Cause with both libertarian freedom and enormous power.

Goff (2019, p. 106) objects that theism incurs a large cost in terms of qualitative parsimony by postulating an immaterial and necessary being which is an addition type of entity to the physical and contingent universe, and it violates the theoretical virtue of having a unified conception of reality by postulating a supernatural God distinct from the natural world.[3] He propose an alternative view (constitutive comopsychism) which postulates that the universe is a conscious subject with a 'basic disposition to form spontaneous mental representations of the complete future consequences of all of the choices available to it' (p. 112). He notes that proponents of KCA have argued that the universe has a timeless, necessarily existent, and personal cause, and argues that the agentive cosmopsychist can accept their conclusions by adopting the following two theses:

- The entity E that is the physical universe exists necessarily and has its spatiotemporal properties contingently (i.e. 'physical universe' is a phase sortal of E as 'adulthood' is a phase sortal of a person), and
- E as a non-spatiotemporal entity caused the Big Bang (i.e. the non-physical phase of E caused its physical phase) (p. 120).

He claims that 'given that physical science tells us nothing of the intrinsic nature of the universe, physical science can give us no grounds for holding that something with such an intrinsic nature is essentially spatiotemporal' (ibid.).

In reply, although parsimony/simplicity is one of the considerations for evaluating the prior probability of hypotheses, it can be defeated by other considerations. Now Swinburne (2004, p. 53) has stated that

> the prior probability of a theory depends on the degree of its fit with background knowledge (an a posteriori matter), and on its simplicity and scope (features internal to the theory and so an a priori matter). A theory fits with our general background knowledge of how the world works in so far as the kinds of entities and laws that it postulates are similar to those that probably (on our evidence) exist and operate in other fields.[4]

The problem with Goff's theory is that it doesn't fit with 'our general background knowledge of how the world works' (Swinburne) and it requires ad hoc postulations in order to make it fit. To illustrate, SETI [Search for Extra-Terrestrial Intelligence] researchers can reasonably conclude that Extra-Terrestrial Intelligent Being exists if they pick up a certain signal under certain circumstances. Suppose someone postulates an alternative hypothesis that the physical universe itself (without the ETI beings) generated the signal. This would be a more parsimonious hypothesis, but it would rightly be rejected because our background knowledge indicates that the physical universe itself (apart from intelligent beings) does not have the capacity to generate such a signal. Thus, the alternative hypothesis has extremely low prior probability. To object to this conclusion by postulating that the physical universe itself might have the capacity to generate such a signal under special circumstances is ad hoc. Likewise, to postulate that the physical universe itself might have intelligence which can set up itself under special conditions is ad hoc. Note that my objection is not question begging because it does not start by assuming that Goff's interpretation of our observation of the universe is wrong. Rather, it starts by observing physical entities and inferring that his postulation of those additional characteristics is ad hoc.

Now, it is not ad hoc to conclude that the universe has a First Cause which is initially changeless, necessarily existent, personal (has libertarian freedom), and intelligent, since this is justified by the reasons and evidences presented in the earlier parts of this book. However, it is ad hoc to postulate that the initial state of the universe is a First Cause which is initially changeless, necessarily existent, personal (has libertarian freedom), and intelligent. The reason is there is no independent evidence that the physical universe which we observe has such properties. On the contrary, all the evidence we have of the universe shows that it does not (for example) freely moves around the planets in ways other than that described by the laws of nature. In other words, our observation of the universe implies that Goff's hypothesis has extremely low prior probability. It is inadequate to respond by saying that our universe does have the property of following the laws of nature which have teleological properties. The reason is because the problem concerning the origin of the universe and fine-tuning does not merely concern the present laws of nature but also the arrangement of the initial conditions. It is like arranging different parts of a factory together (before those parts run according to programmed laws). When we observe the universe it is obvious that it does not have the capacity to bring together different parts of the factory to set up a factory; the laws of nature are unintelligent in that sense. Likewise, it is implausible to think that it could have fine-tuned and set up itself. Consider the analogy of discovering a car factory in a jungle mentioned in Chap. 7. Even if the parts of the factory are faulty for whatever reason (cf. problem of evil Goff mentioned), it is still reasonable to conclude that the factory had an independent designer rather than to think that it designed itself, since it is obvious that the factory is unintelligent and does not have that capacity to set up itself.

Goff might reply by speculating that the universe has a mind and is trying to maximize the good under certain limitations as expressed by the laws of physics.[5]

However, scientific evidence has shown that (regardless of whether the universe has a mind or not) the 'limitations' are very severe. That is, the physical universe behaves in law-like regular ways rather than behaving in ways which indicate that it is capable of arranging things together to form something like a car factory which can set up different systems of an automobile. Therefore, it is unlikely that the universe could have set up itself, or fix its initial conditions in such a way that different systems (e.g. quantum systems, solar systems, biological systems) would eventually form.

Goff might reply by postulating that, because the limitations were broken during the Planck epoch at the beginning of our universe where physical laws break down, the universe might have the capacity to fine-tune itself during that epoch. To illustrate the absurdity of his ad hoc hypothesis, one can postulate that, because the limitations were broken during the Planck epoch at the beginning of our universe where physical laws break down, the universe might have the capacity to generate fine-tuned special signals during that epoch, signals which (because of the fine-tuning and the breaking down of physical laws) cannot be traced back to the Planck epoch but which can be translated as intelligent messages later on. SETI scientists would reject the above hypothesis as ad hoc. They would object that the fact that our current scientific models break down during the Planck epoch does not mean we can postulate anything we want to the universe during the epoch to explain anything we want, even if the resultant hypothesis might be more parsimonious than postulating a universe with aliens. Likewise, scientists ought to reject Goff's hypothesis by arguing that the fact that our current scientific models break down during the Planck epoch does not mean we can postulate any kind of 'theory of everything' to the universe during the epoch to explain anything we want (such as evidence of fine-tuning), even if the resultant hypothesis might be more parsimonious than postulating a universe with the God of traditional theism.

Secondly, the conclusion (justified by the arguments in previous chapters) that this First Cause (A) is initially changeless and (B) has libertarian freedom to initiate or prevent itself from initiating the first event already implies that the First Cause is utterly different from the physical world and not describable by natural laws. Concerning (A), as noted previously, according to quantum physics, physical entities constantly fluctuate (i.e. change) at the quantum level as described by the Heisenberg uncertainty principle (Boddy, Carroll & Pollack 2016). To suggest that our current scientific model collapses in the Planck epoch to such an extent that even the fundamental understanding of physics and of natural law that 'physical entities change' no longer applies seems to be equivalent to postulating a non-physical and 'supernatural' origin, rather than origination by the physical universe itself. Goff might reply by postulating that the universe is not essentially physical, and that he is hypothesizing that a 'non-physical God became the universe'. But how is the change from 'non-physical' to 'physical' not supernatural? Moreover, Goff's hypothesis that 'God became the universe' requires that God must still have been distinct from the universe before 'becoming' the universe. Additionally, his hypothesis is as implausible as suggesting that 'the alien which generated the signal message became the signal'. The 'becoming' involves an (unnecessary) extra step which is less parsimonious. It is simpler to postulate that 'the alien created the signal' without postulating that 'the alien became the signal'. Likewise, it is simpler to postulate that 'God created the fine-tuned universe' without postulating that 'God became the universe and allowed the natural laws to limit himself after the Planck epoch'. Concerning (B), our background knowledge of the scientific evidence indicates that no causal relation found in the hard sciences resemble anything like having the (libertarian) freedom to initiate or prevent itself from initiating an event. Again, this indicates that Goff's hypothesis has extremely low prior probability, and that it is ad hoc for Goff to postulate that the physical universe has this freedom which manifested under special consideration. The point concerning the initial changelessness and libertarian freedom of the First Cause is that we are warranted by the evidence to conclude that there exists an entity with a nature which is distinct from physicality as described by natural laws. The term supernatural is usually used for such an entity. Goff might refuse to use this term, but this does not deny the conclusion that such an entity exists. I have argued above that the conclusion that such an entity is non-identical with the universe is less ad hoc and more parsimonious than his hypothesis that they are identical.

Hence, the First Cause should be regarded as something that is distinct from the physical world. Given this, and given that properties such as being (initially) changeless, necessarily existent, having libertarian freedom and intelligence are contrary to our observation of the physical universe but are what theists traditionally meant by 'God',, who is supposed to be very different from the observed universe, the conclusion of theism and the associated cost of violating qualitative parsimony and unified conception of reality are warranted.

## 8.4 Significance of the Conclusion of KCA-TA

The conclusion that the First Cause is initially changeless as well as immaterial and has libertarian freedom indicates that the First Cause is ontologically distinct from the material universe; this is a hallmark of traditional theism in distinction from pantheism (Forrest 2016). It implies that events describable by physical law have a beginning; that is, there is a first event, which implies that materiality has a beginning, which is consistent with *creatio ex nihilo*.

It is true that the KCA-TA by itself does not prove that this First Cause has other properties which many people associate with God, namely, morally perfect, Triune, and so on. Nevertheless, we still need to consider who is this First Cause of our universe who is immaterial, has libertarian freedom (and hence personal), is intelligent, and enormously powerful (and who might well be morally perfect, Trinity, etc.)? If we do not call this First Cause God, then what shall we call Him? There are good reasons for calling Him God, given that hardly any atheist (a person who affirms that there is no God) would acknowledge that there is such a First Cause and still remain an atheist.

Even if we do not call this First Cause God, we should at least call Him the Creator, given that the First Cause has libertarian freedom and is the designer of our universe. One might seek to find out whether there are evidences which indicate that this Creator had revealed Himself in other ways—for example, through the moral law in human conscience (Baggett and Walls 2016) and His acts in history (Loke 2017, 2020, 2021)—to provide us with additional reasons for thinking that He is indeed morally perfect, and so on, and to reveal to us His ultimate purposes for creation and His plan for our lives.

Robert Lawrence Kuhn (2020) has observed that the question 'Why there is something rather than nothing?' is a question that supersedes all other questions. Against this, Maudlin (2018) claims that this question is 'a silly question which obviously has no satisfactory answer', 'for to "explain" existence you either cite something that exists or you don't. If you do you have begged the question, and if you don't then you haven't provided an explanation.' However, Maudlin fails to note that it is not question begging to cite something with properties which logically terminate the question.

To elaborate, when one asks 'why?', one is looking for an explanation. For example, when one asks 'why is there something called Andrew Loke rather than no Andrew Loke?', the answer is his parents brought him into existence and therefore explain why he exists. Since there cannot be an infinite regress of explanations (see Chap. 5), the series of explanations must terminate in an uncaused First Cause with libertarian freedom, that is, a personal Creator God (Chap. 6). Such a First Cause does not need to be explained, since it is beginningless, unsustained, and necessarily existent (Chap. 3). It would therefore be meaningless to ask why is there a First Cause rather than nothing, because there cannot be an explanation for this First Cause since this First Cause is the terminus to the series of explanations. In other words, this First Cause (God) has properties which logically terminate the question. Therefore, this First Cause is the answer to the question 'Why is there something rather than nothing?' Contrary to Maudlin, this question is not a silly question. Rather, it is one of the most important questions humanity has ever asked, a question which leads humanity to God.

Stephen Hawking (2018, p. 29) has observed that 'it is hard to think of a more important, or fundamental, mystery than what, or who, created and controls the universe'. Albert Einstein has stated that 'everyone who is seriously engaged in the pursuit of science becomes convinced that the laws of nature manifest the existence of a spirit vastly superior to that of men, and one in the face of which we with our modest powers must feel humble' (Jammer 1999, p. 93). Richard Dawkins has acknowledged:

> When I lie on my back and look up at the Milky Way on a clear night and see the vast distances of space and reflect that these are also vast differences of time as well, when I look at the Grand Canyon and see the strata going down, down, down, through periods of time when the human mind can't comprehend, I'm overwhelmingly filled with a sense of, almost worship … it's a feeling of sort of an abstract gratitude that I am alive to appreciate these wonders. When I look down a microscope it's the same feeling: I am grateful to be alive to appreciate these wonders. (Dawkins 2006)

The Teleological and Kalām Cosmological Arguments have shown that there is indeed Someone to worship and to be grateful to, that the universe with its astonishing fine-tuning, amazing mathematical laws of nature, and billions of spectacular stars and galaxies is not the result of 'blind pitiless indifference' (Dawkins 1996, pp. 131–132). Rather, it is the work of a transcendent Ultimate Designer and necessarily existent First Cause who is the Source of these wonders and the 'Maker and Father of all' (Plato, *Laws* 10.893b–899c). It is hard to think of a more important, humbling, and joyful discovery than this, and a more important quest in life than to know the God who created the universe.

## Notes

1. I thank Lucas Giolas for emphasizing this point to me.
2. https://www.youtube.com/watch?v=pGKe6YzHiME.
3. Goff also claims that theism makes false prediction concerning the problem of evil (p. 107). For reply, see Sect. 7.3.
4. Now Swinburne also states that 'a "Theory of Everything" will have no contingent background evidence by which to determine prior probability.

Prior probability must then be determined by purely *a priori* considerations' (2004, p. 60). Swinburne's statement might be explained by the fact that, by the 'theory of everything', he is thinking of an entity (i.e. the God of traditional theism) which is different from the contingent universe and which explains the universe. In which case our contingent background evidence concerning our universe would obviously not apply to such an entity since it only applies to the universe. However, Goff's case is different, since Goff's 'theory of everything' is that the universe itself explains its own fine-tuning. In which case our contingent background evidence concerning our universe does apply. In any case, whether Swinburne himself accepts my objection or not is irrelevant to the soundness of my objection against Goff's theory, which I explain below.

5. I thank Goff for helpful discussion in what follows.

## Bibliography

Baggett, David, and Jerry Walls. 2016. *God and Cosmos: Moral Truth and Human Meaning*. Oxford: Oxford University Press.

Boddy, K.K., S.M. Carroll, and J. Pollack. 2016. De Sitter Space Without Dynamical Quantum Fluctuations. *Found Phys* 46: 702–735.

Bostrom, Nick. 2003. Are We Living in a Computer Simulation? *Philosophical Quarterly* 53 (211): 243–255.

Carrier, Richard. 2003. Fundamental Flaws in Mark Steiner's Challenge to Naturalism in The Applicability of Mathematics as a Philosophical Problem. The Secular Web. https://infidels.org/library/modern/richard_carrier/steiner.html.

Copan, Paul, and William Lane Craig, eds. 2017. *The Kalām Cosmological Argument. 2 Vols*. New York: Bloomsbury Academic.

Craig, William Lane, and James Sinclair. 2009. The Kalām Cosmological Argument. In *The Blackwell Companion to Natural Theology*, ed. William Lane Craig and J.P. Moreland. Chichester: Wiley-Blackwell.

Dawkins, Richard. 1996. *River Out of Eden*. New York: Basic Books.

———. 2006. *The God Delusion*. London: Bantam Press.

Feser, Edward. 2017. *Five Proofs of the Existence of God*. San Francisco: Ignatius Press.

Forrest, Peter. 2016. The Personal Pantheist Conception of God. In *Alternative Concepts of God*, ed. Andrei Buckareff and Yujin Nagasawa. Oxford: Oxford University Press.

Goff, Philip. 2019. Did The Universe Design Itself? *International Journal for Philosophy of Religion* 85: 99–122.

Harrison, E. 1995. The Natural Selection of Universes Containing Intelligent Life. *Quarterly Journal of the Royal Astronomical Society* 36: 193–203.

Haught, John. 2004. Darwin, Design and Divine Providence. In *Debating Design*, ed. William Dembski and Michael Ruse. Cambridge: Cambridge University Press.

Hawking, Stephen. 2018. *Brief Answers to the Big Questions*. New York: Bantam Books.

Heller, M. 2013. Deep Questions on the Nature of Mathematics. *Notices of the American Mathematical Society* 60: 592–594.

Jammer, Max. 1999. *Einstein and Religion*. Princeton: Princeton University Press.

Kuhn, Robert. 2020. Why Anything at All II. Closer to Truth, Episode 1907.

Linford, Dan. 2020. The *Kalām* Cosmological Argument Meets the Mentaculus. *The British Journal for the Philosophy of Science*. https://doi.org/10.1093/bjps/axaa005.

Loke, Andrew. 2017. *The Origins of Divine Christology*. Cambridge: Cambridge University Press.

———. 2020. *Investigating the Resurrection of Jesus Christ: A New Transdisciplinary Approach*. London: Routledge.

———. 2021. Reply to Reviews of Investigating the Resurrection of Jesus Christ. https://www.academia.edu/45588286/Reply_to_reviews_of_Investigating_the_Resurrection_of_Jesus_Christ_Routledge_2020_.

Maudlin, Tim. 2018. Philosophy Has Made Plenty of Progress. *Scientific American*. https://blogs.scientificamerican.com/cross-check/philosophy-has-made-plenty-of-progress/.

Oppy, Graham. 2013a. *The Best Argument Against God*. London: Palgrave Macmillan.

Ratzsch, Del, and Jeffrey Koperski. 2019. Teleological Arguments for God's Existence. In *The Stanford Encyclopedia of Philosophy*, ed. Edward N. Zalta. https://plato.stanford.edu/archives/sum2019/entries/teleological-arguments/.

Sober, Elliott. 2019. *The Design Argument*. Cambridge: Cambridge University Press.

Stoeger, William. 2010. God, Physics and the Big Bang. In *The Cambridge Companion to Science and Religion*, ed. Peter Harrison. Cambridge: Cambridge University Press.
Sudduth, Michael. 2009. *The Reformed Objection to Natural Theology*. London: Routledge.
Swinburne, Richard. 2004. *The Existence of God*. 2nd ed. Oxford: Clarendon.

# Bibliography

Adams, Fred. 2019. The Degree of Fine-Tuning in Our Universe – And Others. *Physics Reports* 807: 1–111. arXiv:1902.03928v1 [astro-ph.CO] (Page Number Follows the arXiv Version).

Adamson, P. 2007. *Al-Kindī*. New York: Oxford University Press.

Aguirre, Anthony. 2007. Eternal Inflation, Past and Future. https://arxiv.org/abs/0712.0571.

Aguirre, Anthony, and John Kehayias 2013. Quantum Instability of the Emergent Universe. Cornell University. https://arxiv.org/abs/1306.3232.

Albrecht, A. 2004. Cosmic Inflation and the Arrow of Time. In *Science and Ultimate Reality*, ed. J. Barrow, P. Davies, and C. Harper. New York: Cambridge University Press.

Ali, Ahmed Farag, and Saurya Das. 2015. Cosmology from Quantum Potential. *Physics Letters B* 741: 276–279.

Almeida, Michael. 2018. *Cosmological Arguments*. Cambridge: Cambridge University Press.

An, Daniel, Krzysztof Meissner, Pawel Nurowski, and Roger Penrose. 2020. Apparent Evidence for Hawking Points in the CMB Sky. https://arxiv.org/abs/1808.01740.

Andersen, H. 2014. A Field Guide to Mechanisms: Part I. *Philosophy Compass* 9: 274–283.



A. Loke, *The Teleological and Kalam Cosmological Arguments Revisited*, Palgrave Frontiers in Philosophy of Religion, https://doi.org/10.1007/978-3-030-94403-2

Anderson, E. 2012. Problem of Time in Quantum Gravity. *Annalen der Physik* 524 (12): 757–786.
Anscombe, Elizabeth. 1974. Whatever has a Beginning of Existence Must Have a Cause': Hume's Argument Exposed. *Analysis* 34: 145–151.
Arkani-Hamed, Nima, and Jaroslav Trnka. 2014. The Amplituhedron. *Journal of High Energy Physics* 2014: 30. https://doi.org/10.1007/JHEP10(2014)030.
Armstrong, David. 1983. *What Is a Law of Nature?* Cambridge: Cambridge University Press.
Aspect, Alain. 2002. Bell's Theorem: The Naive View of an Experimentalist. In *Quantum (Un)speakables*, ed. J. Bell, Reinhold Bertlmann, and A. Zeilinger. Berlin: Springer.
Atkins, Peter. 1995. Science as Truth. *History of the Human Sciences* 8: 97–102.
Audi, Robert, ed. 1999. *The Cambridge Dictionary of Philosophy*. Cambridge: Cambridge University Press.
Bacciagaluppi, G., and A. Valentini. 2009. *Quantum Theory at the Crossroads: Reconsidering the 1927 Solvay Conference*. Cambridge: Cambridge University Press.
Baggett, David, and Jerry Walls. 2016. *God and Cosmos: Moral Truth and Human Meaning*. Oxford: Oxford University Press.
Barbour, J. 1999. *The End of Time*. Oxford: Oxford University Press.
Barnes, Luke. 2012. The Fine-Tuning of the Universe for Intelligent Life. *Publications of the Astronomical Society of Australia* 29: 529–564. https://doi.org/10.1071/AS12015.
———. 2019. A Reasonable Little Question: A Formulation of the Fine-Tuning Argument. *Ergo* 6 (42).
Barr, Stephen. 2012. Modern Cosmology and Christian Theology. In *The Blackwell Companion to Science and Christianity*, ed. Alan Padgett and J. Stump. Chichester: Wiley-Blackwell.
Barrow, J. 1991. *Theories of Everything: The Quest for Ultimate Explanation*. Oxford: Oxford University Press.
Barrow, J., and F. Tipler. 1986. *The Anthropic Cosmological Principle*. Oxford: Oxford University Press.
Baum, L., and P. Frampton. 2007. Turnaround in Cyclic Cosmology. *Physical Review Letters* 98: 071301.
Beebee, Helen, Christopher Hitchcock, and Peter Menzies, eds. 2009. *The Oxford Handbook of Causation*. Oxford: Oxford University Press.
Bell, John. 1964. On the Einstein-Podolsky-Rosen Paradox. *Physics*: 195–200.
———. 1987. *Speakable and Unspeakable in Quantum Mechanics*. Cambridge: Cambridge University Press.

———. 2007. Cosmological Theories and the Question of the Existence of a Creator. In *Religion and the Challenges of Science*, ed. William Sweet and Richard Feist. Aldershot: Ashgate.

Benardete, J. 1964. *Infinity: An Essay in Metaphysics*. Oxford: Oxford University Press.

Bersanelli, Marco. 2011. Infinity and the Nostalgia of the Stars. In *Infinity: New Research Frontiers*, ed. Michael Heller and Hugh Wood. Cambridge: Cambridge University Press.

Bigelow, John, Brian Ellis, and Caroline Lierse. 1992. The World as One of Kind. *British Journal for the Philosophy of Science* 43: 371–388.

Bird, Alexander. 2005. Abductive Knowledge and Holmesian Inference. In *Oxford Studies in Epistemology*, ed. Tamar Gendler and John Hawthorne. Oxford: Oxford University Press.

———. 2007. *Nature's Metaphysics*. Oxford: Oxford University Press.

Bishop, R. 2017. Chaos. In *The Stanford Encyclopedia of Philosophy*, ed. E. Zalta. https://plato.stanford.edu/archives/spr2017/entries/chaos/.

Blackburn, S. 2008. *The Oxford Dictionary of Philosophy*. Oxford: Oxford University Press.

Bliss, Ricki. 2013. Viciousness and the Structure of Reality. *Philosophical Studies* 166 (2): 399–418.

Bliss, Ricki, and Graham Priest. 2018. The Geography of Fundamentality: An Overview. In *Reality and Its Structure: Essays in Fundamentality*, ed. Ricki Bliss and Graham Priest. Oxford: Oxford University Press.

Bliss, Ricki, and Kelly Trogdon 2014. Metaphysical Grounding. *The Stanford Encyclopedia of Philosophy* (Winter 2014 Edition), ed. Edward N. Zalta. http://plato.stanford.edu/archives/win2014/entries/grounding/. Accessed 21 January 2017.

Bobier, Christopher. 2013. God, Time and the Kalām Cosmological Argument. *Sophia* 52: 593–600.

Boddy, K.K., S.M. Carroll, and J. Pollack. 2016. De Sitter Space Without Dynamical Quantum Fluctuations. *Found Phys* 46: 702–735.

Bohn, Einar. 2018. Indefinitely Descending Ground. In *Reality and Its Structure: Essays in Fundamentality*, ed. Ricki Bliss and Graham Priest. Oxford: Oxford University Press.

Bojowald, M., G. Date, and G. Hossain. 2004. The Bianchi IX Model in Loop Quantum Cosmology. *Classical and Quantum Gravity* 21: 3541.

Borde, A., A. Guth, and A. Vilenkin. 2003. Inflationary Spacetimes Are Not Past-Complete. *Physical Review Letters* 90: 151301. Preprint: http://arxiv.org/abs/gr-qc/0110012.

Bostrom, Nick. 2003. Are We Living in a Computer Simulation? *Philosophical Quarterly* 53 (211): 243–255.

Bradley, Darren. 2018. Carnap's Epistemological Critique of Metaphysics. *Synthese* 195, 2247–2265.

Bricmont, Jean. 2017. *Making Sense of Quantum Mechanics*. Cham: Springer Nature.

Brooke, John. 1991. *Science and Religion: Some Historical Perspectives*. Cambridge: Cambridge University Press.

Buchak, L. 2013. Free Acts and Chance. *Philosophical Quarterly* 63: 20–28.

Bussey, Peter. 2013. God as First Cause – A Review of the Kalām Argument. *Science & Christian Belief* 25: 17–35.

Butterfield, Jeremy. 2001. Some Worlds of Quantum Theory. In *Quantum Mechanics: Scientific Perspectives on Divine Action*, ed. Robert Russell et al. Berkeley: Center for Theology and the Natural Sciences.

Butterfield, Jeremy, and C. Isham. 1999. On the Emergence of Time in Quantum Gravity. In *The Arguments of Time*, ed. J. Butterfield. Oxford: Oxford University Press.

Callender, Craig. 2016. Thermodynamic Asymmetry in Time. In *The Stanford Encyclopedia of Philosophy*, ed. Edward Zalta. https://plato.stanford.edu/archives/win2016/entries/time-thermo/.

Cameron, Ross. 2008. Turtles All the Way Down: Regress, Priority and Fundamentality. *The Philosophical Quarterly* 58: 1–14.

———. 2018. Infinite Regress Arguments. In *The Stanford Encyclopedia of Philosophy*, ed. Edward Zalta. https://plato.stanford.edu/archives/fall2018/entries/infinite-regress/.

Cantor, G. 1915. *Contributions to the Founding of the Theory of Transfinite Numbers*, Trans. and Intro. P. Jourdain. New York: Dover.

Cao, ChunJun, Sean M. Carroll, and Spyridon Michalakis. 2017. Space from Hilbert Space: Recovering Geometry from Bulk Entanglement. *Physical Review D* 95: 2.

Capra, Fritjof. 2010. *The Tao of Physics*. Boston, MA: Shambhala.

Carrier, Richard. 2003. Fundamental Flaws in Mark Steiner's Challenge to Naturalism in The Applicability of Mathematics as a Philosophical Problem. The Secular Web. https://infidels.org/library/modern/richard_carrier/steiner.html.

——— 2018. *The Problem with Nothing: Why The Indefensibility of Ex Nihilo Nihil Goes Wrong for Theists*. https://www.richardcarrier.info/archives/14486.

Carroll, John. 2009. Anti-Reductionism. In *The Oxford Handbook of Causation*, ed. Helen Beebee, Christopher Hitchcock, and Peter Menzies. Oxford: Oxford University Press.

Carroll, Sean. 2010. *From Eternity to Here: The Quest for the Ultimate Theory of Time*. New York: Dutton.

———. 2012. In *'Does the universe need God?' In The Blackwell Companion to Science and Christianity*, ed. Alan Padgett and J. Stump. Chichester: Wiley-Blackwell.

———. 2014. *Post Debate Reflections*. http://www.preposterousuniverse.com/blog/2014/02/24/post-debate-reflections/.

———. 2016. *The Big Picture*. London: Oneworld.

———. 2019. *Something Deeply Hidden: Quantum Worlds and the Emergence of Spacetime*. New York: Dutton.

Cartlidge, Edwin. 2018. *New Evidence for Cyclic Universe Claimed by Roger Penrose and Colleagues*. https://physicsworld.com/a/new-evidence-for-cyclic-universe-claimed-by-roger-penrose-and-colleagues/.

Cartwright, Nancy. 1989. *Nature's Capacities and Their Measurement*. Oxford: Clarendon Press.

———. 2004. No God, No Laws. In *God and the Laws of Nature*, ed. Stefano Moriggi and Elio Sindoni. Milan: Angelicum Mondo X.

———. 2016. The Dethronement of Laws in Science. In *Rethinking Order After the Laws of Nature*, 25–52, eds. Cartwright, N. and Ward, K. London: Bloomsbury Academic.

Casati, Roberto, and Achille Varzi. 2010. Events. In *The Stanford Encyclopedia of Philosophy*, ed. Edward Zalta. http://plato.stanford.edu/archives/spr2010/entries/events/.

Chalmers, A. 1999. Making Sense of Laws of Physics. In *Causation and Laws of Nature*, ed. H. Sankey. Dordrecht: Kluwer.

Chan, Man Ho. 2019. Is the History of Our Universe Finite? *Theology and Science* 17 (2): 248–256.

Chan, K., and M. Chan. 2020. A Discussion of Klass Landsman's Criticisms of the Fine-Tuning Argument. *Theology and Science*. https://doi.org/10.1080/14746700.2020.1755544.

Choi, Sungho, and Michael Fara. 2018. Dispositions. In *The Stanford Encyclopedia of Philosophy*, ed. Edward N. Zalta. https://plato.stanford.edu/archives/fall2018/entries/dispositions/.

Churchland, P., and C. Hooker. 1985. *Images of Science*. Chicago: University of Chicago Press.

Clarke, Norris. 1970. A Curious Blindspot in the Anglo-American Tradition of Antitheistic Argument. *The Monist* 54: 181–200.

Clarke, Randolph, and Justin Capes. 2013. Incompatibilist (Nondeterministic) Theories of Free Will. *The Stanford Encyclopedia of Philosophy*, ed. Edward Zalta. http://plato.stanford.edu/archives/spr2013/entries/incompatibilism-theories/. Accessed 20 January 2014.

Cogliati, C. 2010. Introduction. In *Creation and the God of Abraham*, eds. D. B. Burrell, C. Cogliati, J. M. Soskice, and W. R. Stoeger. Cambridge: Cambridge University Press.

Cohoe, Caleb. 2013. There Must Be a First: Why Thomas Aquinas Rejects Infinite, Essentially Ordered, Causal Series. *British Journal for the History of Philosophy* 21: 838–856.

Collins, Robin. 1999. A Scientific Argument for the Existence of God: The Fine-Tuning Design Argument. In *Reason for the Hope Within*, ed. M. Murray. Eerdmans: Grand Rapids.

———. 2009a. God and the Laws of Nature. *Philo* 12: 142–171.

———. 2009b. The Teleological Argument: An Exploration of the Fine-tuning of the Universe. In *Blackwell Companion to Natural Theology*. Oxford: Wiley-Blackwell.

———. 2013. The Fine-Tuning Evidence is Convincing. In *Debating Christian Theism*, ed. J. P. Moreland, Chad V. Meister, Khaldoun Sweis. Oxford: Oxford University Press.

———. 2018. The Argument from Physical Constants. In *Two Dozen (Or So) Arguments for God*, ed. Jerry Walls and Trent Dougherty. Oxford: Oxford University Press.

Copan, Paul, and William Lane Craig, eds. 2017. *The Kalām Cosmological Argument. 2 Vols*. New York: Bloomsbury Academic.

Craig, William Lane. 1979. Kant's First Antinomy and the Beginning of the Universe. *Zeitschrift für Philosophische Forschung* 33: 553–567.

———. 1980. *The Cosmological Argument from Plato to Leibniz*. London: Macmillan.

———. 1990. What Place, Then, for a Creator?: Hawking on God and Creation. *British Journal for the Philosophy of Science* 41: 480–486.

———. 1994. Creation and Big Bang Cosmology. *Philosophia Naturalis* 31: 217–224.

———. 1998. Divine Timelessness and Personhood. *International Journal for Philosophy of Religion* 43: 109–124.

———. 1999. The Ultimate Question of Origins: God and the Beginning of the Universe. *Astrophysics and Space Science* 269–270: 723–740.

———. 2000a. *The Tensed Theory of Time: A Critical Examination*. Synthese Library 293. Dordrecht, the Netherlands: Kluwer Academic Publishers.

———. 2000b. *The Tenseless Theory of Time: A Critical Examination*. Synthese Library 294. Dordrecht, the Netherlands: Kluwer Academic Publishers.

———. 2000c. Naturalism and Cosmology. In *Naturalism: A Critical Analysis*, ed. William Lane Craig and J.P. Moreland. London: Routledge.

———. 2002. Must the Beginning of the Universe Have a Personal Cause? *Faith and Philosophy* 19: 94–105. http://www.reasonablefaith.org/must-the-beginning-of-the-universe-have-a-personal-cause-a-rejoinder#ixzz2YAV5QeN5. Accessed 13 January 2010.

———. 2007. Forming an Actual Infinite by Successive Addition. https://www.reasonablefaith.org/writings/question-answer/forming-an-actual-infinite-by-successive-addition/. Accessed 20 October 2020.

———. 2008. Review of *Philosophical Perspectives on Infinity*, by Graham Oppy. *Philosophia Christi* 10: 201–208.

———. 2009. Q&A #106: Is God Actually Infinite? http://www.reasonablefaith.org/is-god-actually-infinite. Accessed 4 July 2013.

———. 2010a. Reflections on Uncaused Beginnings. *Faith and Philosophy* 27: 72–78.

———. 2010b. Taking Tense Seriously in Differentiating Past and Future: A Response to Wes Morriston. *Faith and Philosophy* 27: 451–456.

———. 2011. #232 The Metric of Time. https://www.reasonablefaith.org/question-answer/P220/the-metric-of-time.

———. 2012. #285 Invasion of the Boltzmann Brains. *Reasonable Faith with William Lane Craig*. https://www.reasonablefaith.org/question-answer/fine-tuned-universe/invasion-of-the-boltzmann-brains.

———. 2013. The Kalām Argument. In *Debating Christian Theism*, ed. J.P. Moreland, Chad V. Meister, and Khaldoun A. Sweis. Oxford: Oxford University Press.

———. 2014. #393 Is the Universe Hostile to Life? *Reasonable Faith with William Lane Craig*. http://www.reasonablefaith.org/is-the-universe-hostile-to-life#ixzz3HKF6ugmn.

———. 2015. Q&A #450: Does the B-theory of Time Exclude Human Freedom? http://www.reasonablefaith.org/does-the-B-theory-of-time-exclude-human-freedom#ixzz4O4einczH. Accessed 15 October 2016.

———. 2015b. God and Cosmology: The Existence of God in Light of Contemporary Cosmology. *Reasonable Faith with William Lane Craig*. https://www.reasonablefaith.org/media/debates/god-and-cosmology-the-existence-of-god-in-light-of-contemporary-cosmol/.

———. 2016. *God Over All: Divine Aseity and the Challenge of Platonism*. Oxford: Oxford University Press.

———. 2018. The Kalām Cosmological Argument. In *Two Dozen (Or So) Arguments for God*, ed. Jerry Walls and Trent Dougherty. Oxford: Oxford University Press.

———. 2020. Explaining the Applicability of Mathematics. https://www.reasonablefaith.org/writings/question-answer/explaining-the-applicability-of-mathematics.

———. 2021. No Trouble: A Reply to Wielenberg. *TheoLogica*. https://doi.org/10.14428/thl.v4i3.58143.

Craig, William Lane, and Sean Carroll. 2015. *God and Cosmology: The Existence of God in Light of Contemporary Cosmology*. Augsburg: Fortress Publishers.

Craig, William Lane, and J.P. Moreland, eds. 2009. *The Blackwell Companion to Natural Theology*. Chichester: Wiley-Blackwell.

Craig, William Lane, and James Sinclair. 2009. The Kalām Cosmological Argument. In *The Blackwell Companion to Natural Theology*, ed. William Lane Craig and J.P. Moreland. Chichester: Wiley-Blackwell.

———. 2012. On Non-Singular Space-times and the Beginning of the Universe. In *Scientific Approaches to the Philosophy of Religion*, ed. Yujin Nagasawa. New York: Palgrave Macmillan.

Craig, William Lane, and Quentin Smith. 1993. *Theism, Atheism and Big Bang Cosmology*. Oxford: Clarendon Press.

Curiel, Erik. 2019. Singularities and Black Holes. In *The Stanford Encyclopedia of Philosophy*, ed. Edward Zalta. https://plato.stanford.edu/archives/spr2019/entries/spacetime-singularities/.

Cushing, James. 2001. Determinism Versus Indeterminism in Quantum Mechanics: A Free Choice. In *Quantum Mechanics: Scientific Perspectives on Divine Action*, ed. Robert J. Russell et al. Berkeley, CA: Center for Theology and the Natural Sciences.

d'Espagnat, Bernard. 2006. *On Physics and Philosophy*. Princeton: Princeton University Press.

Daintith, John, ed. 2009. *A Dictionary of Physics*. Oxford: Oxford University Press.

Davidson, H.A. 1969. John Philoponus as a Source of Medieval Islamic and Jewish Proofs of Creation. *Journal of the American Oriental Society* 89 (2): 357–391.

Davies, Paul. 1983. *God and the New Physics*. New York: Simon & Schuster.

———. 2013. Frozen Accidents: Can the Laws of Physics Be Explained? http://www.pbs.org/wgbh/nova/blogs/physics/2013/12/frozen-accidents-can-the-laws-of-physics-be-explained.

Dawes, Gregory. 2007. What Is Wrong with Intelligent Design. *International Journal for Philosophy of Religion* 61: 69–81.

Dawkins, Richard. 1986. *The Blind Watchmaker*. London: Penguin Books.

———. 1996. *River Out of Eden*. New York: Basic Books.

———. 2006. *The God Delusion*. London: Bantam Press.

———. 2011. *The Magic of Reality: How We Know What's Really True*. New York: Free Press.

Dedekind, Richard. 1963. The Nature and Meaning of Numbers. In *Essays on the Theory of Numbers*, ed. Richard Dedekind and Trans. W. W. Beman. New York: Dover.

Dembski, William. 2002. *No Free Lunch: Why Specified Complexity Cannot Be Purchased Without Intelligence*. Lanham: Rowman & Littlefield.

———. 2006. *The Design Inference: Eliminating Chance Through Small Probabilities*. Cambridge: Cambridge University Press.

Deng, Duen-Min. 2019. A New Cosmological Argument from Grounding. *Analysis*. https://doi.org/10.1093/analys/anz071.

Deutsch, David. 1997. *The Fabric of Reality: The Science of Parallel Universes and Its Implications*. London: Penguin.

Deutsch, David, and M. Lockwood. 1994. The Quantum Physics of Time Travel. *Scientific American* 270: 68–74.

DeWeese, Garrett. 2004. *God and the Nature of Time*. Hampshire: Ashgate.

DeWeese, Garrett, and Joshua Rasmussen. 2005. Hume and the Kalām Cosmological Argument. In *In Defense of Natural Theology*, ed. James Sennett and Douglas Groothuis. Downers Grove, Intervarsity.

Dirac, Paul. 1963. The Evolution of the Physicist's Picture of Nature. *Scientific American* 208: 45–53.

Dorato, Mauro, and Michael Esfeld. 2014. The Metaphysics of laws: dispositionalism vs. Primitivism. In *Metaphysics in Contemporary Physics*, ed. Tomasz Bigaj and Christian Wüthrich. Dordrecht: Kluwer.

Douven, Igor. 2011. Abduction. *The Stanford Encyclopedia of Philosophy* (Spring 2011 Edition), ed. Edward Zalta. http://plato.stanford.edu/archives/spr2011/entries/abduction/.

Dowden, Bradley. 2013a. The Infinite. *Internet Encyclopaedia of Philosophy*. http://www.iep.utm.edu/time/. Accessed 30 January 2014.

———. 2013b. Time. *Internet Encyclopaedia of Philosophy*. http://www.iep.utm.edu/time/. Accessed 5 July 2013.

Dowe, Phil. 2007. Causal Processes. In *The Stanford Encyclopedia of Philosophy*, ed. Edward N. Zalta. http://plato.stanford.edu/entries/causation-process/.

Drees, Willem. 1996. *Religion, Science, and Naturalism*. Cambridge: Cambridge University Press.

———. 2016. The Divine as Ground of Existence and of Transcendental Values: An Exploration. In *Alternative Concepts of God: Essays on the Metaphysics of the Divine*, ed. Andrei Buckareff and Yujin Nagasawa. Oxford: Oxford University Press.

Dretske, Fred. 1965. Counting to Infinity. *Analysis* 25: 99–101.

Duhem, P. 1954. *The Aim and Structure of Physical Theory*, Trans. P. Wiener. Princeton: Princeton University Press.

Dumsday, Travis. 2011. Have the Laws of Nature Been Eliminated? In *Reading the Cosmos: Nature, Science and Wisdom*, ed. Giuseppe Butera. Washington, DC: Catholic University of America Press.

———. 2016. Finitism and Divisibility: A Reply to Puryear. *Australasian Journal of Philosophy* 94: 596–601.

———. 2019. *Dispositionalism and the Metaphysics of Science*. Cambridge: Cambridge University Press.

Dupré, John. 2001. *Human Nature and the Limits of Science*. Oxford: Oxford University Press.

Dürr, Detlef, Sheldon Goldstein, Travis Norsen, Ward Struyve, and Nino Zanghi. 2014. Can Bohmian Mechanics Be Made Relativistic? *Proceedings of the Royal Society A* 470. https://doi.org/10.1098/rspa.2013.0699.

Eagle, Antony. 2019. Chance Versus Randomness. *The Stanford Encyclopedia of Philosophy*, ed. Edward Zalta. https://plato.stanford.edu/archives/spr2019/entries/chance-randomness/.

Earman, John. 1967. On Going Backward in Time. *Philosophy of Science* 34: 211–222.

———. 1995. *Bangs, Crunches, Whimpers, and Shrieks: Singularities and Acausalities in Relativistic Spacetimes*. New York: Oxford University Press.

Earman, John, and Christian Wüthrich. 2010. Time Machines. In *The Stanford Encyclopedia of Philosophy*, ed. Edward Zalta. http://plato.stanford.edu/archives/win2010/entries/time-machine/.

Eddington, A. 1920. *Space, Time and Gravitation*. Reprint edition: Cambridge Science Classics. Cambridge University Press, 1987.

Edwards, Jonathan. 1830. *The Works of Jonathan Edwards*. New York: G. & C. & H. Carvill.

Egg, Matthias, and Michael Esfeld. 2014. Non-local Common Cause Explanations for EPR. *European Journal for Philosophy of Science* 4: 181–196.
Einstein, Albert. 1949. Autobiographical Notes. In *Albert Einstein: Philosopher-Scientist*, ed. P. Schilpp. Peru, IL: Open Court Press, p. 63.
———. 1960. *Ideas and Opinions*, Trans. Sonja Bargmann. New York: Crown publishers.
Ellis, Brian. 2001. *Scientific Essentialism*. Cambridge: Cambridge University Press.
Ellis, George. 2007. Issues in the Philosophy of Cosmology. In *Philosophy of Physics*, ed. J. Butterfield and J. Earman. Amsterdam: Elsevier.
———. 2018. Necessity, Purpose and Chance. In *God's Providence and Randomness in Nature: Scientific and Theological Perspectives*, ed. Robert Russell and Joshua Moritz. West Conshohocken: Templeton Press.
Ellis, George, and R. Maartens. 2004. The Emergent Universe: Inflationary Cosmology with No Singularity. *Classical and Quantum Gravity* 21: 223.
Ellis, George, U. Kirchner, and W. Stoeger. 2004. Multiverses and Physical Cosmology. *Monthly Notices of the Royal Astronomical Society* 347: 921–936.
Erasmus, Jacobus. 2018. *The Kalām Cosmological Argument: A reassessment*. Cham: Springer Nature.
Evans, C. Stephen. 2010. *Natural Signs and Knowledge of God: A New Look at Theistic Arguments*. Oxford: Oxford University Press.
———. 2018. The Naïve Teleological Argument. In *Two Dozen (Or So) Arguments for God*, ed. Jerry Walls and Trent Dougherty. Oxford: Oxford University Press.
Fang, L., and Z. Wu. 1986. An Overview of Quantum Cosmology. In *Quantum Cosmology*, ed. L. Fang and Remo Ruffini. Singapore: World Scientific.
Faye, Jan. 2010. Backward Causation. *The Stanford Encyclopedia of Philosophy*, ed. Edward Zalta. http://plato.stanford.edu/archives/spr2010/entries/causation-backwards/.
Feferman, Solomon. 1998. *In the Light of Logic*. Oxford: Oxford University Press.
Feser, Edward. 2013. *Aristotle on Method and Metaphysics*. London: Palgrave.
———. 2017. *Five Proofs of the Existence of God*. San Francisco: Ignatius Press.
Feynman, Richard. 1967. *The Character of Physical Law*. Cambridge, MA: MIT Press.
———. 1998. *The Meaning of It All*. New York: Basic Books.
Fine, Kit. 1994. Essence and Modality. *Philosophical Perspectives* 8: 1–16.
———. 2002. The Varieties of Necessity. In *Conceivability and Possibility*, ed. Tamar Gendler and John Hawthorne. Oxford: Clarendon.

Fitelson, Brandon, Christopher Stephens, and Elliot Sober. 1999. How Not to Detect Design. *Philosophy of Science* 66: 472–488.

Ford, Kenneth. 2011. *101 Quantum Questions*. Cambridge, MA: Harvard University Press.

Forrest, Peter. 2016. The Personal Pantheist Conception of God. In *Alternative Concepts of God*, ed. Andrei Buckareff and Yujin Nagasawa. Oxford: Oxford University Press.

Foster, John. 2004. *The Divine Lawmaker*. Oxford: Oxford University Press.

Frederick, Danny. 2013. A Puzzle About Natural Laws and the Existence of God. *International Journal for the Philosophy of Religion* 73: 269–283.

Friederich, Simon. 2018. Fine-Tuning. In *The Stanford Encyclopedia of Philosophy*, ed. Edward Zalta. https://plato.stanford.edu/archives/spr2018/entries/fine-tuning/.

Frisch, Matthias. 2009. The Most Sacred Tenet'?, Causal Reasoning in Physics. *British Journal for the Philosophy of Science* 60: 459–474.

Frisch, Mathias. 2012. No Place for Causes? Causal Skepticism in Physics. *European Journal for Philosophy of Science* 2: 313–336.

———. 2014. *Causal Reasoning in Physics*. Cambridge: Cambridge University Press.

Gale, G. 1990. Cosmological Fecundity: Theories of Multiple Universes. In *Physical Cosmology and Philosophy*, ed. J. Leslie. New York: Macmillan.

Gambini, Rodolfo, and Jorge Pullin. 2013. Loop Quantization of the Schwarzschild Black Hole. *Physical Review Letters* 110: 211301.

Gendler, Tamar, and John Hawthorne, eds. 2002. *Conceivability and Possibility*. Oxford: Oxford University Press.

Giberson, Karl, and Mariano Artigas. 2007. *Oracles of Science: Celebrity Scientists Versus God and Religion*. New York: Oxford University Press.

Gingerich, O. 2006. *God's Universe*. Cambridge, MA: Harvard University Press.

Glass, David. 2012a. Darwin, Design and Dawkins' Dilemma. *SOPHIA* 51: 31–57.

———. 2012b. Can Evidence for Design Be Explained Away? In *Probability in the Philosophy of Religion*, ed. V. Harrison and J. Chandler. Oxford: Oxford University Press.

Gödel, Kurt. 1949. A Remark about the Relationship between Relativity Theory and Idealistic Philosophy. In *In Albert Einstein: Philosopher-Scientist*, ed. P. Schilpp. La Salle: Open Court.

Goff, Philip. 2019. Did The Universe Design Itself? *International Journal for Philosophy of Religion* 85: 99–122.

———. 2021. Our Improbable Existence Is No Evidence for a Multiverse. https://www.scientificamerican.com/article/our-improbable-existence-is-no-evidence-for-a-multiverse/.

Goldman, R. 1997. *Einstein's God.* Northvale, NJ: Joyce Aronson Inc.

Goldstein, Sheldon. 2013. Bohmian Mechanics. In *The Stanford Encyclopedia of Philosophy*, ed. Edward Zalta. http://plato.stanford.edu/archives/spr2013/entries/qm-bohm/.

Gott, Richard I.I.I. 1991. Closed Timelike Curves Produced by Pairs of Moving Cosmic Strings – Exact Solutions. *Physical Review Letters* 66: 1126.

Gott, Richard I.I.I., and Li-Xin Li. 1998. Can the Universe Create Itself? *Physical Review D* 58: 023501–023501.

Gould, Paul, ed. 2014. *Beyond the Control of God? Six Views on the Problem of God and Abstract Objects*. London: Bloomsbury.

Greenberger, Daniel, and Karl Svozil. 2005. Quantum Theory Looks at Time Travel. In *Quo Vadis Quantum Mechanics?* ed. A. Elitzur, S. Dolev, and N. Kolenda. Berlin: Springer.

Greiner, Walter, and Joachim Reinhardt. 2009. *Quantum Electrodynamics*. Berlin: Springer.

Griffiths, David. 1989. *Introduction to Electrodynamics*. New Jersey: Prentice Hall.

Grünbaum, Adolf. 1989. The Pseudo-Problem of Creation in Physical Cosmology. *Philosophy of Science* 56: 373–394.

———. 1991. Creation as a Pseudo-Explanation in Current Physical Cosmology. *Erkenntnis* 35: 233–254.

———. 1994. Some Comments on William Craig's 'Creation and Big Bang Cosmology'. *Philosophia Naturalis* 31: 225–236.

———. 1998. Theological Misinterpretations of Current Physical Cosmology. *Philo* 1: 15–34.

———. 2000. A New Critique of Theological Interpretations of Physical Cosmology. *The British Journal for the Philosophy of Science* 51: 1–43.

———. 2004. The Poverty of Theistic Cosmology. *The British Journal for the Philosophy of Science* 55: 561–614.

———. 2005. Rejoinder to Richard Swinburne's 'Second Reply to Grünbaum. *The British Journal for the Philosophy of Science* 56: 927–938.

———. 2009. Why Is There a Universe AT ALL, Rather Than Just Nothing? *Ontology Studies* 9: 7–19.

Guth, Alan. 2000. Inflation and Eternal Inflation. *Physics Reports* 333: 555–574.

Hagar, Amit. 2014. *Discrete Or Continuous? The Quest for Fundamental Length in Modern Physics*. Cambridge: Cambridge University Press.

Hájek, Alan. 2019. Interpretations of Probability. In *The Stanford Encyclopedia of Philosophy*, ed. Edward Zalta. https://plato.stanford.edu/archives/fall2019/entries/probability-interpret/.

Halper, Phillip. 2021. The Kalām Cosmological Argument: Critiquing a Recent Defence. *Think* 20: 153–165.

Halvorson, Hans. 2018. A Theological Critique of the Fine-Tuning Argument. In *Knowledge, Belief, and God*, ed. Matthew Benton, John Hawthorne, and Dani Rabinowitz. Oxford: Oxford University Press.

Halvorson, Hans, and Helge Kragh. 2011. Cosmology and Theology. In *The Stanford Encyclopedia of Philosophy*, ed. Edward Zalta. http://plato.stanford.edu/archives/win2011/entries/cosmology-theology/.

Hanley, Richard. 2004. No End in Sight: Causal Loops in Philosophy, Physics and Fiction. *Synthese* 141: 123–152.

Harrison, E. 1995. The Natural Selection of Universes Containing Intelligent Life. *Quarterly Journal of the Royal Astronomical Society* 36: 193–203.

Hartle, James, and Stephen Hawking. 1983. Wave Function of the Universe. *Physical Review D* 28: 2960–2975.

Haught, John. 2004. Darwin, Design and Divine Providence. In *Debating Design*, ed. William Dembski and Michael Ruse. Cambridge: Cambridge University Press.

Hawking, Stephen. 1987. Quantum Cosmology. In *Three Hundred Years of Gravitation*, ed. S. Hawking and W. Israel. Cambridge: Cambridge University Press.

———. 1988. *A Brief History of Time*. London: Bantam.

———. 1992. Chronology Protection Conjecture. *Physical Review D* 46: 603.

———. 1993. *Black Holes and Baby Universes and Other Essays*. New York: Bantam.

———. 1997. The Objections of an Unashamed Positivist. In *The Large, the Small, and the Human Mind*, ed. R. Penrose and M. Longair. New York: Cambridge University Press.

———. 2003. *Cosmology from the Top Down*. http://arxiv.org/abs/astro-ph/0305562.

———. 2005. *The Origin of the Universe*. http://www.hawking.org.uk/the-origin-of-the-universe.html.

———. 2018. *Brief Answers to the Big Questions*. New York: Bantam Books.

Hawking, Stephen, and Leonard Mlodinow. 2010. *The Grand Design*. New York: Bantam Books.

Hawthorne, J. 2000. Before-effect and Zeno Causality. *Noûs* 34: 622–633.

Hawthorne, John, and Yoaav Isaacs. 2017. Misapprehensions About the Fine-Tuning Argument. *Royal Institute of Philosophy Supplement 81*: 133–155.

Hawthorne, J., and Y. Isaacs. 2018. Fine-tuning Fine-Tuning. In *Knowledge, Belief, and God*, ed. M. Benton, J. Hawthorne, and D. Rabinowitz. Oxford: Oxford University Press.

Heisenberg, Werner. 1989. *Physics and Philosophy*. London: Penguin Books.

Helen De Cruz, and Johan De Smedt. 2016. Naturalizing Natural Theology. *Religion, Brain & Behavior* 6 (4): 355–361.

Heller, M. 2013. Deep Questions on the Nature of Mathematics. *Notices of the American Mathematical Society* 60: 592–594.

Heller, Michael, and Hugh Woodin, eds. 2011. *Infinity: New Research Frontiers*. Cambridge: Cambridge University Press.

Hilbert, David. 1964. On the Infinite. In *Philosophy of Mathematics*, ed. P. Benacerraf and H. Putnam. Englewood Cliffs: Prentice-Hall.

Hitchcock, Christopher. 2012. Probabilistic Causation. In *The Stanford Encyclopedia of Philosophy*, ed. Edward Zalta. http://plato.stanford.edu/archives/win2012/entries/causation-probabilistic/.

Hoefer, Carl. 2010. Causal Determinism. In *The Stanford Encyclopedia of Philosophy*, ed. Edward Zalta. http://plato.stanford.edu/archives/spr2010/entries/determinism-causal/.

Holder, Rodney. 2004. *God, the Multiverse, and Everything*. Aldershot: Ashgate.

———. 2012. Quantum Theory and Theology. In *The Blackwell Companion to Science and Christianity*, ed. Alan Padgett and J. Stump. Chichester: Wiley-Blackwell.

Hossenfelder, Sabine. 2019. Screams for Explanation: Finetuning and Naturalness in the Foundations of Physics. *Synthese*. https://doi.org/10.1007/s11229-019-02377-5.

Huemer, Michael. 2018. *Paradox Lost*. Cham: Springer.

Huggett, Nick, and Christian Wüthrich. 2013. Emergent Spacetime and Empirical (in)coherence. *Studies in History and Philosophy of Modern Physics* 44: 276–285.

Huggett, Nick, and Christian Wuthrich. forthcoming. *Out of Nowhere: The Emergence of Spacetime in Quantum Theories of Gravity*. Oxford: Oxford University Press.

Hume, David. 1739/1978. *A Treatise of Human Nature*, ed. L. A. Selby-Bigge. 2nd ed. Oxford: Oxford University Press.

———. 1748/1999. *An Enquiry Concerning Human Understanding*, ed. Tom L. Beauchamp. New York: Oxford University Press.

———. 1779/1993. Dialogues Concerning Natural Religion. In *Dialogues and Natural History of Religion*, ed. J.A.C. Gaskin. Oxford: Oxford University Press.

———. 1932. In *The Letters of David Hume*, ed. J.Y.T. Greig, vol. 1. Oxford: Clarendon.

Hunter, Joel. 2013. Time Travel. *Internet Encyclopedia of Philosophy*. http://www.iep.utm.edu/timetrav/. Accessed 18 June 2013.

Isham, C. J. 1997. Creation of the universe as a quantum process. In *Physics, philosophy, and theology: A common quest for understanding* (pp. 375–408) (3rd ed.), ed. R. J. Russell, W. R. Stoeger, & G. V. Coyne. Vatican City State: Vatican Observatory.

Issacs, A., ed. 2000. *Oxford Dictionary of Physics*. 4th ed. Oxford: Oxford University Press.

Jammer, Max. 1999. *Einstein and Religion*. Princeton: Princeton University Press.

Jantzen, Benjamin. 2014. *An Introduction to Design Arguments*. New York: Cambridge University Press.

Jastrow, Robert. 2000. *God and the Astronomers*. New York: W.W. Norton.

Jow, Dylan, and Douglas Scott. 2020. Re-evaluating Evidence for Hawking Points in the CMB. *Journal of Cosmology and Astroparticle Physics* 3. https://doi.org/10.1088/1475-7516/2020/03/021.

Kant, Immanuel. 1965. *Critique of Pure Reason*, Trans. Norman Kemp-Smith. London: Macmillan.

Kim, Y.-H., et al. 2000. A Delayed Choice Quantum Eraser. *Physical Review Letters* 84: 1–5.

Klement, Kevin. 2020. Russell's Paradox. *Internet Encyclopedia of Philosophy: A Peer-Reviewed Academic Resources*. www.iep.utm.edu/par-russ/.

Kojonen, Rope. 2021. *The Compatibility of Evolution and Design*. Palgrave Frontiers in Philosophy of Religion Series. Cham: Springer Nature.

Koons, Robert. 1997. A New Look at the Cosmological Argument. *American Philosophical Quarterly* 34: 193–211.

———. 2000. *Realism Regained: An Exact Theory of Causation, Teleology, and the Mind*. New York: Oxford University Press.

———. 2014. A New Kalām Argument: Revenge of the Grim Reaper. *Noûs* 48: 256–267.

———. 2019. The Universe Has a Cause. In *Contemporary Debates in Philosophy of Religion*, ed. Michael Peterson and Raymond Van Arragon. Chichester: Wiley Blackwell.

Koons, Robert and George Bealer, eds. 2010. *The Waning of Materialism*. Oxford: Oxford University Press.

Koons, Robert, and Timothy Pickavance. 2015. *Metaphysics: The Fundamentals*. Chichester: Wiley-Blackwell.

Krauss, Lawrence. 2012. *A Universe from Nothing*. New York: Free Press.
Kripke, Saul. 1980. *Naming and Necessity*. Oxford: Blackwell.
Kuhn, Robert. 2020. Why Anything at All II. Closer to Truth, Episode 1907.
Ladyman, James, Don Ross, David Spurrett, and John Collier. 2007. *Every Thing Must Go: Metaphysics Naturalized*. Oxford: Oxford University Press.
Lange, Marc. 2009. *Laws and Lawmakers. Science, Metaphysics, and the Laws of Nature*. Oxford: Oxford University Press.
———. 2016. *Because Without Cause: Non-Causal Explanations in Science and Mathematics*. Oxford: Oxford University Press.
Laudan, Larry. 1981. A Confutation of Convergent Realism. *Philosophy of Science* 48: 19–49.
———. 1990. Demystifying Underdetermination. In *Scientific Theories*, ed. C. Wade Savage. Minneapolis: University of Minnesota Press.
Leftow, Brian. 2005. Eternity and Immutability'. In *The Blackwell Guide to the Philosophy of Religion*, ed. W. Mann. Malden, Blackwell.
Leon, Felipe. 2011. Moreland on the Impossibility of Traversing the Infinite: A Critique. *Philo* 14: 32–42.
Leslie, John. 1982. Anthropic Principle, World Ensemble, Design. *American Philosophical Quarterly* 19 (2): 141–151.
———. 1989. *Universes*. London: Routledge.
———. 2016. A Way of Picturing God. In *Alternative Concepts of God*, ed. Andrei Buckareff and Yujin Nagasawa. Oxford: Oxford University Press.
Levy, Neil, and Michael McKenna. 2009. Recent Work on Free Will and Moral Responsibility. *Philosophy Compass* 4: 96–133.
Lewis, David. 1976. The Paradoxes of Time Travel. *American Philosophical Quarterly* 13: 145–152.
———. 1986. *On the Plurality of Worlds*. New York: Basil Blackwell.
Lewis, Peter. 2016. *Quantum Ontology*. Oxford: Oxford University Press.
Lewis, Geraint, and Luke Barnes. 2016. *A Fortunate Universe: Life in a Finely Tuned Cosmos*. Cambridge: Cambridge University Press.
Liddle, A., and J. Loveday. 2009. *The Oxford Companion to Cosmology*. Oxford: Oxford University Press.
Linde, Andrei. 1994. The Self-Reproducing Inflationary Universe. *Scientific American* 271: 48–55.
Linford, Dan. 2020. The *Kalām* Cosmological Argument Meets the Mentaculus. *The British Journal for the Philosophy of Science*. https://doi.org/10.1093/bjps/axaa005.
Linnemann, N. 2020. On Metaphysically Necessary Laws from Physics. *European Journal for Philosophy of Science* 10: 23.

Livio, Mario. 2009. *Is God a Mathematician?* New York: Simon & Schuster.

Loke, Andrew. 2010. Divine Omnipotence and Moral Perfection. *Religious Studies* 46: 525–538.

———. 2012a. Is an Uncaused Beginning of the Universe Possible? A Response to Recent Naturalistic Metaphysical Theorizing. *Philosophia Christi* 14: 373–393.

———. 2012b. Is an Infinite Temporal Regress of Events Possible? *Think* 11: 105–122.

———. 2014a. A Modified Philosophical Argument for a Beginning of the Universe. *Think* 13: 71–83.

———. 2014b. No Heartbreak at Hilbert's Hotel: A Reply to Landon Hedrick. *Religious Studies* 50: 47–50.

———. 2014c. The Benefits of Studying Philosophy for Science Education. *Journal of the NUS Teaching Academy* 4: 27–35.

———. 2014d. *A Kryptic Model of the Incarnation*. London: Routledge.

———. 2016a. On Finitism and the Beginning of the Universe: A Reply to Stephen Puryear. *Australasian Journal of Philosophy* 94: 591–595.

———. 2016b. On Beginningless Past, Endless Future, God, and Singing Angels: An Assessment of the Morriston-Craig Dialogue. *Neue Zeitschrift für Systematische Theologie und Religionsphilosophie* 58: 57–66.

———. 2016e. On the Infinite God Objection: A Reply to Jacobus Erasmus and Anné Hendrik Verhoef. *Sophia* 55: 263–272.

———. 2017a. *God and Ultimate Origins: A Novel Cosmological Argument. Palgrave Frontiers in Philosophy of Religion Series*. Cham, Switzerland: Springer Nature.

———. 2017b. *The Origins of Divine Christology*. Cambridge: Cambridge University Press.

———. 2017c. A Reply to Peter Lyth on Whether an Infinite Temporal Regress of Events Is Possible. *Think* 16: 77–81.

———. 2018. Review of Timothy Pawl's in Defense of Conciliar Christology: A Philosophical Essay. *Faith and Philosophy* 34: 114–119.

———. 2019. Theological Critiques of Natural Theology: A Reply to Andrew Moore. *Neue Zeitschrift für Systematische Theologie und Religionsphilosophie* 61: 207–222.

———. 2020. *Investigating the Resurrection of Jesus Christ: A New Transdisciplinary Approach*. London: Routledge.

———. 2021a. The Kalām Cosmological Argument and divine omniscience: an evaluation of recent discussions in Sophia. *Sophia* 59: 651–656.

———. 2021b. Reply to Reviews of Investigating the Resurrection of Jesus Christ. https://www.academia.edu/45588286/Reply_to_reviews_of_Investigating_the_Resurrection_of_Jesus_Christ_Routledge_2020_.

———. 2022a. *Evil, Suffering and Christian Theism*. London: Routledge.

———. 2022b. *The Origin of Humanity and Evolution: Science and Scripture in Conversation*. London: T & T Clark.

Loose, Jonathan, Angus Menuge, and J.P. Moreland. 2018. *The Blackwell Companion to Substance Dualism*. Oxford: Wiley Blackwell.

Lowe, E.J. 2002. Kinds, Essence, and Natural Necessity. In *Individuals, Essence and Identity*. Heidelberg: Springer.

———. 2005. Necessary and Sufficient Conditions. In *The Oxford Companion to Philosophy*, ed. Ted Honderich. New York: Oxford University Press.

———. 2008. *Personal Agency: The Metaphysics of Mind and Action*. Oxford: Oxford University Press.

Luna, Laureano, and Jacobus Erasmus. 2020. A Philosophical Argument for the Beginning of Time. *Prolegomena* 19 (2): 161–176.

Mackenzie, Ruari, et al. 2017. Evidence Against a Supervoid Causing the CMB Cold Spot. *Monthly Notices of the Royal Astronomical Society* 470 (2): 2328–2338.

Mackie, Penelope. 2005. Causality. In *The Oxford Companion to Philosophy*, 2nd ed. New York: Oxford University Press. http://www.oxfordreference.com/views/ENTRY.html?subview=Main&entry=t116.e356.

Malpass, Alex. 2019. *Successive Addition*. https://useofreason.wordpress.com/2019/02/21/successive-addition/. Accessed 14 April 2020.

———. 2020. Counting Forever. https://useofreason.wordpress.com/2020/07/06/counting-forever/#comments. Accessed 6 July 2020.

———. 2021. All the Time in the World. *Mind*, fzaa086. https://doi.org/10.1093/mind/fzaa086.

Malpass, Alex, and Wes Morriston. 2020. Endless and Infinite. *Philosophical Quarterly*. https://doi.org/10.1093/pq/pqaa005.

Manson, Neil A. 2003. *God and Design: The Teleological Argument and Modern Science*. London: Routledge.

Manson, N. 2020. How Not to Be Generous to Fine-Tuning Sceptics. *Religious Studies* 56 (3): 303–317.

Martin, C.B. 2008. *The Mind in Nature*. Oxford: Clarendon Press.

Maudlin, Tim. 2007. *The Metaphysics Within Physics*. Oxford: Oxford University Press.

———. 2018. Philosophy Has Made Plenty of Progress. *Scientific American.* https://blogs.scientificamerican.com/cross-check/philosophy-has-made-plenty-of-progress/.

Mawson, T. 2011. Explaining the Fine Tuning of the Universe to Us and the Fine Tuning of Us to the Universe. *Royal Institute of Philosophy Supplement* 68 (2011): 25–50.

McDonnell, N. 2018. Transitivity and Proportionality in Causation. *Synthese* 195: 1211–1229.

McGrath, Alister. 2009. *A Fine-tuned Universe: The Quest for God in Science and Theology*. Louisville: Westminster John Knox Press.

———. 2018. *The Territories of Human Reason: Science and Theology in an Age of Multiple Rationalities*. Oxford: Oxford University Press.

McGrew, Timothy. 2013. Miracles. In *The Stanford Encyclopedia of Philosophy*, Edward N. Zalta (ed.). Available at http://plato.stanford.edu/archives/spr2013/ entries/miracles/.

McGrew, Timothy, Lydia McGrew and Eric Vestrup. 2001. Probabilities and the fine-tuning argument: A sceptical view. *Mind* 110 (440): 1027–1038.

Miksa, Rad. 2020. Deny the Kalam's Causal Principle, Embrace Absurdity. *Philosophia Christi* 22: 239–255.

Monton, Bradley. 2010. Design Inferences in an Infinite Universe. In *Oxford Studies in Philosophy of Religion, Volume II*. Oxford: Oxford University Press.

Moore, A.W. 2001. *The Infinite*. London: Routledge.

Moreland, J.P. 1997. Agent Causation and the Craig/Grünbaum Debate About Theistic Explanation of the Initial Singularity. *American Catholic Philosophical Quarterly* 71: 539–554.

———. 1998. The Explanatory Relevance of Libertarian Agency. In *Mere Creation*, ed. William Dembski. Downer's Grove: Intervarsity.

———. 2001. The Kalām Cosmological Argument. In *Philosophy of Religion: Selected Readings*, ed. Michael Peterson, William Hasker, Bruce Reichenbach, and David Basinger. New York: Oxford University Press.

———. 2017. Libertarian Agency and the Craig/Grünbaum Debate About Theistic Explanation of the Initial Singularity. In *The Kalām Cosmological Argument*, ed. Paul Copan and William Lane Craig. New York: Bloomsbury Academic.

Moreland, J.P., and William Lane Craig. 2003. *Philosophical Foundations for a Christian Worldview*. Downers Grove: InterVarsity Press.

Morganti, M. 2015. Dependence, Justification and Explanation: Must Reality Be Well-Founded? *Erkenntnis* 80: 555–572.

Morrison, M. 2015. *Reconstructing Reality. Models, Mathematics, and Simulations*. Oxford: Oxford University Press.

Morriston, Wes. 2000. Must the Beginning of the Universe Have a Personal Cause? *Faith and Philosophy* 17: 149–169.

———. 2002a. Craig on the Actual Infinite. *Religious Studies* 38: 147–166.

———. 2002b. Causes and Beginnings in the *Kalām* Argument: Reply to Craig. *Faith and Philosophy* 19: 233–244.

———. 2003. Must Metaphysical Time Have a Beginning? *Faith and Philosophy* 20: 296–301.

———. 2010. Beginningless Past, Endless Future, and the Actual Infinite. *Faith and Philosophy* 27: 439–450.

———. 2013. Doubts About the Kalām Argument. In *Debating Christian Theism*, ed. J.P. Moreland, Chad V. Meister, and Khaldoun A. Sweis. Oxford: Oxford University Press.

———. 2021. Infinity, Time, and Successive Addition. *Australasian Journal of Philosophy*. https://doi.org/10.1080/00048402.2020.1865426.

Mullins, R.T. 2015. *The End of the Timeless God*. Oxford: Oxford University Press.

———. 2020. The Divine Timemaker. *Philosophia Christi* 22: 207–233.

Mumford, Stephen. 2004. *Laws in Nature*. London: Routledge.

———. 2009. Causal Powers and Capacities. In *The Oxford Handbook of Causation*, ed. Helen Beebee, Christopher Hitchcock, and Peter Menzies. Oxford: Oxford University Press.

Nagel, Thomas. 2004. Review of Bede Rundle's Why there is Something. Rather than Nothing, Times Literary Supplement, May 7.

Newton, William. 2014. A Case of Mistaken Identity: Aquinas's Fifth Way and Arguments of Intelligent Design. *New Blackfriars* 95: 569–578.

Ney, A. 2016. Microphysical Causation and the Case for Physicalism. *Analytic Philosophy* 57: 141–164.

Nomura, Yasunori. 2012. Static Quantum Multiverse. *Physical Review* 86: 083505.

Norton, John. 2003. Causation as Folk Science. *Philosophers' Imprint* 3 (4): 1–22. http://www.philosophersimprint.org/003004/3.

O'Connor, Timothy. 2000. *Persons and Causes: The Metaphysics of Free Will*. Oxford: Oxford University Press.

———. 2004. And This All Men Call God. *Faith and Philosophy* 21: 417–435.

———. 2016. Probability and Freedom. *Res Philosophica* 93: 289–293.

Oaklander, Nathan. 2004. *The Ontology of Time*. New York: Prometheus Books.

Oderberg, David. 2001. The Kalām Cosmological Argument Neither Bloodied nor Bowed. *Philosophia Christi* 3: 193–196.

———. 2002. Traversal of the Infinite, the 'Big Bang,' and the *Kalām* Cosmological Argument. *Philosophia Christi* 4: 303–334.

———. 2006. Instantaneous Change without Instants. In *Analytical Thomism: Traditions in Dialogue*, ed. C. Paterson and M.S. Pugh. Aldershot: Ashgate.

———. 2010. 'Whatever Is Changing Is Being Changed by Something Else': A Reappraisal of Premise One of the First Way. In *Mind, Method and Morality*, ed. J. Cottingham and P. Hacker. Oxford: Oxford University Press.

Oppy, Graham. 1991. Craig, Mackie, and the Kalām Cosmological Argument. *Religious Studies* 27: 189–197.

———. 2006a. *Philosophical Perspectives on Infinity*. Cambridge: Cambridge University Press.

———. 2006b. *Arguing About Gods*. Cambridge: Cambridge University Press.

———. 2009. Cosmological Arguments. *Nous* 43: 31–48.

———. 2010. Uncaused Beginnings. *Faith and Philosophy* 27: 61–71.

———. 2012. Science, Religion, and Infinity. In *The Blackwell Companion to Science and Christianity*, ed. Alan Padgett and J. Stump. Chichester: Wiley-Blackwell.

———. 2013a. *The Best Argument Against God*. London: Palgrave Macmillan.

———. 2013b. Ultimate Naturalistic Causal Explanations. In *The Puzzle of Existence*, ed. Tyron Goldschmidt. New York: Routledge.

———. 2014. *Describing Gods*. Cambridge: Cambridge University Press.

———. 2015. Uncaused Beginnings Revisited. *Faith and Philosophy* 32: 205–210.

———. 2019a. The Universe Does Not Have a Cause. In *Contemporary Debates in Philosophy of Religion*, ed. Michael Peterson and Raymond Van Arragon. Chichester: Wiley Blackwell.

———. 2019b. Review of Paul Copan and William Lane Craig, eds., The Kalām Cosmological Argument. *Philosophia Christi* 21: 445–449.

———. 2019c. Review of P. Copan and Craig, W. The Kalām Cosmological Argument Volume Two: Scientific Evidence for the Beginning of the Universe. *European Journal for Philosophy of Religion* 11: 225–229.

Oriti, Daniele. 2014. Disappearance and Emergence of Space and Time in Quantum Gravity. *Studies in the History and Philosophy of Modern Physics* 46: 186–199.

Padgett, Alan. 1992. *God, Eternity, and the Nature of Time*. New York: St. Martin's.

Paley, William. 1794/1859. *A View of the Evidences of Christianity*. London: John W. Parker and Son.

Parsons, Keith. 1999. Defending Objectivity. *Philo* 2: 84.

Pearce, Kenneth. 2017. Foundational Grounding and the Argument from Contingency. In *Oxford Studies in Philosophy of Religion*, vol. 8. Oxford: Oxford University Press.

———. 2021a. Monistic Metaphysical Rationalism Requires Grounding Indeterminism. Presentation at Global Philosophy of Religion Conference, Birmingham University.

———. 2021b. Foundational Grounding and Creaturely Freedom. *Mind*, fzab024. https://doi.org/10.1093/mind/fzab024.

Peckham, John. 2018. *Theodicy of Love: Cosmic Conflict and the Problem of Evil*. Grand Rapids: Baker.

Peels, Rick. 2018. Does Evolution Conflict with God's Character? *Modern Theology* 34 (4) https://doi.org/10.1111/moth.12435.

Pelczar, Michael. 2015. *Sensorama: A Phenomenalist Analysis of Spacetime and Its Contents*. Oxford: Oxford University Press.

Penrose, Roger. 1989. *The Emperor's New Mind: Concerning Computers, Minds and the Laws of Physics*. Oxford: Oxford University Press.

———. 2004. *The Road to Reality: A Complete Guide to the Laws of the Universe*. London: Random House.

———. 2010. *Cycles of Time: An Extraordinary New View of the Universe*. London: Bodley Head.

Penrose, Roger, and William Lane Craig. 2019. The Universe: How Did It Get Here & Why Are We Part of It? [Video File]. https://www.youtube.com/watch?v=9wLtCqm72-Y. Accessed 22 April 2020.

Perrine, Timothy, and Stephen Wykstra. 2017. Skeptical Theism. In *The Cambridge Companion to the Problem of Evil*, ed. Chad Meister and Paul Moser. Cambridge: Cambridge University Press.

Pettersson, Thomas, and P. Lefèvre 1995. *The Large Hadron Collider. CERN Document Server*. http://cdsweb.cern.ch/record/291782.

Philipse, Herman. 2012. *God in the Age of Science?* Oxford: Oxford University Press.

Philoponus, John. c.530–34/1987. *Against Aristotle on the Eternity of the World*, Trans. C. Wildberg. London: Duckworth.

———. c.529/2004. *Against Proclus On the Eternity of the World*, Trans. Michael Share, 1–5. London: Duckworth.

Pitts, Brian. 2008. Why the Big Bang Singularity Does Not Help the Kalām Cosmological Argument for Theism. *The British Journal for the Philosophy of Science* 59: 675–708.

Plantinga, Alvin. 2011. *Where the Conflict Really Lies: Science, Religion, and Naturalism*. Oxford: Oxford University Press.

Polkinghorne, John. 1998. *Believing in God in the Age of Science*. New Haven: Yale University Press.

———. 2000. *Faith, Science, and Understanding*. New Haven: Yale University Press.

———. 2005. *Exploring Reality: The Intertwining of Science and Religion*. New Haven: Yale University Press.

———. 2006a. Christianity and Science. In *The Oxford Handbook of Religion and Science*, ed. Philip Clayton and Zachary Simpson. Oxford: Oxford University Press.

———. 2006b. Space, Time, and Causality. *Zygon* 41: 975–983.

———. 2007. *Quantum Physics and Theology: An Unexpected Kinship*. New Haven: Yale University Press.

———. 2011a. *Science and Religion in Quest of Truth*. New Haven: Yale University Press.

———. 2011b. Mathematical Reality. In *Meaning in Mathematics*, ed. John Polkinghorne. Oxford: Oxford University Press.

Poplawski, N.J. 2010. Radial Motion into an Einstein-Rosen Bridge. *Physics Letters B* 687: 110–113.

Pratt, Bill. 2012. Does Quantum Mechanics Invalidate the Law of Non-Contradiction. Part 2. https://www.toughquestionsanswered.org/2012/01/04/does-quantum-mechanics-invalidate-the-law-of-non-contradiction-part-2/.

Priest, Graham, Francesco Berto, and Zach Weber. 2018. Dialetheism. *The Stanford Encyclopedia of Philosophy* (Fall 2018 Edition), ed. Edward N. Zalta. https://plato.stanford.edu/archives/fall2018/entries/dialetheism/.

Prior, Arthur. 1968. Limited Indeterminism. In *Papers on Time and Tense*. Oxford: Clarendon Press.

Pruss, Alexander. 2009. The Leibnizian Cosmological Argument. In *The Blackwell Companion to Natural Theology*, ed. William Lane Craig and J.P. Moreland. Chichester: Wiley-Blackwell.

———. 2016. Divine Creative Freedom. *Oxford Studies in the Philosophy of Religion* 7: 213–238.

———. 2018. *Infinity, Causation, and Paradox*. Oxford: Oxford University Press.

Psillos, Stathis. 2009. Regularity Theories of Causation. In *Beebee, Hitchcock, and Menzies*, 131–157. Oxford: Oxford University Press.

Puryear, Stephen. 2014. Finitism and the Beginning of the Universe. *Australasian Journal of Philosophy* 92: 619–629.

———. 2016. Finitism, Divisibilty, and the Beginning of the Universe: Replies to Loke and Dumsday. *Australasian Journal of Philosophy*. https://doi.org/10.1080/00048402.2016.1194443.

Putnam, Hilary. 2005. A Philosopher Looks at Quantum Mechanics (Again). *British Journal for Philosophy of Science* 56: 615–634.

Quine, W.V.O. 1951. Two Dogmas of Empiricism. In *Reprinted in From a Logical Point of View*, 2nd ed., 20–46. Cambridge, MA: Harvard University Press.

Qureshi-Hurst, E., and A. Pearson. 2020. Quantum Mechanics, Time and Theology: Indefinite Causal Order and a New Approach to Salvation. *Zygon* 55: 663–684.

Rasmussen, Joshua. 2010. Cosmological Arguments from Contingency. *Philosophy Compass* 5: 806–819.

———. 2018. Review of God and Ultimate Origins. *European Journal for Philosophy of Religion* 10: 189–194.

Rasmussen, Joshua, and Felipe Leon. 2018. *Is God the Best Explanation of Things: A Dialogue*. Cham: Springer Nature.

Ratzsch, Del, and Jeffrey Koperski. 2019. Teleological Arguments for God's Existence. In *The Stanford Encyclopedia of Philosophy*, ed. Edward N. Zalta. https://plato.stanford.edu/archives/sum2019/entries/teleological-arguments/.

Reichenbach, Bruce. 2021. Cosmological Argument. In *The Stanford Encyclopedia of Philosophy*, ed. Edward Zalta. https://plato.stanford.edu/archives/win2016/entries/cosmological-argument/.

Reid, Thomas. 1983. In *Inquiry and Essays*, ed. Ronald E. Beanblossom and Keith Lehrer. Indianapolis: Hackett.

Reutlinger, A. 2017. Explanation Beyond Causation? New Directions in the Philosophy of Scientific Explanation. *Philosophy Compass*. https://onlinelibrary.wiley.com/doi/abs/10.1111/phc3.12395.

Rodriguez-Pereyra, Gonzalo. 2019. Nominalism in Metaphysics. *The Stanford Encyclopedia of Philosophy* (Summer 2019 Edition), ed. Edward N. Zalta. https://plato.stanford.edu/archives/sum2019/entries/nominalism-metaphysics/.

Rosen, Gideon. 2020. Abstract Objects. In *The Stanford Encyclopedia of Philosophy*, ed. Edward Zalta. https://plato.stanford.edu/archives/spr2020/entries/abstract-objects/.

Rosenberg, Alex. 2005. *Philosophy of Science: A Contemporary Introduction*. 2nd ed. London: Routledge.

Rovelli, C. 2018a. Physics Needs Philosophy. Philosophy Needs Physics. *Foundations of Physics* 48 (5): 481–491.

———. 2018b. Creath, Richard, "Logical Empiricism", The Stanford Encyclopedia of Philosophy (Summer 2017 Edition), Edward N. Zalta (ed.), https://plato.stanford.edu/archives/sum2017/entries/logical-empiricism/.

Rowe, William. 2003. Reflections on the Craig-Flew Debate. In *Does God Exist? The Craig-Flew Debate*, ed. Stan Wallace. Aldershot: Ashgate.

Rundle, Bede. 2004. *Why There Is Something Rather Than Nothing*. Oxford: Oxford University Press.

Russell, Bertrand. 1918. On the Notion of Cause. In *Mysticism and Logic and Other Essays*. New York: Longmans Green and Co.

———. 1957. *Why I Am Not a Christian*. London: Unwin Paperbacks.

———. 1969. *History of Western Philosophy*. London: Allen and Unwin.

Russell, Bertrand, and Frederick Copleston. 1964. Debate on the Existence of God. In *The Existence of God*, ed. John Hick. New York: Macmillan.

Schaffer, Jonathan. 2008. The Metaphysics of Causation. *The Stanford Encyclopedia of Philosophy* (Fall 2008 Edition), ed. Edward N. Zalta. http://plato.stanford.edu/archives/fall2008/entries/causation-metaphysics/.

———. 2010. Monism: The Priority of the Whole. *Philosophical Review* 119: 31–76.

———. 2016. Grounding on the Image of Causation. *Philosophical Studies* 173: 49–100.

Shoemaker, S. 1969. Time Without Change. *Journal of Philosophy* 66: 363–381.

Sidelle, Alan. 2002. On the Metaphysical Contingency of Laws of Nature. In *Conceivability and Possibility*, ed. Tamar Gendler and John Hawthorne. Oxford: Oxford University Press.

Sider, Theodore. 2002. Time Travel, Coincidences and Counterfactuals. *Philosophical Studies* 110: 115–138.

Siegel, Ethan. 2018. Why Haven't We Bumped into Another Universe Yet. https://medium.com/starts-with-a-bang/ask-ethan-why-havent-we-bumped-into-another-universe-yet-fa15b45b0ce9.

Simon, Jonathan. 2015. Review of Michael Pelczar's *Sensorama. Notre Dame Philosophical Reviews: An Electronic Journal*. https://ndpr.nd.edu/news/sensorama-a-phenomenalist-analysis-of-spacetime-and-its-contents/.

Sinhababu, N. 2016. Divine Fine-Tuning vs Electrons in Love. *American Philosophical Quarterly* 53: 423–432.

Skow, B. 2016. *Reasons Why*. Oxford: Oxford University Press.

Smith, Quentin. 1991. Atheism, Theism and Big Bang Cosmology. *Australasian Journal of Philosophy* 69: 48–66.

———. 1996. Causation and the Logical Impossibility of a Divine Cause. *Philosophical Topics* 24: 169–191.

———. 1998. Why Stephen Hawking's Cosmology Precludes a Creator. *Philo* 1: 75–93.

———. 2002. Time Was Created by a Timeless Point. In *God and Time*, ed. Gregory E. Ganssle and David M. Woodruff. Oxford: Oxford University Press.

———. 2007. Kalām Cosmological Arguments for Atheism. In *The Cambridge Companion to Atheism*, ed. Michael Martin. Cambridge: Cambridge University Press.

Smith, Sheldon. 2013. Causation in Classical Mechanics. In *The Oxford Handbook of Philosophy of Physics*, ed. R. Batterman. Oxford: Oxford University Press.

Smith, Quentin, and Nathan Oaklander. 1995. *Time, Change and Freedom*. London: Routledge.

Smolin, Lee. 1997. *The Life of the Cosmos*. New York: Oxford University Press.

Sobel, Jordan. 2004. *Logic and Theism: Arguments for and Against Beliefs in God*. Cambridge: Cambridge University Press.

Sober, Elliot. 2003. The Argument from Design. In *God and Design*, ed. N. Manson (pp. 25–53). Routledge. Reprinted in *The Blackwell Guide to Philosophy of Religion*, 2004, ed. W. Mann (pp. 117–147).

———. 2019. *The Design Argument*. Cambridge: Cambridge University Press.

Sorabji, R. 1983. *Time, Creation and the Continuum*. Ithaca, NY: Cornell University Press.

———. 2006. *Time, Creation and the Continuum*. Ithaca, NY: Cornell University Press.

Stanley, Matthew. 2009. Myth 21: That Einstein Believed in a Personal God. In *Galileo Goes to Jail: and Other Myths About Science and Religion*, ed. Ronald Numbers. Cambridge, MA: Harvard University Press.

Steiner, Mark. 1998. *The Applicability of Mathematics as a Philosophical Problem*. Cambridge, MA: Harvard University Press.

Steinhardt, P., and N. Turok. 2005. The Cyclic Model Simplified. *New Astronomy Reviews* 49: 43–57. Preprint: http://arxiv.org/abs/astro-ph/0404480.

Stenger, Victor. 2000. Natural Explanations for the Anthropic Coincidences. *Philo* 3 (2): 50–67.

———. 2013. The Universe Shows No Evidence of Design. In *Debating Christian Theism*, ed. J.P. Moreland, Chad V. Meister, and Khaldoun Sweis. Oxford: Oxford University Press.

Stenmark, Mikael. 2003. Scientism. In *Encyclopedia of Science and Religion*, ed. J. Wentzel Vrede van Huyssteen, 2nd ed. New York: Macmillan Reference.

Stoeger, William. 2001. Epistemological and Ontological Issues Arising from Quantum Theory. In *Quantum Mechanics: Scientific Perspectives on Divine Action*, ed. Robert Russell et al. Berkeley: Center for Theology and the Natural Sciences.

———. 2010. God, Physics and the Big Bang. In *The Cambridge Companion to Science and Religion*, ed. Peter Harrison. Cambridge: Cambridge University Press.

Strawson, Galen. 1989. *The Secret Connexion*. Oxford: Clarendon.

Suarez, Francisco. 2002. *On Creation, Conservation, and Concurrence*, Trans. Alfred Freddoso. South Bend.: St. Augustine's Press.

Sudduth, Michael. 2009. *The Reformed Objection to Natural Theology*. London: Routledge.

Susskind, Leonard. 2012. Was There a Beginning?. https://arxiv.org/abs/1204.5385.

Swinburne, Richard. 1993. God and Time. In *Reasoned Faith*, ed. Eleonore Stump. Ithaca: Cornell University Press.

———. 1996. *Is There a God?* New York: Oxford University Press.

———. 2004. *The Existence of God*. 2nd ed. Oxford: Clarendon.

———. 2005a. Second Reply to Grünbaum. *The British Journal for the Philosophy of Science* 56: 923–924.

———. 2005b. The Value and Christian Roots of Analytical Philosophy of Religion. In *Faith and Philosophical Analysis*, ed. H. Harris and C. Insole. Aldershot: Ashgate.

———. 2012. Arguments to God from the Observable Universe. In *The Blackwell Companion to Science and Christianity*, ed. Alan Padgett and J. Stump. Chichester: Wiley-Blackwell.

Tahko, Tuomas E., and E. Jonathan Lowe. 2020. Ontological Dependence. *The Stanford Encyclopedia of Philosophy* (Fall 2020 Edition), ed. Edward N. Zalta. https://plato.stanford.edu/archives/fall2020/entries/dependence-ontological/.

Tegmark, Max. 2008. The Mathematical Universe. *Foundations of Physics* 38: 101–150.

———. 2015. Infinity Is a Beautiful Concept – And It's Ruining Physics. https://www.discovermagazine.com/the-sciences/infinity-is-a-beautiful-concept-and-its-ruining-physics.

Tooley, Michael. 1987. *Causation: A Realist Approach*. Oxford: Oxford University Press.

Topham, J. 2010. Natural Theology and the Sciences. In *The Cambridge Companion to Science and Religion*, ed. Peter Harrison. Cambridge: Cambridge University Press.

Towler, Mike. 2009a. *Pilot Wave Teory, Bohmian Metaphysics, and the Foundations of Quantum Mechanics Lecture 7*. www.tcm.phy.cam.ac.uk/~mdt26/PWT/lectures/bohm7.pdf. Accessed 20 January 2017.

———. 2009b. *De Broglie-Bohm Pilot-Wave Theory and the Foundations of Quantum Mechanics*. http://www.tcm.phy.cam. ac.uk/~mdt26/pilot_waves.html. Accessed 20 January 2017.

Trigg, Roger. 1993. *Rationality and Science: Can Science Explain Everything?* Oxford: Blackwell.

Trogdon, Kelly. 2018. Inheritance Arguments for Fundamentality. In *Reality and Its Structure: Essays in Fundamentality*, ed. Ricki Bliss and Graham Priest. Oxford: Oxford University Press.

Van Bendegem, Jean Paul. 2011. The Possibility of Discrete Time. In *The Oxford Handbook of Philosophy of Time*, ed. Craig Callender. Oxford: Oxford University Press.

Veneziano, G., and M. Gasperini. 2003. The Pre Big Bang Scenario in String Cosmology. *Physics Reports* 373: 1–212.

Vilenkin, Alexander. 1982. Creation of Universes from Nothing. *Physics Letters B* 117: 25.

———. 2006. *Many Worlds in One*. New York: Hill and Wang.

———. 2015. The Beginning of the Universe. *Inference: International Review of Science* 1 (4) https://inference-review.com/article/the-beginning-of-the-universe. Accessed 22 April 2015.

Vilenkin, Alexander, and Max Tegmark. 2011. The Case for Parallel Universes. *Scientific American*. https://www.scientificamerican.com/article/multiverse-the-case-for-parallel-universe/.

Voltaire. 1764/1901. Philosophical Dictionary. In *The Works of Voltaire*, Vol. 11. New York: E. R. DuMont.

Wall, Aron. 2013a. The Generalized Second Law Implies a Quantum Singularity Theorem. *Classical and Quantum Gravity* 30 (16): 165003. Preprint: arXiv:1010.5513v4 [gr-qc].

———. 2013b. Corrigendum: The Generalized Second Law Implies a Quantum Singularity Theorem. *Classical and Quantum Gravity* 30 (19): 199501.

Wall, Aron C. 2013c. A Discrete, Unitary, Causal Theory of Quantum Gravity. *Classical and Quantum Gravity* 30: 115002.

Wall, Aaron. 2014. Did the Universe Begin? IV: Quantum Eternity Theorem. http://www.wall.org/~aron/blog/did-the-universe-begin-iv-quantum-eternity-theorem/. Accessed 20 January 2017.

Waters, Ben. 2013. Methuselah's Diary and the Finitude of the Past. *Philosophia Christi* 15 (2): 463–469.

Weaver, C. 2016. Yet Another New Cosmological Argument. *International Journal for Philosophy of Religion* 80 (1): 11–31.

———. 2019. *Fundamental Causation: Physics, Metaphysics, and the Deep Structure of the World*. London: Routledge.

Wenmackers, Sylvia. 2016. Children of the Cosmos. In *Trick or Truth?: The Mysterious Connection Between Physics and Mathematics*, ed. Anthony Aguirre, Brendan Foster, and Zeeya Merali. Springer Nature: Cham.

Wheeler, John, and Richard Feynman. 1949. Classical Electrodynamics in Terms of Direct Interparticle Action. *Reviews of Modern Physics* 21: 425–434.

Wielenberg, Erik. 2020. Craig's Contradictory Kalām: Trouble at the Moment of Creation. *TheoLogica*. https://doi.org/10.14428/thl.v4i3.55133.

Wigner, Eugene. 1960. The Unreasonable Effectiveness of Mathematics in the Natural Sciences. *Communications on Pure and Applied Mathematics* 13 (1): 1–14.

Williamson, Jon. 2009. Probabilistic Theories of Causality. In *The Oxford Handbook of Causation*, ed. Helen Beebee, Christopher Hitchcock, and Peter Menzies. Oxford: Oxford University Press.

Wilson, Alastair. 2020. *The Nature of Contingency: Quantum Physics as Modal Realism*. Oxford: Oxford University Press.

Wolff, J. 2013. Are Conservation Laws Metaphysically Necessary? *Philosophy of Science* 80 (5): 898–906.

Wu, Y.L. 2003. Symmetry Principle Preserving and Infinity Free Regularization and Renormalization of Quantum Field Theories and the Mass Gap. *International Journal of Modern Physics A* 18: 5363–5420.

Zarepour, Mohammad. 2020. Infinite Magnitudes, Infinite Multitudes, and the Beginning of the Universe. *Australasian Journal of Philosophy*. https://doi.org/10.1080/00048402.2020.1795696.

# Index[1]



[1] Note: Page numbers followed by 'n' refer to notes.


A. Loke, *The Teleological and Kalam Cosmological Arguments Revisited*, Palgrave Frontiers in Philosophy of Religion, https://doi.org/10.1007/978-3-030-94403-2

D

The manufacturer's authorised representative in the EU is Springer Nature Customer Service Centre GmbH, Europaplatz 3, 69115 Heidelberg, Germany. If you have any concerns regarding our products, please contact ProductSafety@springernature.com

Printed and bound by CPI Group (UK) Ltd, Croydon, CR0 4YY
30/09/2026
02225499-0016